S0-AWV-379

Emerging Technologies in Hazardous Waste Management

ACS SYMPOSIUM SERIES 422

Emerging Technologies in Hazardous Waste Management

D. William Tedder, EDITOR
Georgia Institute of Technology

Frederick G. Pohland, EDITOR
University of Pittsburgh

Developed from a symposium sponsored
by the Division of Industrial and Engineering Chemistry, Inc.,
of the American Chemical Society
at the Industrial and Engineering Chemistry Winter Symposium,
Atlanta, Georgia,
May 1–4, 1989

American Chemical Society, Washington, DC 1990

CHEM SEP/ae

Library of Congress Cataloging-in-Publication Data

Emerging Technologies in Hazardous Waste Management
D. William Tedder and Frederick G. Pohland, editors

p. cm.—(ACS Symposium Series, ISSN 0097–6156; 422).

"Developed from a symposium sponsored by the Division of Industrial and Engineering Chemistry, Inc., of the American Chemical Society at the Industrial and Engineering Chemistry Winter Symposium, Atlanta, Georgia, May 1–4, 1989."

Includes bibliographical references.

ISBN 0–8412–1747–5

1. Hazardous wastes—Management—Congresses.
2. Sewage—Purification—Congresses.

I. Tedder, D. W. (Daniel William), 1946– . II. Pohland, F. G. (Frederick George), 1931– . III. American Chemical Society. Division of Industrial and Engineering Chemistry. IV. American Chemical Society. Division of Industrial and Engineering Chemistry. Winter Symposium (1989: Atlanta, GA.). V. Series.

TD1020.E44 1990
628.4'2—dc20 90–20
CIP

The paper used in this publication meets the minimum requirements of American National Standard for Information Sciences—Permanence of Paper for Printed Library Materials, ANSI Z39.48–1984.

∞

PRINTED IN THE UNITED STATES OF AMERICA

ACS Symposium Series

M. Joan Comstock, *Series Editor*

Foreword

The ACS SYMPOSIUM SERIES was founded in 1974 to provide a medium for publishing symposia quickly in book form. The format of the Series parallels that of the continuing ADVANCES IN CHEMISTRY SERIES except that, in order to save time, the papers are not typeset but are reproduced as they are submitted by the authors in camera-ready form. Papers are reviewed under the supervision of the Editors with the assistance of the Series Advisory Board and are selected to maintain the integrity of the symposia; however, verbatim reproductions of previously published papers are not accepted. Both reviews and reports of research are acceptable, because symposia may embrace both types of presentation.

Contents

INDEXES

Preface

IN MAY 1989, the Division of Industrial and Engineering Chemistry, Inc., held a special symposium with the theme "Emerging Technologies for Hazardous Waste Treatment." Approximately 70 papers, covering a wide range of topics, were presented during the four-day meeting in Atlanta.

The symposium organizers gave the definition of "hazardous" wastes the broadest possible interpretation, and participants included a mix of politicians, government regulators, and interveners. Two sessions were conducted simultaneously. One focused on water technologies. The other considered topics relating to the management of solids and residues. Speakers addressed problems and issues ranging from municipal wastes, radionuclide control, and cytotoxic and biologically active hazards, to chemical contamination and degradation, and species migration in water, soils, and residues.

Although no single volume can do justice to a subject as broad as hazardous waste management, we believe that the present collection will be of general interest. No topic is afforded comprehensive treatment, but there is virtually something for everyone. Research topics cover the gamut from fundamental studies, the evaluation of significant analytical problems and species degradation mechanisms, to semipilot-scale testing of innovative management and treatment technologies. Not surprisingly, the authors hail from varied backgrounds and interests. They include physicists, chemists, and biologists, along with mechanical, civil, environmental, nuclear, and chemical engineers.

Manuscripts submitted during the symposium were peer reviewed and subsequently revised. The final selection of edited papers included as chapters in this volume was based on scientific review and merit, the editors' perceptions of their lasting value or innovative aspects, and the general applicability of either the technology itself or the scientific methods advocated by the authors. Dominant themes brought from the symposium to the book include reactive techniques for the destruction of chlorinated species and innovative methods for economically detoxifying large quantities of contaminated soils.

One goal of the meeting was to promote interaction between investigators who seldom cross paths and, as readers will see, we were largely successful in this regard. We hope this volume is as successful in helping promote better solutions to a number of longstanding problems.

Acknowledgments

The symposium was supported by several outstanding organizations which are committed to solving waste problems and reducing environmental pollution. Their financial support demonstrates their interest and recognition that research is essential to finding the best solutions to these problems. Their generosity was essential to the overall success of the symposium, and we gratefully recognize it here.

Our 1989 special supporters were Merck and Company, Inc., of Rahway, New Jersey; Monsanto Company of St. Louis, Missouri; Nalco Chemical Company of Naperville, Illinois; Petroleum Research Fund of Washington, D.C.; and Texaco Inc., of Beacon, New York.

We also gratefully acknowledge the generous support of several organizations which were event sponsors. These are the Ansul Company of Marinette, Wisconsin; Hoechst-Celanese of Chatham, New Jersey; ICI Americas, Inc., of Wilmington, Delaware; and Merck and Company, Inc., of Rahway, New Jersey.

D. WILLIAM TEDDER
Georgia Institute of Technology
Atlanta, GA 30332–0100

FREDERICK G. POHLAND
University of Pittsburgh
Pittsburgh, PA 15261–2294

December 11, 1989

Chapter 1

Emerging Technologies for Hazardous Waste Management

An Overview

D. William Tedder[1] and Frederick G. Pohland[2]

[1]School of Chemical Engineering, Georgia Institute of Technology, Atlanta, GA 30332–0100
[2]Department of Civil Engineering, University of Pittsburgh, Pittsburgh, PA 15261–2294

Since World War II, waste management problems have been exacerbated by two factors—quantity and toxicity. Waste quantities have steadily increased due to population growth and, perhaps more significantly, because the specific waste generation rates (i.e., per man, woman or child) continue to increase, especially in the U.S. Americans are affluent consumers who demand the best and recycle little. Increasing requirements for consumer products, coupled with rapidly expanding technological innovations to make these products affordable, have placed high-technology products in the hands of many. General access to the personal computer, for example, was unthinkable only a few years ago. However, the manufacturers of televisions, automobiles, and personal computers, as well as the creators of countless other consumer items, also produce industrial wastes that must be managed. Disposable plastic objects, containers, and old newspapers are among the more obvious contributors to increasing municipal waste quantities, but hazardous wastes also result from their manufacture (1).

Increasing waste toxicities are relatively new phenomena. In those bygone days when real horsepower pulled our vehicles, wastes were more visible but less toxic. Horse manure is easily recycled as fertilizer and is biodegradable. Today we travel by automobiles and their exhausts are less visible than horse manure, less readily biodegradable, but usually more hazardous. Additional auto-related wastes, such as battery acids, paints and metal processing wastes, and residues from oil refining and mining operations, are both visible and highly toxic, but only slightly biodegradable.

Toxic and hazardous wastes are increasingly prevalent and exotic. Radioactive wastes, for example, were virtually nonexistent prior to World War II. Similarly, industrial wastes from solvent use, electrochemical applications, fertilizers, and pesticides, just to name a few sources, have all come from relatively recent, but highly beneficial, technological innovations. It should not be surprising that tradeoffs exist; useful technologies may also have less than desirable side-effects.

The benefits from our lifestyle today are undeniable and we need not consider the waste problems insurmountable (2). With effort and commitment, they can be solved. Moreover, the best solutions will both avoid future problems and reduce

0097–6156/90/0422–0001$06.00/0

costs simultaneously (3). In the meantime, better technologies are needed to detoxify and isolate existing hazardous wastes from the environment. Within the technical community, these needs are widely recognized. Literature focusing on wastewater (4,5,6), hazardous waste treatment (7,8,9), land disposal technologies (10), and the related legislative and legal issues (11) continues to grow exponentially.

Our present contribution focuses on selected technologies that are currently under development. Since these are *emerging* technologies and management techniques, neither process safety nor economic considerations are discussed in detail. While both factors are important, additional research and development are needed before these aspects can be assessed with an acceptable level of confidence.

Wastewater Management Technologies

Traditional concerns with industrial water quality have focused on process use (e.g., pretreatment for cooling tower and boiler water) rather than wastewater treatment (12). This latter issue has been of more concern to municipalities rather than to industries, but this situation is rapidly changing. With the widespread use of chemicals and increasingly limited water supplies, reuse and waste minimization, rather than discharge, are emerging as dominant strategies for industrial wastewater management.

Groundwater pollution (e.g., from landfill and land disposal leachates) is now a more frequent industrial concern, especially in light of tighter regulations and long-term liabilities (11,13,14). It is increasingly difficult for generators to abandon wastes, and this reality is changing perceptions and emphases. Short-term benefits are being weighed more carefully against long-term liabilities. Less tangible considerations, such as public and community relations, are also galvanizing corporate sensitivities. More commonly, the prudent decision is to minimize waste generation and provide in-house stabilization, rather than discharge minimally treated wastes to municipal or private waste handlers and thereby assume greater long-term liabilities.

Nonetheless, existing landfill, migration, and surface runoff problems persist. The resulting polluted waters usually contain only trace levels of restricted species, but management is still required. Moreover, traditional methods are often ineffective and new treatment methods are needed to economically manage these wastes. Chemical or biological oxidation or reduction techniques are often possibilities, although partially degraded species from inadequate treatment may be more noxious than the parent compounds. Thus, these technologies must be thoroughly understood before deployment can be safely initiated.

A related issue concerns the ability to model species degradation and fate at infinite dilution in water, based on limited experimental data. Regardless of the degradation technology, the chemistry is complex because of multiple reaction paths and solutes, species competition, and other interactions. Considerable efforts have already been expended toward the development of predictive techniques for estimating species migration from landfills, land disposal sites, etc. (e.g., Valocchi (15)).

Water purification by solute separation and retention is another important option. The main problems with such technologies relate to the extremely low pollutant concentrations, the relatively high decontamination factors that are required, and the ability to either regenerate or stabilize spent separating agents. Modeling capabilities are also needed to guide research and the selection of alternatives.

In Chapter 2, Tseng and Huang examine the fundamental kinetics of phenol oxidation in the presence of TiO_2 and UV radiation. They find that the oxidation rate is affected by pH, the TiO_2 and dissolved oxygen concentrations, as well as those of phenol. Oxygen plays an important role, and the reaction sequence is likely initiated by several highly reactive oxygen radicals. Their kinetic expression has $\sim 2/3$ concentration dependence on phenol. The TiO_2 concentration dependence is the classic form described by Hougen and Watson (16) which suggests chemisorption with catalytic reaction on the particle surface.

Pacheco and Holmes (see Chapter 3) also present results on oxidation in the presence of TiO_2 and UV, but with salicylic acid as the reactive solute. Their emphasis is on the engineering design of solar driven reactors, rather than fundamentals, and they compare the relative efficiencies of falling-film versus glass-tube reactors. They confirm Tseng and Huang's observations and discuss some of the design limitations. While increased TiO_2 concentrations accelerate reaction rates, higher TiO_2 concentrations also increase stream turbidity and reduce UV absorption. Since both UV irradiation and the catalyst are important, competition results and an optimal TiO_2 concentration must exist for a particular waste stream. It also helps to maintain high dissolved oxygen concentrations in the reactor and, while poisoning effects are not discussed, they are doubtless important in many instances.

A third study of heterogeneous catalysis for water purification takes a broader view. In Chapter 4, Bull and McManamon discuss its use to enhance oxidation rates for several different solutes and catalyst preparations. They find that many pollutants can be effectively destroyed including: sulfide, thiosulfate, mercaptans and cyanide species. Their catalysts include various transition metals on aluminosilicate and zeolite supports. Several catalytic systems exhibit encouraging results over a range of pHs. Hence a wider class of streams may be effectively treated with reduced levels of pH adjustment. They examine the destruction of three key inorganic pollutants; thiosulfate, sulfide, and cyanide. In their experiments, H_2O_2 and aqueous solute were mixed and pumped upflow through jacketed beds packed with metal-supported catalysts. Iron on mordenite and copper on synthetic zeolites were found particularly effective. The practical operating range was $\sim 4 \leq pH \leq 12$, but it depended on the pollutant. For example, cyanide with $pH \leq 6$–8 liberated HCN which is usually unacceptable. However, pH did not appear to affect catalytic activity and the relatively large catalytic particle sizes used permitted an efficient packed bed design.

Castrantas and Gibilisco (see Chapter 6) present results where UV and H_2O_2 are used without a heterogeneous catalyst or other additives to promote the destruction of phenol and substituted phenols. They point out that this approach is somewhat simpler than methods using soluble additives (e.g., Fenton's reagent), and it is more effective under a wider pH range. Another advantage is that H_2O_2, being unstable, does not require subsequent removal as would be the case with iron soluble salts, for example. Compared to UV alone, the UV and H_2O_2 system gives almost a factor of 20 increase in reaction rates.

Homogeneous phase reaction with UV catalyzed oxidation is a classical method which has been studied for many years. Most investigators, however, have examined reactivities in mixtures that initially contained only one oxidizable solute. With respect to wastewaters, this situation seldom exists and Sundstrom et al. present results in Chapter 5 which suggest that solute interactions can inhibit oxidation. In particular, they find that mixtures of two priority pollutants, benzene and trichloroethylene, oxidize more slowly than the individual species. While this result is partly due to competition for the oxidant and mass action, benzene and its derivatives apparently inhibit trichloroethylene oxidation. Consequently, a reactor's volume requirements may be seriously underestimated if its design is based on pure component oxidation rates.

Coincidentally, Sundstrom, Weir, and Redig's work has implications for Desai et al.'s study of chemical structure and biodegradability relationships in Chapter 9. In both cases, most available data describe solute destruction rates in relatively simple systems. Although the reactive pathways are quite different, unfavorable interactions are plausible in each. The complexity of most wastewater streams, and the need to design for a wide range of applications, suggest that solute interactions and mixture reaction rates will require much more study before reliable design and scale-up procedures can be defined a priori.

The use of catalysts, H_2O_2, etc., to promote oxidation are advanced technologies compared to current practice. Such systems generate radical species in sufficient quantities that they become the primary detoxifying agents. Although many of these technologies have been available for several decades, their complex chemistry is still only partially understood. Consequently, modeling has been difficult and the results of limited value. Nonetheless, modeling is needed and in Chapter 7, Peyton presents results from his efforts to describe advanced oxidation systems involving methanol, formaldehyde, and formic acid as pollutants. He concludes that lower oxidant dose rates are more efficient for the destruction of dilute contaminants and that, in the absence of promoters, H_2O_2 and UV radiation may be the most appropriate combination in many situations.

Solute species may also be destroyed using anaerobic reduction rather than oxidation chemistry. One possibility for inducing such chemistry lies in freshwater lake sediments. In Chapter 8, Wiegel et al. discuss aspects of using isolates of methanogenic sediments to degrade 2,4-dichlorophenol. They conclude that each of the probable sequential degradation steps is catalyzed by a different microorganism. Dechlorination is apparently rate limiting. Nonetheless, they were able to obtain a fast growing, dehalogenating, and stable enrichment culture.

A wastewater stream typically contains many species that must be controlled. Currently, experimental measurements are essential to characterize degradation behavior (e.g., using a biological process), and such studies are often time-consuming and expensive. On the other hand, a generalized, theoretical procedure for estimating degradation rates may be achievable. Desai et al. present one possible technique for characterization using respirometry data and molecular structure in Chapter 9. They present Monod kinetic parameters for the substituted phenols.

Wastewater treatment by purification is a clear possibility for many streams. Obstacles to the use of a separation technology often relate to costs and the degree to which a pollutant can be removed. Inorganic fluorine species are often present in industrial effluents, and they are increasingly prevalent. In Chapter 10, Nomura et al. describe the use of hydrous cerium oxide adsorbent for reducing F^- concentrations to ~1 mg/L. Their system has higher selectivity and capacity than many current alternatives.

Strontium contamination in wastewater is less generally problematic. Nonetheless, the removal techniques discussed by Watson et al. are important (see Chapter 11). They present a separation method which is often overlooked—the synergistic combination of microorganisms and polymeric adsorbents. The same bioaccumulation factors that compel regulatory agencies to issue very low release limits can be used to advantage, and this contribution is a case in point. The authors explain how relatively inexpensive biological materials can be incorporated into particulate forms, such as gel beads, and used to effectively remove dissolved metal ions from wastewaters.

Affinity dialysis is another promising method for removing metal ions from wastewater. Hu et al. present experimental results and a theoretical model for describing such processes in Chapter 12. Their contribution is a useful introduction to this subject.

Eyal et al. present a review of mixed ionic, or coupled, extractants and their use in Chapter 13. They focus on acid recovery from aqueous streams and, compared to ion exchange, this technology offers recovery options with reduced dilution factors. Mixed solvents, comprised of amines and organic acids for example, can be used to recover inorganic acids and salts simultaneously. These authors emphasize a flowsheet they have developed for coextracting H_2SO_4 and $ZnSO_4$. Their solvent is a kerosene diluent blended with an extractant mixture of tricapryl amine, Alamine 336, and di(2-ethylhexyl)phosphoric acid.

Soils, Residues, and Recycle Techniques

The historical practice of solid waste disposal in near-surface facilities (e.g., landfills, land disposal sites, low-level radwaste burial grounds, etc.) becomes more expensive and difficult as available land and high-quality groundwater resources continue to diminish. Extensive site characterization (17) and monitoring (18,19) efforts are now required, but they do little to mitigate the public controversy surrounding any new site selection process. Increasingly, generators are finding it more beneficial to minimize waste quantities while maximizing recycle and reuse.

Historical practices, either deliberate or accidental, have left numerous site legacies that must be managed. While much of the public concern rightly focuses on species migration and groundwater contamination, more effective remediation and control may result from soil management and on-site treatment practices. In some instances, modified site operations, as well as chemical and biological treatment technologies, will have useful roles in solids management.

For many contaminated soils, high-temperature calcining (i.e., incineration) is perhaps the most obvious treatment. While attractive for some relatively concentrated liquid wastes (20), other alternatives should be considered for soil remediation. Low-temperature technologies, either chemical or biological treatment, for example, can achieve similar results although at somewhat slower rates (21).

Significant quantities of soils, contaminated with various chlorinated species, have accumulated since World War II. These wastes include soils contaminated with highly toxic species such as polychlorinated dibenzo-*p*-dioxins, dibenzofurans, and biphenyls. In Chapter 14, Tiernan et al. describe a dechlorination process where polyethylene glycol and solid KOH are mixed and heated gently to form a highly reactive potassium alkoxide (KPEG) species which then forms an ether and KCl upon contact with aliphatic and certain aromatic chlorinated species. Liquid wastes can also be treated in this manner and toxicity is often substantially reduced.

Creosote and pentachlorophenol-contaminated soils are also significant waste problems. These wastes have resulted largely from the operation of wood-treating plants. While many of these plants are no longer in operation, virtually all sites have contaminated soil problems. Borazjani et al. (see Chapter 15) describe site specific studies where pollutant migration and bioremediation (composting with chicken manure) were evaluated. They find that bioremediation is feasible and that the degradation rates (measured as half-lives) can be related to molecular structure. Moreover, composting does not appear to accelerate pollutant migration. In fact, the added organic matter appears to retard migration compared to species movement in untreated soils.

Many industrial sludges contain lead, mercury, and other toxic metal species. Although current regulations require that toxic metal wastes be disposed of in hazardous waste landfills, their codisposal with municipal solid wastes is a historical practice which, in certain cases, still continues. In Chapter 16, Gould et al. examine factors that influence the mobility of toxic metals in landfills operated with leachate recycle. They find that reducing conditions, induced by anaerobic bacteria, decrease mobility through a variety of mechanisms. Under these conditions, landfill disposal of municipal solid waste retards the migration of many toxic metal species.

In Chapter 17, Reinhart et al. present results of a similar study in which the fates of several chlorinated organic species were monitored. They find that municipal solid waste has an affinity for many such species and thereby reduces their concentrations in leachates. Hence, municipal landfills have a finite assimilative capacity for industrial wastes contaminated with such species. With extended contact in landfills, biodegradation by naturally occurring anaerobic species is enhanced.

Municipal sewage sludges are commonly dewatered and sent to landfills. In Chapter 18, Diallo et al. describe an alternative where ~75 mass% of a organic sewage sludge is dissolved by acidification with concentrated H_2SO_4. Their technique yields significant sludge mass and volume reductions, and the potential for generating organic liquids which may be usable as fuels or solvents. They also point out that acid recycle, a significant economic factor in their flowsheet, may be accomplished using the solvent extraction technology discussed by Eyal and Baniel in Chapter 13.

Shieh et al. (see Chapter 19) suggest an alternative for disposing of incinerator ash. Left untreated, many incinerator ashes are hazardous wastes because of their toxic metal content. In particular, ash residues from coal and municipal solid waste incineration must often be controlled. These authors point out, however, that such ashes may be fabricated into concrete blocks which are then used to form reef barriers. They report leachability and stability test results suggesting that this option is environmentally acceptable. Their studies further indicate that reef blocks fabricated from such incinerator ashes have maintained their physical and chemical integrity in the ocean environment without adversely affecting marine life.

In Chapter 20, Bostick et al. describe treatment and disposal options for ^{99}Tc. This radioisotope exhibits relatively high bioaccumulation factors when released to the environment. Although it is a synthetic isotope resulting from the operation of light-water nuclear power reactors, its behavior shares a common feature with more common toxic species such as Pb and Hg. Namely, it is immobilized by reducing conditions. In this case, iron filing are useful in removing both Hg and ^{99}Tc from a dilute aqueous solution. Iron filings also reduce species leaching rates from grout waste forms.

The last two contributions are concerned with recycle. Campbell et al. discuss the use of fluorine peroxide to separate Pu from ash residues in Chapter 21. They describe a low-temperature volatility process that offers improved capabilities for decontaminating transuranic wastes, particularly incinerator ash. It also has potential applications at sites producing UO_2 light water reactor fuels from UF_6 where uranium-contaminated ash and CaF_2 are generated.

Solvent wastes are also a significant hazard. Too often, used solvents are illegally disposed as orphan wastes in landfills or unmarked burial grounds. In Chapter 22, Joao et al. describe a simple, but effective, method for recycling toxic solvent wastes that are generated in peptide synthesis. Their contribution emphasizes the importance of recycle to minimize waste production and disposal costs, as well as solvent costs. The best waste management techniques are not always the most obvious or the simplest approaches. However, the best methods always reduce both costs and waste quantities.

Summary

While the contributors to this volume address different problems, there are common threads. Anaerobic bacteria, for example, apparently have an important role to play in decreasing toxicity in soil systems through species degradation. These bacteria may also retard the migration of many organic and inorganic species through their maintenance of a chemically reducing environment. Microorganisms may also play important roles in water purification through their immobilization and the exploitation of their naturally occurring bioaccumulative properties.

Our a priori ability to predict species degradation rates in wastewaters may be limited in the foreseeable future. The chemistry of dilute wastewaters is usually complex and, while we can predict general trends, meaningful quantitative analysis without experimentation is more elusive. Group contribution theory is probably one useful key, but considerably more work is needed and unexpected interactions between solutes, affecting their degradation rates in mixtures, are likely.

Waste stabilization and degradation are both important, especially for existing wastes. Recycle, however, will doubtless become increasingly important, especially in the chemical industries and for the management of future wastes. Increased efforts are needed in all of these areas and in the development of integrated strategies for modifying management approaches in the waste producing industries. In many instances, acceptable methods must be found for either reducing or eliminating hazardous waste production altogether. In cases where waste production cannot be eliminated, intraprocess recycle may prove essential to the economical production of environmentally acceptable waste forms.

Literature Cited

1. Schofield, A. N. A long term perspective of burial grounds. In *Land Disposal of Hazardous Waste*, Ellis Horwood Limited, Chichester, England, 1988.

2. Boraiko, A. A. Storing up trouble ...hazardous waste. *National Geographic*, 167(3), 318–351, March 1985.

3. McArdle, J. L., Arozarena, M. M., and Gallagher, W. E. *Treatment of Hazardous Waste Leachates: Unit Operations and Costs.* Noyes Data Corp., Park Ridge, NJ, 1988.

4. Abel, P. D. *Water Pollution Biology.* Halsted Press, New York, 1989.

5. Vigneswaran, S. and Aim, R. B., editors. *Water, Wastewater, and Sludge Filtration.* CRC Press, Boca Raton, FL, 1989.

6. U.S. National Research Council, Water Science and Technology Board. *Hazardous Waste Site Management: Water Quality Issues Report on a Colloquium: February 19–20, 1987*, National Academy Press, Washington, DC, 1988.

7. Travis, C. C. and Cook, S. C. *Hazardous Waste Incineration and Human Health.* CRC Press, Boca Raton, FL, 1989.

8. Kokoszka, L. C. and Flood, J. W. *Enivronmental Management Handbook: Toxic Chemical Materials and Waste.* Marcel Dekker, Inc., New York, 1989.

9. Conway, R. A., editor. *Waste Minimization Practice: the 8th Symposium on Hazardous and Industrial Solid Waste Testing and Disposal, 12–13 November 1987, Clearwater, FL.* Volume 1043 of *Special Technical Publications*, American Society of Testing and Materials, 1989.

10. Kittel, J. H., editor. *Near-Surface Land Disposal.* Volume 1 of *Radioactive Waste Management Handbook*, Harwood Academic Publishers, New York, 1989.

11. Wentz, C. A. *Hazardous Waste Management.* McGraw-Hill, New York, 1989.

12. *Betz Handbook of Industrial Water Conditioning.* Betz Laboratories, Inc., Trevose, PA, 8th edition, 1980.

13. *RCRA Orientation Manual.* EPA/530-SW-86-001, U.S. Environmental Protection Agency, Washington, DC, 1986.

14. Krueger, R. F. and Severn, D. J. Regulation of pesticide disposal. In Krueger, R. F. and Seiber, J. N., editors, *Treatment and Disposal of Pesticide Wastes*, American Chemical Society, Washington, DC, 1984.

15. Valocchi, A. J. Mathematical modelling of the transport of pollutants from hazardous waste landfills. In *Land Disposal of Hazardous Waste*, Ellis Horwood Limited, Chichester, England, 1988.

16. Hougen, O. A. and Watson, K. M. *Chemical Process Principles, Part Three, Kinetics and Catalysis.* John Wiley & Sons, Inc., New York, 1947.

17. Cartwright, K. Site characterization. In Kittel, J. H., editor, *Near-Surface Land Disposal*, Harwood Academic Publishers, New York, 1989.

18. Williams, G. M. Integrated studies into groundwater pollution by hazardous waste. In *Land Disposal of Hazardous Waste*, Ellis Horwood Limited, Chichester, England, 1988.

19. Golchert, N. W. Environmental monitoring. In Kittel, J. H., editor, *Near-Surface Land Disposal*, Harwood Academic Publishers, New York, 1989.

20. Ferguson, T. L. and Wilkinson, R. R. Incineration of pesticide wastes. In Krueger, R. F. and Seiber, J. N., editors, *Treatment and Disposal of Pesticide Wastes*, American Chemical Society, Washington, DC, 1984.

21. Chakrabarty, A. M. *Biodegradation and Detoxification of Environmental Pollutants.* CRC Press, Inc., Boca Raton, FL, 1982.

RECEIVED December 7, 1989

WASTEWATER MANAGEMENT TECHNOLOGIES

Chapter 2

Mechanistic Aspects of the Photocatalytic Oxidation of Phenol in Aqueous Solutions

Jesseming Tseng and C. P. Huang[1]

Environmental Engineering Program, Department of Civil Engineering, University of Delaware, Newark, DE 19716

A photo-catalytic oxidation process using photo-catalyst TiO_2 and ultra violet light to decompose phenol in aqueous solution was studied. The photo-catalyst was thoroughly characterized for specific surface area (BET), surface charge (zeta potential), surface morphology (scanning electron microscopy, SEM) and surface composition (x-ray energy dispersion analysis, EDAX). Parameters such as oxygen, temperature, pH, concentration of photocatalyst, and phenol concentration that may affect the oxidation reaction were thoroughly examined. Results show that oxygen plays one of the most important roles in phenol photo-oxidation. Trace amounts of chloride can inhibit the phenol decomposition reaction while aluminum impurity enhances it. At high phenol concentrations, e.g. ca > 5 $\times 10^{-2}$ M, hydroquinone and a biphenyl dimer are the major intermediates of the oxidation reaction. Free radicals such as hydroxyl and super oxide are keys to photo-oxidation of phenol.

Pollution by hazardous organic waste has become an important environmental issue due to their high toxicity potential. The U. S. Environmental Protection Agency (EPA) has identified over one hundred organic chemicals as priority pollutants(1). The adverse effects of these hazardous chemicals are well documented. Phenol and its related compounds are commonly found in various industrial effluents and have been reported in hazardous waste sites around the country(2). According to Jones (3) phenol can be found at concentrations ranging from 10 to 100 mg/L in the industrial wastewater of thermal and catalytic cracking of petroleum products.

[1]Address correspondence to this author.

0097–6156/90/0422–0012$08.00/0

Photocatalytic oxidation process involving the use of semiconductors has gained recent attention in the field of innovative treatment of hazardous organic wastes. Upon irradiation, a semiconductor generates electron/hole pairs with free electrons produced in the nearly empty conduction band (cb) and positive holes remaining in the valence band (vb)(4,5). The holes, acting as strong oxidizing agents, migrate to the semiconductor surface and react with organic compounds. Figure 1 shows the typical reaction scheme of a n-type semiconductor such as TiO_2. Depending on the ambient conditions, the lifetime of an electron/hole separation process can be from a few nano-seconds to a few hours(6). The recombination of electron/hole pairs can take place either between energy bands or on the surface. As a result the photocatalytic efficiency is reduced. To impede the recombination process, conducting materials such as noble metals, can be incorporated into the semiconductor to facilitate electron transfer and prolong the lifetime of the electron/hole separation process(7,8). Very recently, researchers have investigated the production of fuel such as hydrogen from water using semiconductor and light. This type of work includes photocatalytic dissociation of liquid or vapor water on the surface of powdered semiconductors and single crystals(9,10). Izumi et al (11,12) have used platinized (10% by weight) TiO_2 to decompose benzoic acid and adipic acid. They have proposed a photo-Kolbe type reaction mechanism for the oxidation process. Kawai and Sakata (13-16) have studied the photocatalytic oxidation of various substrates such as chlorine and nitrogen compounds using TiO_2 deposited with 5% platinum and reported that organic decomposition and hydrogen production occurs simultaneously and that thermal effect contributes little to the oxidation. Kawai and Sakada (13,15) have also studied the photocatalytic oxidation of natural materials such as glucose, ethanol, cellulose and lignin; food stuffs such as potato, fatty oil, and herbs; wood such as cherry wood, white dutch clover and water hyacinth; green algae, dead animals and excrement using TiO_2 under a xenon lamp. They have also found that nitrogen and chlorine are converted to NH_3 and HCl, respectively without producing any residual organic chemicals.

Fujihira et al.(17-19), however, have reported an incomplete oxidation of benzene and toluene. Neither Kawai and Sakada nor Fujihira have provided any details of the kinetics and the nature of the reaction. All experiments were conducted in neutral or in NaOH (5 M) solution using TiO_2 catalyst and a 500 watt xenon lamp.

Kruatler and Bard (20) have reported the production of alkane, carbon dioxide and hydrogen from mixtures of organic acids. Pavlik and Tatayanon (21) have demonstrated that lactams can be oxidized to imides following steps similar to those of the Fenton solution. Barbeni et al. (22,23) have studied the degradation of chlorinated hydrocarbons, 2,4,5-trichlorophenoxyacetate and 2,4,5-trichlorophenol, using TiO_2 as the photocatalyst and proposed the formation of ·OH free radicals from reactions between the water molecules and the positive holes. The ·OH free radicals can subsequently participate in a series of reactions with the chlorinated organic.

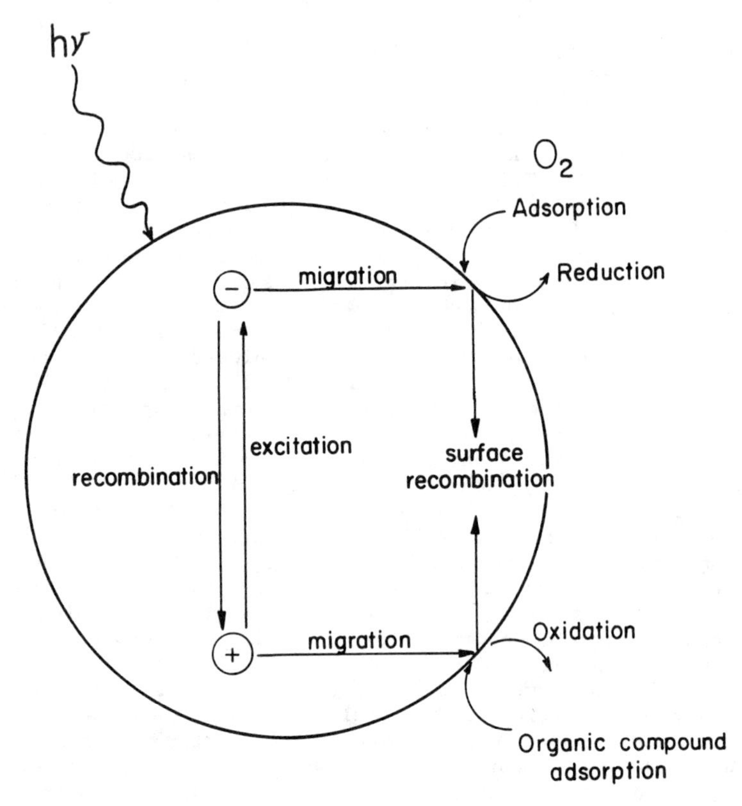

Figure 1. Schematic Redox Reactions of an n-type Semiconductor.

Ollis(24,25) has examined the photo-oxidation of chlorinated hydrocarbons with TiO_2 and reported that the extent of conversion increases in the order: chloroolefins > chloroparaffins > chloroacetic acids. Matthews(26-28) and Gratzel (29) have used thin films of TiO_2 to study the oxidation of some organic impurities in water and reported that the kinetics of photo-oxidation reactions can be described by a Langmuir type adsorption equation.

While there is general agreement among researchers that photocatalyst TiO_2 is promising in decomposing organic substances, a systematic study on the various parameters affecting the photo-oxidation reaction is lacking. The major objective of this study was therefore to investigate the decomposition of organic matter exemplified by phenol using titanium dioxide as the photocatalyst and ultra violet as the light source for illumination. Various factors such as pH, oxygen content, phenol concentration, light intensity and concentration of photo-catalyst were studied. A mechanism for phenol decomposition is proposed.

Methods and Materials

Preparation of Photocatalyst

Titanium dioxide, commercial grade, provided by the Du Pont Company (Wilmington, Delaware) was used. The oxide was pretreated by rinsing 15 g of the sample with 1 liter of $HClO_4$ (1 M) solution followed by washing with distilled water several times until the conductivity of the supernatant was ca 10 μmho/cm. Samples were centrifuged at 10,000 rpm for 30 minutes, using a centrifuge (model RC5 Sorval-Du Pont Company). The titanium dioxide was then dried overnight at 105 °C and ground to fine powder prior to any experiments. During the early phase of this study, the untreated TiO_2 was used. The results show that the phenol decomposition with untreated TiO_2 was extremely slow. Results from EDAX analysis indicated that untreated sample contains 1.145 % (by weight) chloride. Bickley has reported that the photoadsorption activity of oxygen on the surface of TiO_2 is remarkably decreased in presence of F^- or PO_4^{3-} ions(30). In order to improve the oxidation efficiency, the TiO_2 was treated for chloride removal. The treated TiO_2 was thus used throughout the remaining study.

Surface Characterization of Photocatalyst

The surface charge of the titanium dioxide was characterized by zeta potential measurements using a Laser-zee Meter (Pen-ken Inc.). The pH_{zpc} is 9.0 and 9.3 in $NaClO_4$ solutions for the treated and the untreated TiO_2 samples, respectively. The specific surface area of TiO_2 was determined by the BET method (Quantasorb-Model QS-7, Quanta Chrome Co). The impurities of both the treated and the untreated TiO_2 were determined by scanning electron microscopy (SEM) and energy dispersive x-ray (EDAX). The particle size and shape were determined by transmission electron microscopy (TEM). Figure 2 shows the size and shape of the treated TiO_2 under 60,000x and 22,000x magnifications. Table I summarizes the major properties of the treated and the untreated TiO_2.

Table I. Major Properties of Titanium Dioxide

Properties	Untreated	Treated
Structure	Rutile	Rutile
pH_{zpc}	9.3	9.0
Surface area (m^2/g)	6.6	6.1
Band gap energy (eV)	3.2	3.2#
Impurities (% by wt.)	0.729 (Al)	0.708 (Al)
	1.145 (Cl)	
Size	-	10.60 (a)*
(μm)		4.72 (b)*

* *a* represents the length of the particles and *b* denotes the diameter of the particles

\# Ref (29)

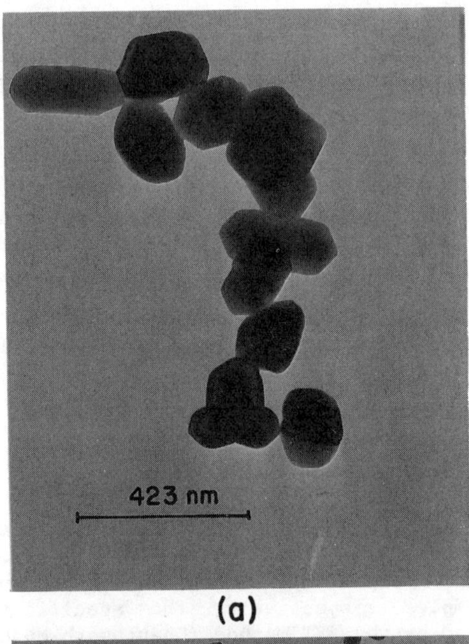

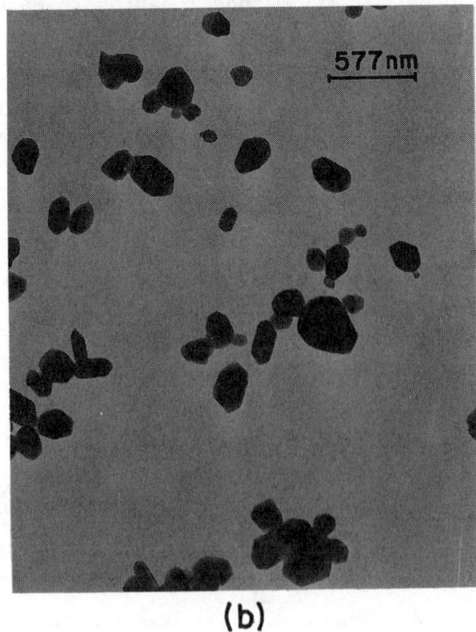

Figure 2. Transmission Electron Micrograph of the TiO_2 Particles After Pretreatment. (a): 60,000x Magnification; (b): 22,000x Magnification.

Light Source

The light source was a 1,600 watt medium pressure mercury vapor discharge lamp (American Ultraviolet Co.). The spectral irradiance for the UV lamp (260 watts/m^2) ranges from 228 and 420 nm at a distance of the 1 meter from the light source according to infortion provided by the maker(31). A typical spectrum for a mercury lamp is shown in Figure 3(32). The light intensity was recorded with a radiometer (Model 65A, YSI- Kettering Co.).

Test Tube Reactors

Pyrex glass tubes were used. The transmittance capability of pyrex glass tubes was scanned from 200 through 800 nm wavelength. The results show that any light having a wavelength longer than 325 nm can completely pass through the pyrex glass. All photo-catalytic oxidation experiments were performed in pyrex tubes containing 15 mL of 10^{-3} M phenol solution and 0.15 g of TiO_2. In order to study the effect of oxygen, different atmospheres were created by introducing pure nitrogen or oxygen into the system. An Ascarite II trap was used to remove carbon dioxide. A pyrogallol solution was used to remove oxygen prior to nitrogen bubbling. Dissolved oxygen was monitored by a dissolved oxygen (DO) meter calibrated by the modified Winkler method (33). Generally, 0.3 ppm oxygen was found in solution under nitrogen bubbling. In the pure oxygen atmosphere, 31.5 ppm dissolved oxygen can be obtained. Samples were shaken over a reciprocal shaker (American Optical Co.) at 180 strokes per minute to insure complete mixing. The temperature of the sample was

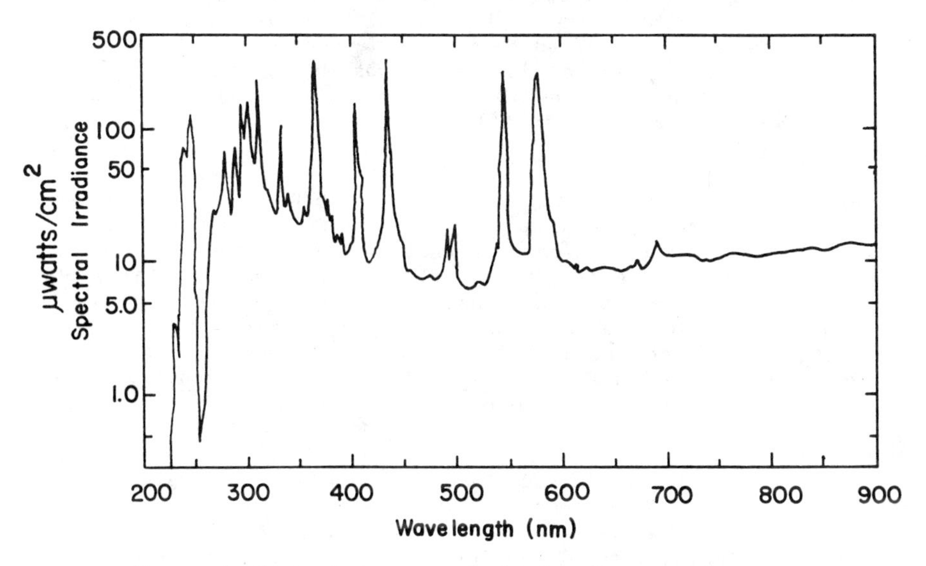

Figure 3. Typical Spectra of the Low Pressure Mercury Lamp.

controlled by a thermostat pump in conjunction with a cooler. At the end of a given reaction time, samples were filtered with 0.45 μm microfilters (Gelman, Supor-450, 25mm). The residual concentration of phenol was measured with an uv-vis spectrophotometer (Hitachi/ Perkin-Elmer, model 139) at a wavelength of 271 nm. Figure 4 shows the schematic set up for the test tube experiments. Test tubes were tightened with teflon septa air-tight caps. The phenol solution was bubbled with nitrogen or oxygen prior to experiments.

For the gas phase studies, a pressure lok gas syringe was used to withdraw gas from the test tube. The compositions of the gas phase in the head space were detected by a GC/MS (Hewlett Packard, Model 5890/5970) under an isothermal condition.

Study of Intermediates

Samples (500 μL) were dried at room temperature in a desiccator to evaporate water. After drying, 500 μL of benzene was added to extract phenol from the samples. The residual solid was dissolved in 10 μL methanol. Phenol and its intermediate compounds were determined by GC/FID and GC/MS techniques.

Results and Discussion

Oxygen/Nitrogen Atmosphere

Figure 5 shows that oxygen plays a significant role in phenol oxidation. In a nitrogen atmosphere the percentage of phenol removal is small compared to that of an oxygen atmosphere. Figure 5 also shows that phenol removal is insignificant in the absence of TiO_2. It has been speculated that ozone may be generated under the oxygen atmosphere. However, this is not possible since the short UV wavelengths that are responsible for ozone production are totally absorbed by the pyrex glass tube(34,35).

Temperature Effect

Results from our previous studies have shown that the photocatalytic oxidation reaction can be complete within 4 hours under 90 ^{o}C(35). Figure 6 shows the oxidation of phenol over a temperature ranging from 20 to 50 ^{o}C. The results clearly indicate the importance of temperature. As expected, increases in temperature greatly raise the rate of phenol decomposition. However the temperature does not exhibit any effect on phenol decomposition under a nitrogen atmosphere. Oxygen appears to be much important than temperature in photocatalytic oxidation reactions.

Effect of pH

The optimum pH for phenol oxidation is at a neutral pH region, e.g. 5-9; a slightly small removal at pH 3 was observed (Figure 7). These results agree well with those reported by Matthews(28). Oxidation of 4-chlorophenol, measured by carbon dioxide yield, was the maximum at neutral pH values and decreased or increased as pH values shifted towards the acidic or the alkaline region. The oxidation of benzoic acid, ethanol, propan-2-ol and methanol is dramatically reduced at pH 3. The photocatalytic oxidation of the surfactant DBS is favored in neutral solutions(36). Ollis and

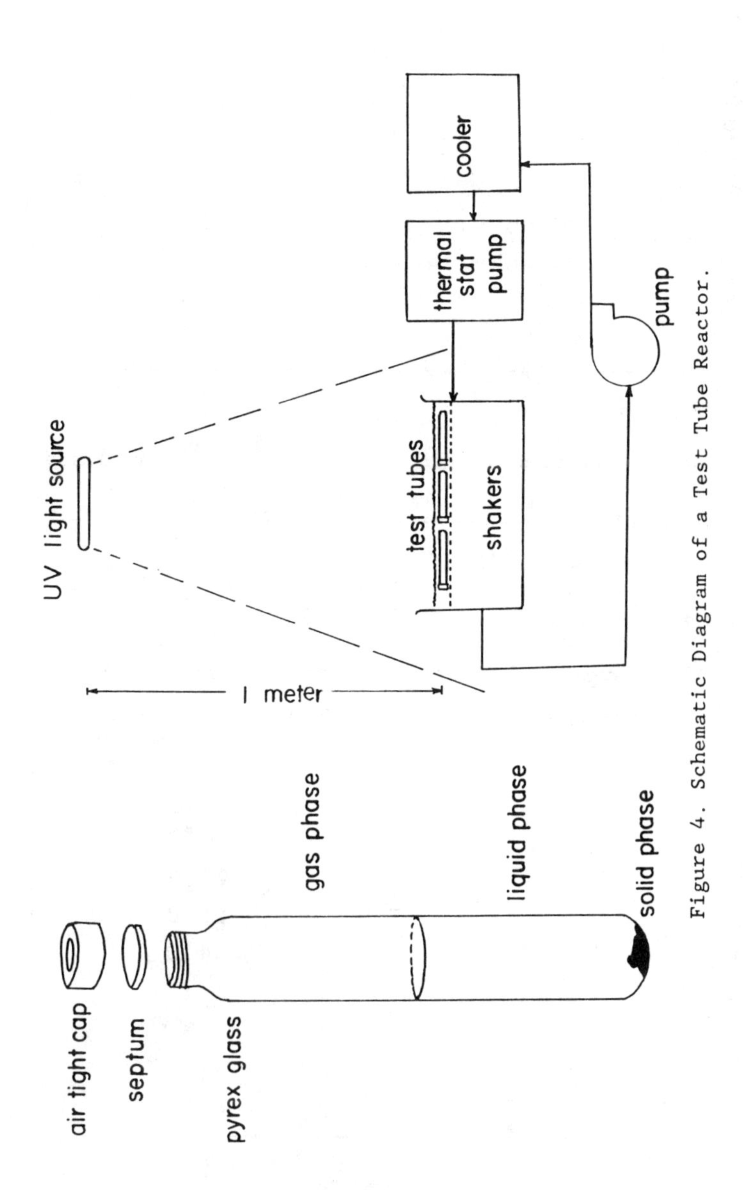

Figure 4. Schematic Diagram of a Test Tube Reactor.

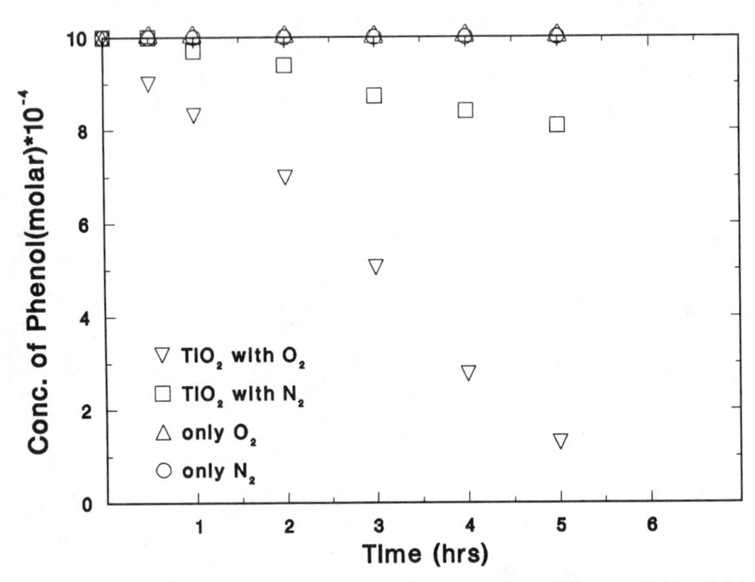

Figure 5. Effect of Oxygen on the Decomposition of Phenol.

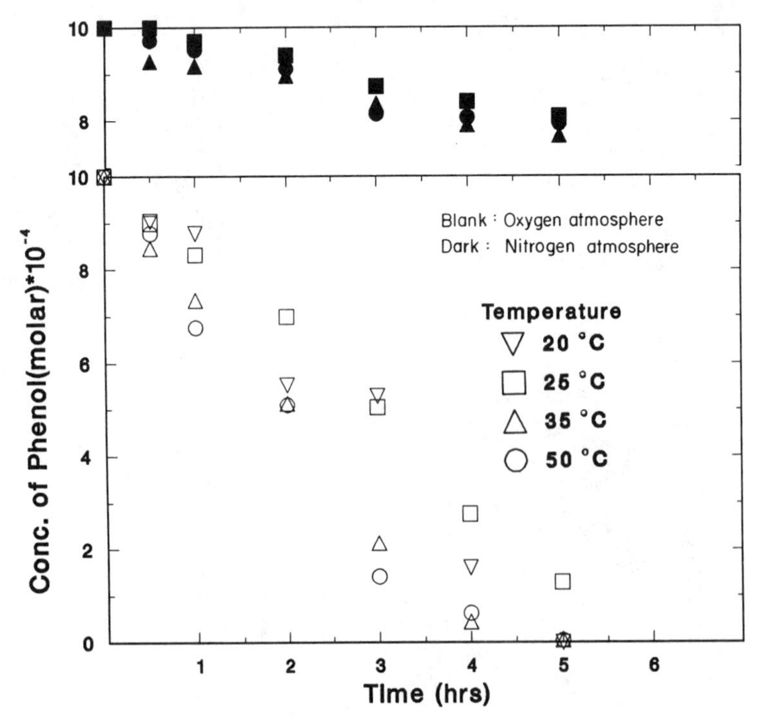

Figure 6. Effect of Temperature on the Decomposition of Phenol.

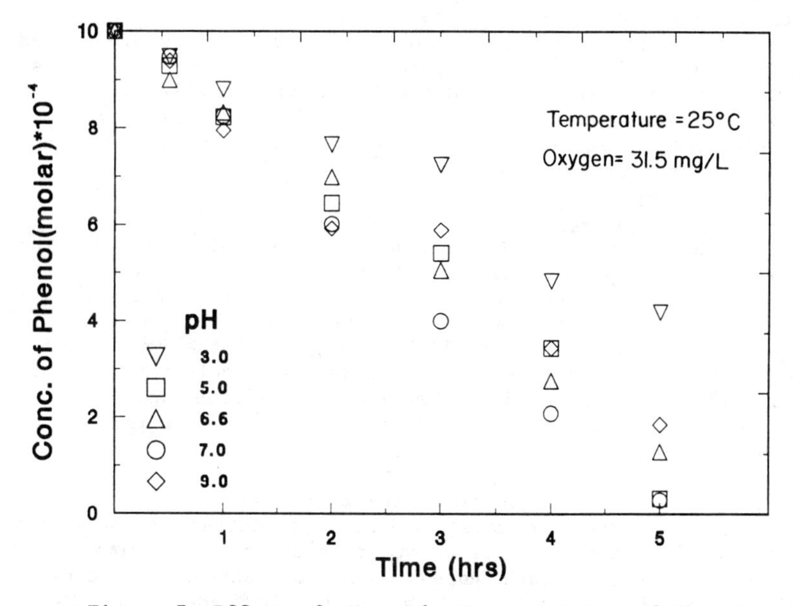

Figure 7. Effect of pH on the Decomposition of Phenol.

Pruden(24) have reported that protons inhibit chloroform degradation on illuminated TiO_2. Kraeutler and Bard(20) have found the rate of CO_2 formation from acetic acid/sodium acetate on undoped anatase to be a maximum at pH 3.6. Matsumura et al.(37) have reported the rates of gas evolution on illuminated CdS to be the highest at low pH for formic acid and the greatest at high pH for methanol. Gas evolution rate for the photo-oxidation of formaldehyde was essentially pH independent, indicating higher (oxidizing) energy is needed for lower pH values.

Oxidation has not been found to be significant for phenol at low pH values e.g. 3.0. Since protons are potential-determining ions for TiO_2, the surface charge development is affected by pH. As organic adsorption is an important step in photocatalytic oxidation and the TiO_2 surface is positively (or protonated) charged at low pH, the rate of organic oxidation may be hindered. Moreover, a highly protonated surface can not provide hydroxyl groups which are needed for hydroxyl radical formation. Therefore a neutral pH is the most favorable for photocatalytic oxidation reactions. Similarly, in the alkaline pH region, the TiO_2 surface is deprotonated. The adsorption of anionic phenolated ions is inhibited. The extent of phenol oxidation is therefore reduced.

Concentration of Photocatalyst

Figure 8 shows the phenol decomposition as affected by the concentration of TiO_2. The concentration of TiO_2 ranged from 0.5 to 10.0 g/L. The results show that at 1.0 g/L TiO_2 can perform as well as at 10 g/L. A high catalyst concentration may create a turbidity thereby blocking the passage of light. In general the reaction can be completed in 5 hours under our experimental conditions. A zero order reaction was found for phenol decomposition which is in disagreement with others who have reported a first order reaction(24,25).

It was also noted that light intensity can determine the reaction order, i.e. zero order or first order. By decreasing the light intensity to 310 watts/m^2, a first order reaction was observed(38). A plot of the initial rate versus solid concentration yields a parabolic curve (Figure 9). It is likely that adsorption takes place between phenol and the solid under ultra violet illumination. A similar experiment was there conducted under darkness. Figure 10 shows that phenol adsorption was insignificant after 30 hours of experiment. Under ultra violet light illumination the surface property of TiO_2 can be altered. Apparently hydroxyl radicals are formed under illumination. It is speculated that hydroxyl radicals can trap the organic compound and bring about photocatalytic oxidation reaction (12,25,26,28,39). By plotting the reciprocal of the initial rate versus the ceciprocal of the amount of catalyst, a linear relationship is obtained. The relationship between the rate (M/hr) and the concentration of photocatalyst is as follows:

$$r = \frac{2.33 \times 10^{-5}\ [S]}{1 + 1.11\ [S]}$$

where [S] is the solid concentration (g/L).

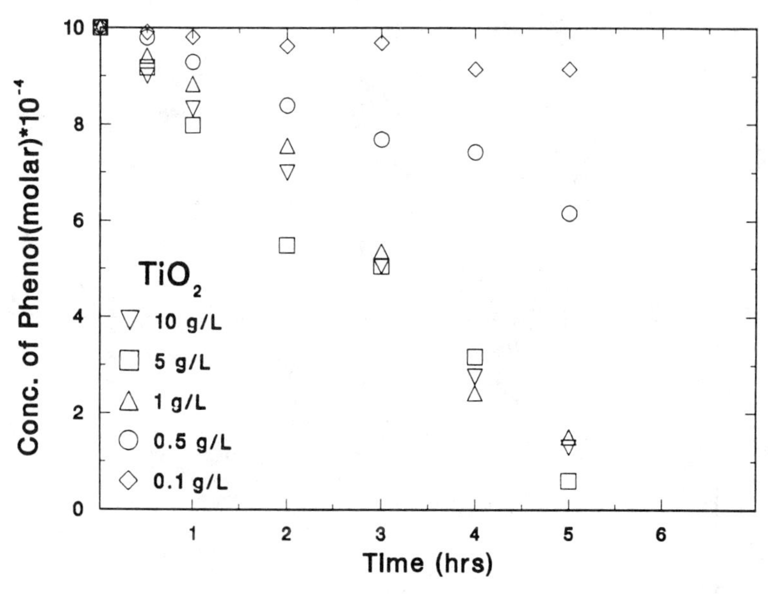

Figure 8. Effect of Catalyst Dose on the Decomposition of Phenol.

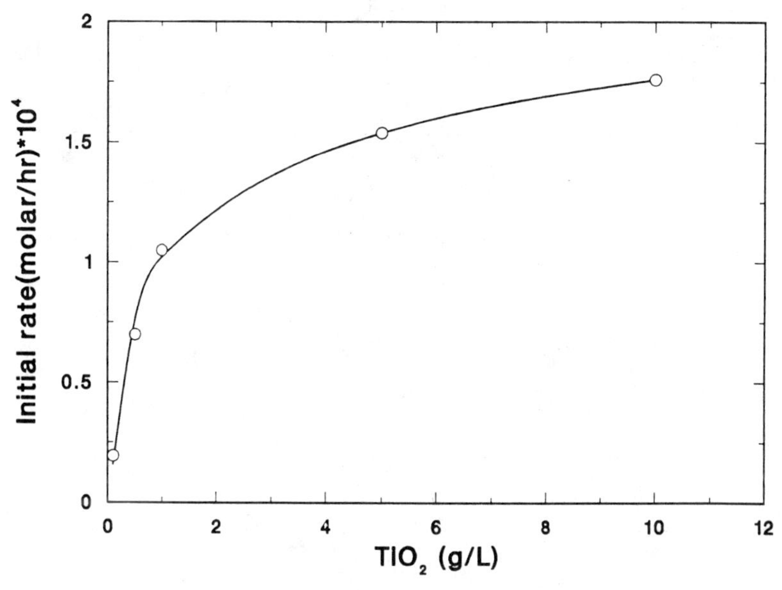

Figure 9. Initial Phenol Oxidation Rate as a Function of Solid Concentration.

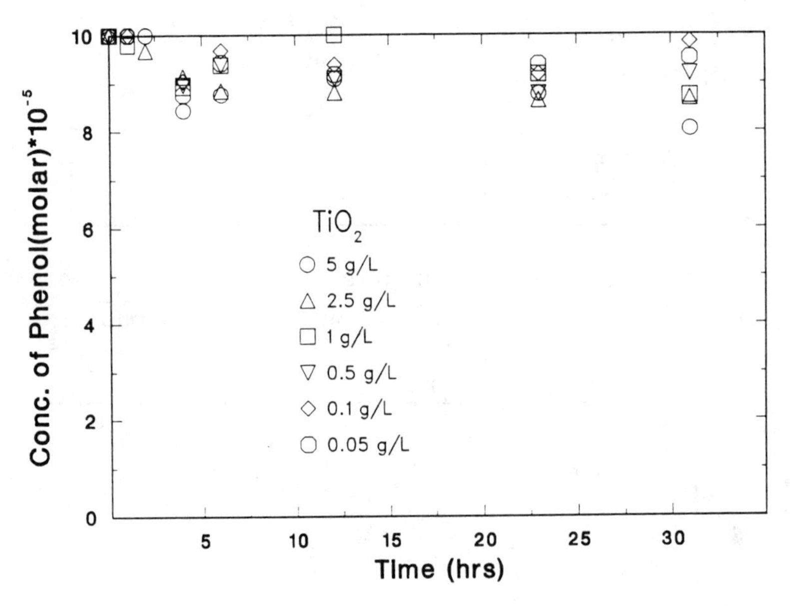

Figure 10. Adsorption of Phenol onto TiO_2 under Dark Condition.

Concentration of Phenol

The effect of phenol concentration is presented in Figure 11. The lower the concentration, the higher the efficiency of phenol decomposition. At a concentration of 10^{-4} M (9.4 mg/L), phenol can be totally oxidized within 3 hours. The results clearly demonstrate that the photocatalytic oxidation process is promising. By plotting the initial reaction rate (M/hr) versus phenol concentration (M), a straight line results (Figure 12). From Figure 12 the reaction rate in terms of phenol concentration is obtained:

$$r = 1.98\ [C]^{0.6837}$$

where C is the concentration of phenol (M).

Oxygen Effect

As shown above, it is clear that oxygen is crucial to photocatalytic oxidation. A series of experiments were then conducted using different solution volumes to create different volumes of head space in the test tube while keeping the concentration of TiO_2 constant at 10 g/L. The solution was saturated with oxygen for 4 hours prior to any experiments. Figure 13 show that oxidation reaction is slow in the absence of head space. In the absence of a head space, the concentration of oxygen in the sample was 3.2 x 10^{-5} moles (in 32 mL solution). At the end of 6 hours, the theoretical demand of oxygen for phenol oxidation is 2.2 x 10^{-5} moles. This is equivalent to a 68.8 % oxygen utilization. As the volume of solution decreases (or head space increases), the oxygen content in the liquid phase is not sufficient to oxidize the phenol present. A small head space not only impedes oxygen transfer between the gas and the liquid phase but also limits the total oxygen supply. Although it has been suggested that it is possible to form oxygen from water under photo-irradiation (40-42), this oxygen photo-generation process is too slow to match up the oxygen consumption rate during phenol decomposition. Therefore, as the volume of gas increases the reaction rate increases.

Carbon Dioxide Production

In order to verify the completeness of the photocatalytic oxidation process, the production of carbon dioxide was monitored. A 1 mL volume of gas sample was withdrawn from the test tube by a pressure lok syringe and then injected into a GC/MS with an isothermal program. An ion chrotomagraph program was used to analyze the mass spectrum at 44.0 a.m.u. for CO_2. GC/MS results are presented in Figure 14. Pure carbon dioxide (grade 5, 99.999%) was used as standard for calibration. Figure 15 gives the amount of carbon dioxide produced in moles as phenol is oxidized. The production of carbon dioxide is small during the first 3 hours when intermediates are being produced. Phenol can not be oxidized directly to carbon dioxide in this process. However at prolonged reaction time, the intermediates can be broken down to yield carbon dioxide. The carbon dioxide production rate increases during the next 3 hours. A comparison of the theoretical and experimental CO_2 production is listed in Table II.

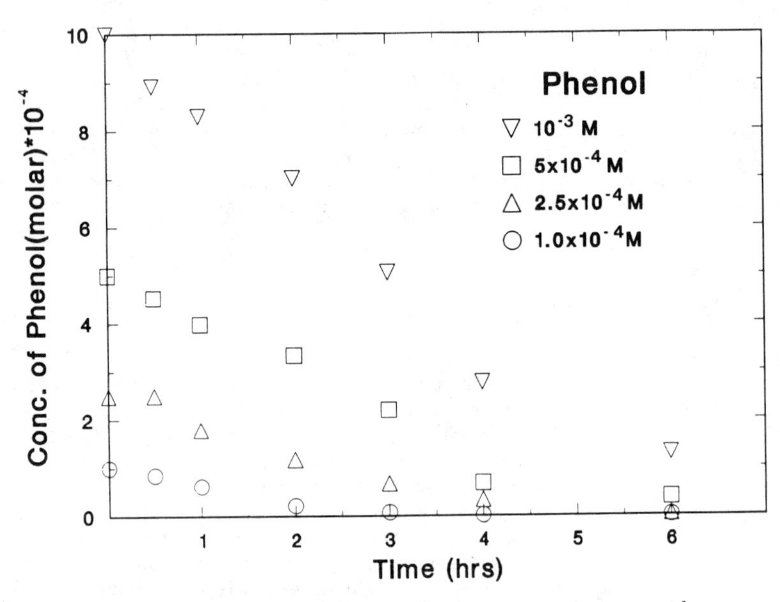

Figure 11. Effect of Phenol Concentrations on the Decomposition of Phenol.

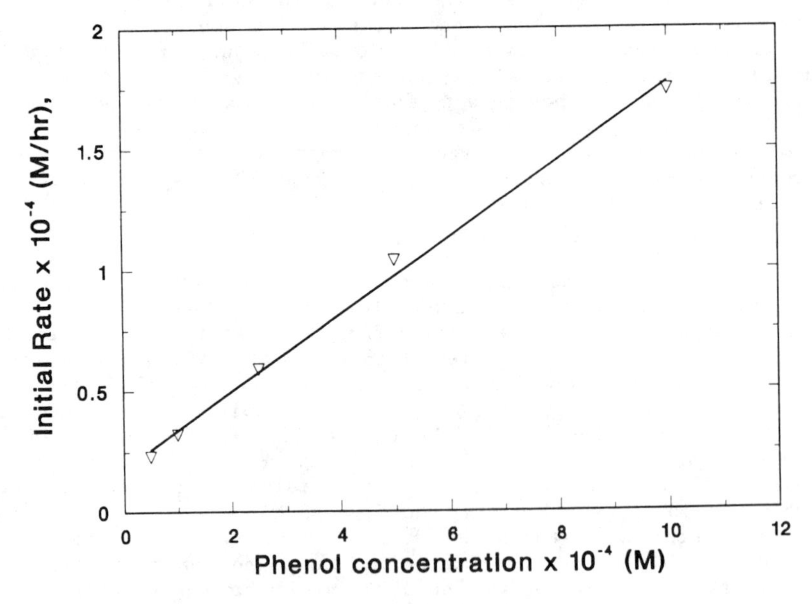

Figure 12. Initial Phenol Oxidation Rate as a Function of Phenol Concentration.

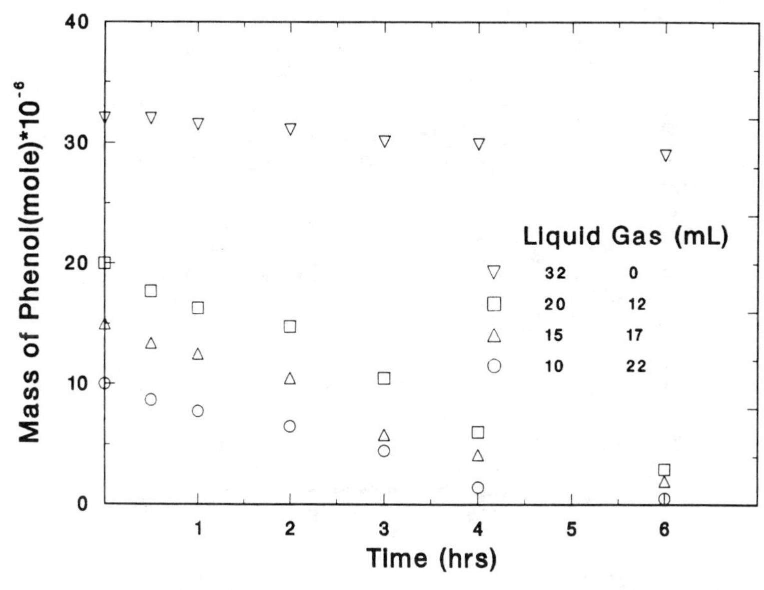

Figure 13. Effect of Gas Phase Oxygen on the Decomposition of Phenol.

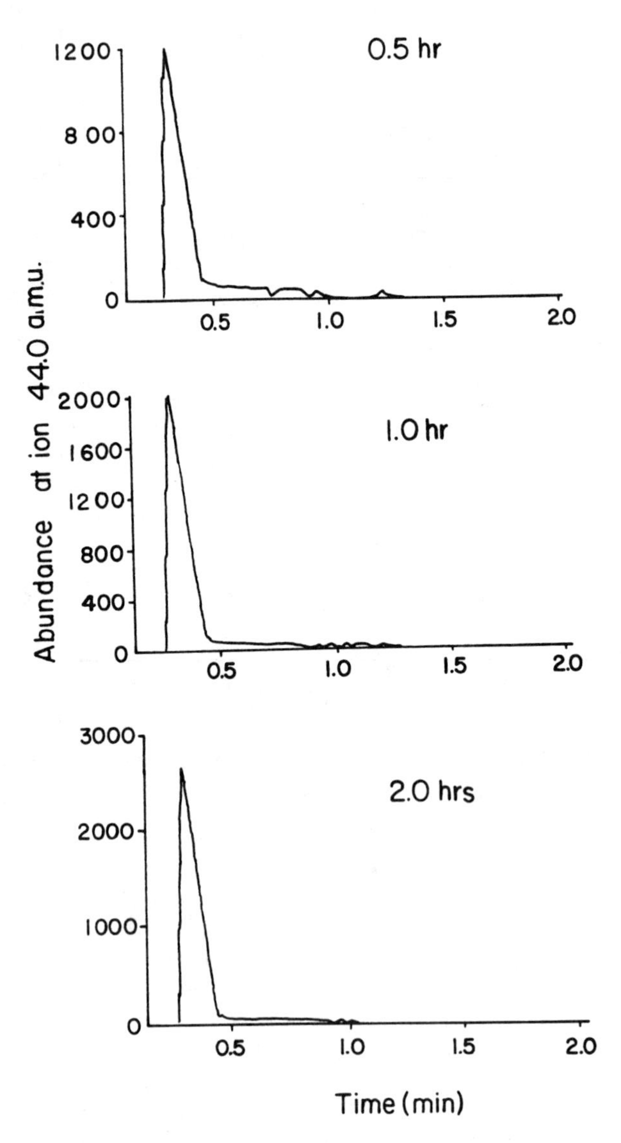

Figure 14. CO_2 Production as Detected by GC/MS.

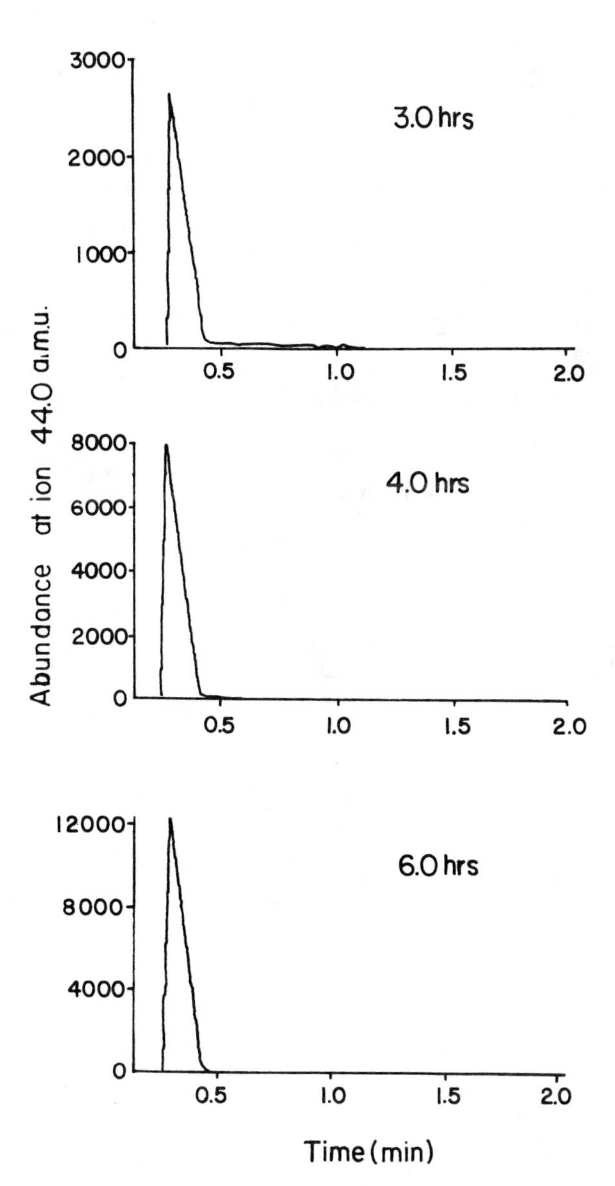

Figure 14. *Continued.*

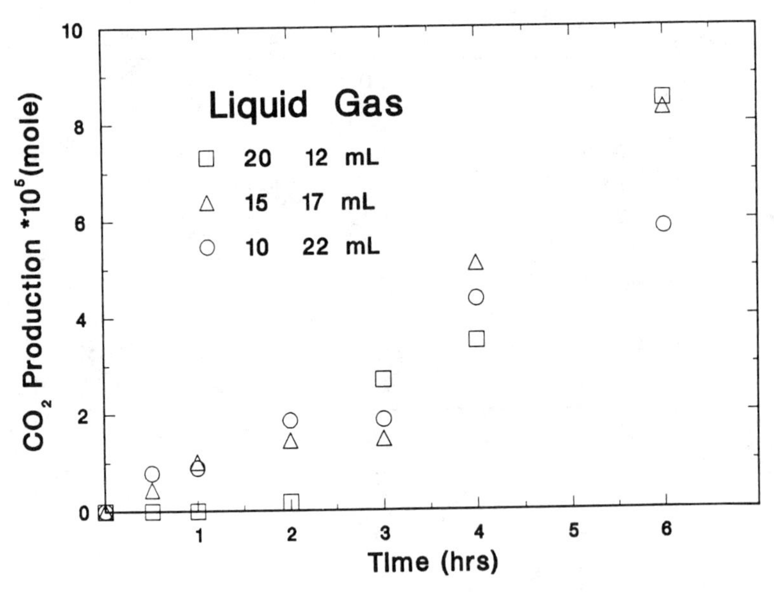

Figure 15. CO_2 Production During Phenol Decomposition.

Table II. Carbon Dioxide Production in Phenol Decomposition[a]

Time (hours)	Phenol removal (%)	CO_2 produced(10^{-5} moles) theoretical	experimental
0.5	10.8	0.97	0.44
1.0	17.0	1.53	1.01
2.0	29.9	2.69	1.44
3.0	49.5	4.46	1.47
4.0	72.4	6.52	5.10
6.0	87.1	7.84	8.29

a.15 mL phenol solution.

The result clearly suggests that phenol can not be oxidized directly to carbon dioxide during the early phases of the photocatalytic oxidation.

Intermediate Analysis

Two intermediates, hydroquinone and phenol dimer were found in a 5 x 10^{-2} M phenol solution in the course of the reaction. Figure 16 shows the intensity of phenol and intermediates with respect to time by the GC/MS technique. After 2 hours the phenol concentration remains constant which is indicative of a coating of intermediates on the surface of TiO_2. This obstruction leads to the deactivation of surface sites for photocatalytic oxidation. It must be mentioned that these intermediates are not detectable at dilute phenol concentrations, i.e. $< 10^{-4}$ M.

Mechanisms of Phenol Decomposition

Based on the results presented above, a mechanism of the photocatalytic oxidation of phenol is proposed:

$$TiO_2 + h\nu \longrightarrow e_{cb}^- + h_{vb}^+ \quad (1)$$
(excitation; electron/hole pair production)

$$e_{cb}^- \longrightarrow e_{tr}^- \quad (2a)$$
(electron migration to the surface)

$$h_{vb}^+ \longrightarrow h_{tr}^+ \quad (2b)$$
(hole migration to the surface)

$$H_2O \longrightarrow H^+ + OH^- \quad (3)$$
(water dissociation)

$$O_2 + e_{tr}^- \longrightarrow \cdot O_2^- \quad (4)$$

$$H^+ + \cdot O_2^- \longrightarrow \cdot HO_2 \quad (5)$$

$$\cdot HO_2 \longrightarrow 1/2\ O_2 + \cdot OH \quad (6)$$

$$2\ OH^- + 2\ h_{tr}^+ \longrightarrow 2\ \cdot OH \quad (7)$$

$$\cdot O_2^- + e_{tr}^- \longrightarrow \cdot O_2^{2-} \quad (8)$$

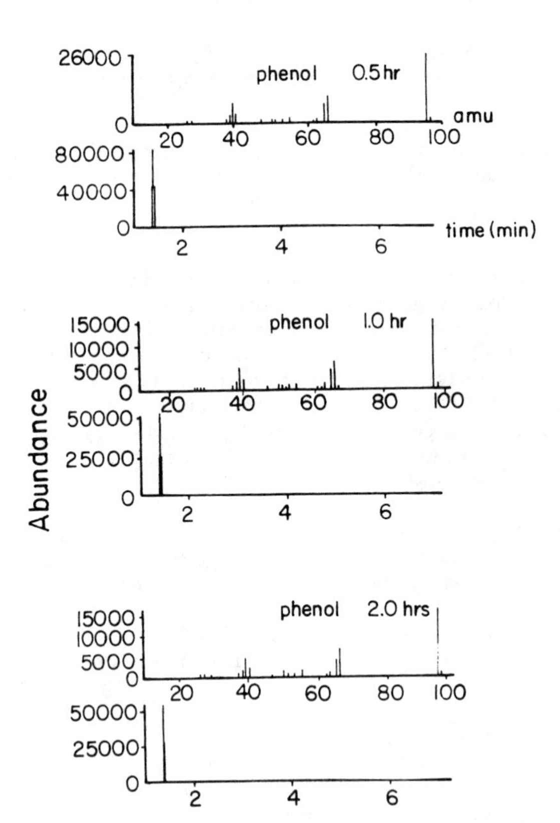

Figure 16. Intermediates of Phenol Decomposition as Detected by GC/MS.

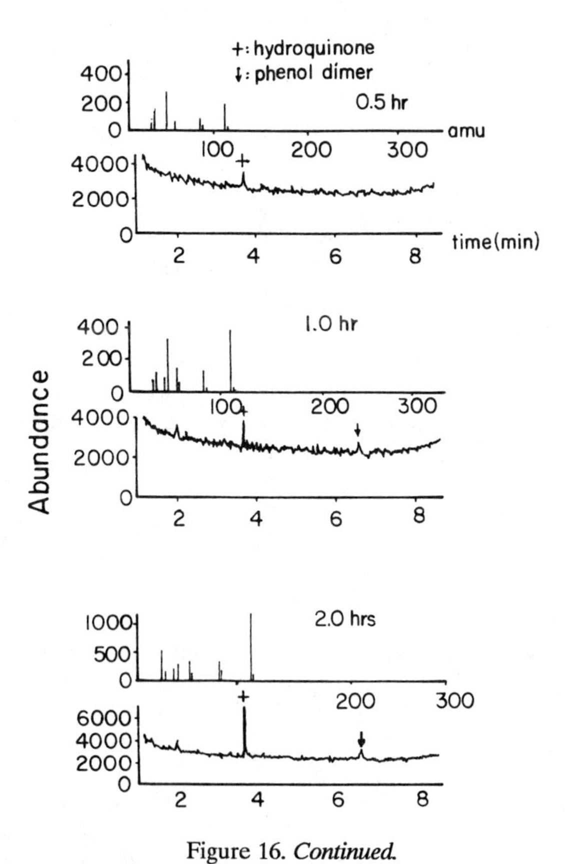

Figure 16. *Continued.*

$$C_6H_5OH \;+\; \cdot OH \;\longrightarrow\; HO{-}C_6H_4{-}OH \qquad (9)$$

$$HO{-}C_6H_4{-}OH \;+\; \cdot O_2^{\,2-} \;\longrightarrow\; O{=}C_6H_4{=}O \;+\; 2\ OH^- \qquad (10)$$

$$O{=}C_6H_4{=}O \;+\; 2\ \cdot OH \;\longrightarrow\; HOOC{-}CH{=}CH{-}COOH \qquad (11)$$

$$HOOC{-}CH{=}CH{-}COOH \;\longrightarrow\; HO - \underset{\underset{O}{\|}}{C} - \overset{\overset{H}{|}}{C} = \overset{\overset{H}{|}}{C} - \underset{\underset{O}{\|}}{C} - OH \qquad (12)$$

$$HO - \overset{\overset{H}{|}}{\underset{\underset{O}{\|}}{C}} - \overset{\overset{H}{|}}{C} = C - \underset{\underset{O}{\|}}{C} - OH \;\longrightarrow\; \overset{\overset{H}{|}}{\underset{\underset{O}{\|}}{C}} - \underset{\underset{O}{\|}}{C} - OH + HO - \underset{\underset{O}{\|}}{C} - CH \qquad (13)$$

$$\overset{\overset{H}{|}}{\underset{\underset{O}{\|}}{C}} - \underset{\underset{O}{\|}}{C} - OH \;+\; H_2O \;\longrightarrow\; HO - \underset{\underset{O}{\|}}{C} - \underset{\underset{O}{\|}}{C} - OH \;+\; H_2 \qquad (14)$$

$$HO - \underset{\underset{O}{\|}}{C} - \underset{\underset{O}{\|}}{C} - OH \;\longrightarrow\; H - \underset{\underset{O}{\|}}{C} - OH \;+\; CO_2 \qquad (15)$$

$$\mathrm{H-\overset{\displaystyle O}{\overset{\|}{C}}-OH} + 1/2\ O_2 \longrightarrow CO_2 + H_2O \qquad (16)$$

$$\mathrm{HO-\overset{\displaystyle O}{\overset{\|}{C}}-CH_3} + 2\ O_2 \longrightarrow 2\ CO_2 + 2\ H_2O \qquad (17)$$

Bickley et al.(43) have studied photoadsorption of oxygen at the surface of titanium dioxide (rutile) and proposed a dissociation chemisorption at a surface $Ti^{4+}O^{2-}$ pair leading to OH^- being retained as a surface species. These OH^- groups can be trapped by positive holes with the formation of hydroxyl radical, i.e. $OH^- + h^+ \longrightarrow \cdot OH$. The formation of hydroxyl radical has also been proposed by Izumi(11), Matthews(26) and Hashimoto(44). According to our observation, the oxygen content changed slightly in the gas phase. Therefore, the reaction step (Eq.6) is proposed to explain the generation of oxygen as a means to maintain the oxygen content in the system.

There are two possible sources for the generation of hydroxyl radicals , i. e. Eqs (6) and (7). Equation (7) shows that the consumption of hydroxyl ion results in a pH drop. Our results do show that the pH drops during the first 2 hours then remains constant for the rest of reaction period due to the generation of OH^- ions as indicated in Eq (10). Oxidation of hydroquinone by the O_2^{2-} radical leads to a more alkaline environment. Contrary to the ozonation process in which catechol and resorcinol are commonly reported, hydroquinone is the sole species observed in the photo-oxidation of phenol (Eq. 9). In order to break down the C = C bond, hydroxyl radicals must react with the o-quinone to produce muconic acid which also has been found in the ozonation process. From the mechanism proposed, it expected to find the following intermediates such as muconic acid, maleic acid, oxalic acid, formic acid and acetic acid in the system. However within the time scale of and under the present experimental conditions, it is difficult to observe these intermediates. It appears that these intermediates are relatively unstable under a photocatalytic oxidation process.

Summary

Photocatalytic oxidation of phenol is affected by pH, solid concentration, oxygen concentration and concentration of phenol. Phenol oxidation is most favorable under the neutral pH region. An empirical equation can be obtained based on solid and phenol concentration as follows:

$$r = 1.98\ [C]^{0.6837}\ (2.147 \times 10^{-4}\ [S]/\ (1 + 1.11\ [S]))$$

The rate of phenol oxidation increases with the phenol concentration. Results show that oxygen plays the most important role. Reaction between oxygen molecules and electrons with the production of oxygen radicals, e. g. O_2^- and O_2^{2-}. These highly active radicals initiate a sequence of reactions. It is also suggested that hydroxyl radicals, from reaction between positive holes and OH^- and with an oxidation potential of (45), play the major role in breaking the C = C bonds of the phenol compound.

Photocatalytic oxidation can be an effective process in degrading hazardous organic substances such as phenol. Toxic organics can be broken into an environmentally acceptable chemicals such as CO_2. TiO_2 is an inexpensive semiconductor(46) and should be affordable to many water and wastewater treatment utilities. Moreover, TiO_2 is stable upon photo-irradiation.

Acknowledgments

This research on which this paper is based was supported in part by the United States Department of the Interior as authorized by the Water Research and Development Act of 1978 (P.L. 95.467). Contents of this publication do not necessarily reflect the views and policies of the State Department of the Interior, nor mention of trade names and commercial products constitute their endorsement by the U. S. Government. We would like to thank Earl Buckley of the Mechanical Engineering Department, University of Delaware for assistance on TEM and SEM work.

Literature Cited

1. U.S., Office of Technology Assessment, Technologies of Management Strategies for Hazardous Waste Control, 1983, Washington, D. C. Government Printing Office.
2. Ellis, W. D., Payne, J. R., Tafuri, A. N. and Freestone, F. J., "The Develpoment of Chemical Countermeasures for Hazardous Waste Contaminated Soil" in 1984 Hazardous Material Spill Conference, Edited by Ludwigson, J. 1984, p118, Government Institutes, Inc. MD.
3. Jones, H. R., Environmental Control in the Organic and Petrochemical Industries, Noyes Data Corporation, Park Ridge, N. J., 1971.
4. Schiavello, M., "Basic Concepts in photocatalysis", in Photocatalysis and Environment- Trends and Applications, edited by M. Schiavello, Kluwer Academic Publishers, The Netherlands, 1987, pp 351-360.
5. Bard, A. J., "Photoelectrochemistry and Heterogeneous Photocatalysis at Semiconductors", J. Photochemistry, 1979, 10, pp 59-75.
6. Pleskov, Y. V. and Gurevich, Y. Y., Semiconductor Photoelectrochemistry, Consultants Bureau, N. Y., 1986, p29.
7. Izumi, I., Dunn W. W., Wilbourn, K. O., Fan, F. R. F. and Bard, A. J., "Heterogeneous Photocatalytic Oxidation of Hydrocarbons on Platinized TiO_2 Powders", J. Phys. Chem., 1980, 84, pp3207-3210.
8. Stadler, K. H. and Boehm, H.P., "Catalysis of the Photooxidation of Sulfurous Acid by Photodeposited Platinum, Palladium or Ruthenium on a Titanium Dioxide Carrier", Zeitschrift fur Physikalische Chemie Neue Folge, Bd. 144, 1985, S. pp 9-19.

9. Herrmann, J. and Pichat, P., Hydrogen Production by Water Photoelectrolysis with a Powdered Semiconductor Anode, Chem. Phys. Letter, 1981, 84(3), p555.
10. Pichat, P., Disdier, J. and Hermmann, T., "Photocatalytic Oxidation of Liquid (or Gaseous) 4-Tert-Butyltoluene to 4-Tert-Butylbenzaidehyde by O_2 (or Air) over TiO_2", New J. Chemistry, 1986, Vol. 10, No.10, pp545-551.
11. Izumi, I., Dunn W. W., Wilbourn, K. O., Fan, F. R. F. and Bard, A. J., "Heterogeneous Photocatalytic Oxidation of Hydrocarbons on Platinized TiO_2 Powders", J. Phys. Chem., 1980, **84**, pp3207-3210.
12. Izumi, I., Fan, Fu-Ren F. and Bard, A. J., "Heterogeneous Photocatalytic Decomposition of Benzoic Acid and Adipic Acid on Platinized TiO_2 Powder. The Photo-Kolbe Decarboxylative Route to the Breakdown of the Benzene Ring and to the Production of Butane", J. Phys. Chem., 1981, **85**(3), pp218-223.
13. Kawai, T. and Sakata, T., "Photocatalytic Decomposition of Gaseous Water Over TiO_2 and TiO_2-RuO_2 Surface", Chem. Phys. Letter, 72(1),1980, p87.
14. Kawai, T. and Sakata T., "Conversion of Carbohydrate into Hydrogenfuel by a Photocatalytic Process", Nature, 1980, **286**, 474-476.
15. Kawai, T. and Sakata, T., "Photocatalytic Hydrogen Production from Water by the Decomposition of Poly-vinylchloride, Protein, Algae, Dead Insects, and Excrement", Chemistry Letters, Chemical Society of Japan, 1981, pp81-84.
16. Kawai, T., "Examples for Photogeneration of Hydrogen and Oxygen from Water", in Energy Resource Through Photochemistry and Catalysis, (Ed Gratzel, M.) Academic Press, 1983, pp297-330.
17. Fujihira, M. et al, "Heterogeneous Photocatalytic Oxidation of Aromatic Compounds on Semiconductor Materials. The Photo-fent-on Type Reaction", Chem. Letter (Chem. Soc. Japan), 1981, p1053.
18. Fujihira, M. et al, Heterogeneous Photo-oxidation of Aromatic Compounds by Semiconductor Materials", Nature, 293, 1981, p206.
19. Fujihira, M. et al, "Heterogeneous Photocatalytic Reaction on Semiconductor Materials, J. Electroanal. Chem., 1981, **126**, p277.
20. Kraeutler, B., & Bard, A. J., "Heterogeneous Photocatalytic Decomposition of Saturated Carboxylic Acids on TiO_2 Powder. Decarboxylative Route to Alkanes", J. Amer. Chem. Soc., 1978, 100, pp 5985-5992.
21. Pavlik, J. W. and Tantayanon, S., "Photocatalytic Oxidations of Lactams and N-Acylamines", J. Am. Chem. Soc., 1981, 103, p6755.
22. Barbeni, M., Morello, M., Pramauro, E., Pelizelli, E., Vincenti, M., Borgarello, E. and Serpone, N., "Sunlight Photodegradation of 2,4,5- trichlorophenoxy Acetate and 2,4,5-trichlorophenol on TiO_2", Chemosphere, 1987, **16**, p 1165.
23. Barbeni, M., Pramauro, E., Pelizelli, E., Vincenti, M., Borgarello, E. and Serpone, N., "Photodegradation of Pentachlorophenol Catalyzed by Semiconductor Particles", Chemosphere, 1985, **14**, p195.
24. Ollis, D. F. and Pruden, A. L., "Degradation of Chloroform by Photoassisted Heterogeneous Catalysis in Dilute Aqueous Suspension of Titanium Dioxide", Environ. Sci. Technol., 1983, **17**, 628-631.

25. Ollis, D. F., Hsiao, C. Y., Budiman, L., and Lee, C. L., "Heterogeneous Photoassisted Catalysis: Conversions of Perchloroethylene, Dichloroethane, Chloroacetic Acids, and Chlorobenzenes", J. Catalysis, 88, 1984, 89-96.
26. Matthews, R. W. "Hydroxylation Reactions Induced by Near Ultraviolet Photolysis of Aqueous Titanium Dioxide Suspension", J. Chem. Soc. Faraday Trans. I, **80**, 1984, 457-471.
27. Matthews, R. W., "Photocatalytic Oxidation of Chlorobenzene in Aqueous Suspensions of Titanium Dioxide", J. Catal., 1986, **97**, p565.
28. Matthews, R. W., "Photo-oxidation of Organic Material in Aqueous Suspensions of Titanium Dioxide", Water Res., 1986, **20**, 569-578.
29. Gratzel, M. "Photocatalysis with Colloidal Semiconductors and Polycrystalline Films", in Photocatalysis and Environment-Trends and Applications, (Ed M. Schiavello) Kluwer Academic Publishers, The Netherlands, 1987, p 369.
30. Bickley, R. I., "Photoadsorption and Photodesorption at the Gas-solid Interface- Part II. Photo-electronic Effect Relating to Photochromic Changes and to Photosorption", in Photocatalysis and Environment- Trends and Applications, (Ed M. Schiavello) Kluwer Academic Publishers, The Netherlands, 1987, pp 223-232.
31. American Ultraviolet Co., Tecnical data for porta-cure irradiators., N. J., 1987.
32. Oriel Co. Light Sources Monochromators Detection Systems, vol.II, 1985, p43.
33. Standard Methods, 421 B. Azide Modification, Ed 15th, 1985, p390.
34. Phillips, R., Sources and Applications of Ultraviolet Radiation, Academic Press. N. Y. 1985.
35. Huang, C. P., A. P. Davis and J. M. Tseng, "The Removal of Some Toxic Organic Chemicals from Contaminated Groundwaters by Photocatalytic Oxidation", Project annual report to DOI, Grant number:1408001-G1216, 1987.
36. Hidaka, H., Kubota, H., Grätzel, M., Pelizzetti, E., & Serpone, N., "Photodegradation of Surfactants II: Degradation of Sodium Dodecylbenzene Sulphonate Catalysed by Titanium Dioxide Particles", J. Photochem., 1986, 35, pp 219-230.
37. Matsumura, M., Hiramoto, M., Iehara, T., & Tsubomura, H., "Photocatalytic and Photoelectrochemical Reactions of Aqueous Solutions of Formic Acid, Formaldehyde, and Methanol on Platinized CdS Powder and at a CdS electrode", J. Phys. Chem., 1984, 88, pp 248-250.
38. Tseng, J. M. and C. P. Huang, "Photocatalytic Oxidation of Phenol by Al_2O_3-TiO_2", to be summitted to Water Research, 1989.
39. Okamoto, K., Yamamoto, Y., Tanaka, H., Tanaka, M. and Itaya, A., "Heterogeneous Photocatalytic Decomposition of Phenol over TiO_2 power", Bull. Chem. Soc. Japan., 1985, 58, pp 2015-2022.
40. Gratzel, M. "Molecular Engineering in Photoconversion Systems", in Energy Resource Through Photochemistry and Catalysis, (Ed Gratzel, M.) Academic Press, 1983, pp 71-98.

41. Zamaraev, K. I. and Parmon, V. N., "Development of Molecular Photocatalytic Systems for Solar Energy Conversion: Catalysts for Oxygen and Hydrogen Evolution from Water", in Energy Resources Through Photochemistry and Catalysis, (Ed Gratzel, M.) Academic Press, 1983, pp 123-162.
42. Kiwi, J., "Examples for Photogeneration of Hydrogen and Oxygen from Water", in Energy Resource Through Photochemistry and Catalysis, (Ed Gratzel, M.) Academic Press, 1983, pp 297-330.
43. Bickley, R. I. and Stone, F. S., "Photoadsorption and Photocatalysis at Rutile Surface. I. Photoadsorption of Oxygen", J. of Catal., 1973, 31, pp389-397.
44. Hashimoto, K. Kawai, T. and Sakata, T., "Hydrogen Production with Visible Light by Using Dye-Sensitized TiO_2 powder", Nouveau J. De Chimie, 1987, 7, 249-253.
45. Rice, R. G. and Gomez-Taylor, "Oxidation Byproducts from Drinking Water Treatment", in Treatment of Drinking Water for Organic Contaminants, (Ed Huck, P. M. and Toft) P. Pergamon Press, 1986, p 109.
46. Key Chemicals & Polymers, 7th Ed., Chemical & Engineering News, American Chemical Society, Washington D. C., 1985, p43.

RECEIVED December 1, 1989

Chapter 3

Falling-Film and Glass-Tube Solar Photocatalytic Reactors for Treating Contaminated Water

James E. Pacheco and John T. Holmes

Solar Thermal Collector Technology Division 6216, Sandia National Laboratories, Albuquerque, NM 87185

Photocatalytic destruction of a model compound has been demonstrated with bench scale and with two engineering scale experiments. For the engineering scale experiments two solar concentrating hardware systems were used: a glass-tube/trough system and a falling-film/heliostat system. The ultraviolet (UV) power on the reactors of the bench scale, glass-tube/trough and falling film/heliostat systems were 0.15, 1600, and 5200 watts of UV. In the bench scale experiments, the use of hydrogen peroxide as an oxidant with a titanium dioxide catalyst under ultraviolet illumination yielded reaction rates significantly greater than without the oxidant. The thermal performances of both solar systems are favorable for photocatalytic processes, since destruction of the contaminant can take place before boiling occurs. The trough / glass tube system was evaluated in clear and partly cloudy weather, and the amount of time to attain the same level of destruction was proportional to the total incident solar energy. With the heliostat / falling-film system, experiments were performed at two power levels. The concentration of the model compound was reduced below detectable limits in less than 15 seconds of exposure. Preliminary data indicate initial reaction rates are linear with ultraviolet intensity, though further experimental studies need to be performed to verify this relation.

As much as 1% of the usable groundwater in the U.S. may be contaminated according to estimates by the Environmental Protection Agency (EPA) (1). With over 200 million tons of hazardous waste in the form of contaminated water being generated each year, the Department of Energy (DOE) has designated groundwater detoxification as its number one hazardous waste priority (2). Much of the contaminated groundwater has aqueous pollutants that are in low concentration (parts per million range) but still exceed EPA's contamination limits by orders of magnitudes.

Studies (2-13) have demonstrated destruction of dilute organic chemicals in water by using low intensity ultraviolet (UV) light with titanium dioxide catalysts. When photons of sufficient energy are absorbed on the catalyst surface, electrons and holes are produced which form very powerful oxidizing agents: peroxide and hydroxyl radicals. These radicals have been found to completely mineralize a large number of organic compounds including solvents, PCB's, dioxins, pesticides, dyes, etc., resulting in only carbon dioxide, water, and dilute acids as the products (4-6).

$$C_xH_yO_zCl_w + O_2 \xrightarrow[TiO_2]{Solar\ UV} CO_2 + H_2O + HCl$$

Most of these studies have concentrated on using **low intensity** UV light with either titanium dioxide or oxidizing agents (11) (e.g., H_2O_2 and/or O_3) but little attention has been paid to the combination of titanium dioxide with an oxidizing agent such as hydrogen peroxide along with **concentrated solar** UV light.

A potential application of this process is decontamination of groundwater. The advantage of this process over others is it provides a method for actually destroying the pollutant and not just transforming it from one phase to another as is the case of some conventional methods, such as air stripping, that move the contaminant from the water to air.

The objective of our program, a joint effort between Sandia National Laboratories and the Solar Energy Research Institute (SERI), is to develop an efficient and economical technology to detoxify groundwater or dilute waste streams utilizing solar energy. We determined photocatalytic reaction rates of a model compound for two engineering scale solar concentrating hardware systems, a **trough / glass-tube** system and a **heliostat / falling-film** system. We also conducted bench scale experiments with low-intensity ultraviolet light to guide the engineering scale experiments. Titanium dioxide (Degussa-P25, anatase form) was used as a catalyst since it has demonstrated high photocatalytic activity in previous studies (13).

Experiment.

At the Solar Thermal Test Facility at Sandia National Laboratories in Albuquerque, NM, two engineering scale hardware configurations were used to perform solar detoxification experiments in addition to bench scale experiments. For the bench scale experiments we used a Spectronics ENF-240C ultraviolet lamp (365nm) with an output of about 0.15 W centered over a 10 cm diameter Pyrex brand petri dish with 75 ml of solution on top of a stirrer/hotplate. The UV intensity was approximately 2 watts of UV per liter. We varied the loading of catalyst, the oxidant, and temperature.

The first solar system used line focusing parabolic troughs, having a 90 degree rim angle and aluminized reflective surfaces which reflect 69% of the ultraviolet. The troughs have a width of 2.1 meters, and a concentration of 51 "suns". The length of the trough is 36.4 meters for an aperature area of 77.5 square meters (15). The reactor is a 3.2 cm diameter borosilicate glass pipe mounted at the focus of the trough. Borosilicate glass transmits about 90% of the ultraviolet. For this trough system the calculated ultraviolet power was 1600 W of UV for an intensity of 38 watts-UV per unit of reactor volume accounting for losses in the reflective film and the borosilicate glass. We circulated 1100 liters (300 gallons) of water at a flowrate of 130 liters/minute (34 gpm). The residence time in the trough was 20 seconds. Figure 1 is a diagram of this system.

The second system uses large tracking mirrored structures, 37 square meters each, called heliostats to reflect and concentrate solar radiation onto the falling-film reactor, mounted in the solar tower 37 meters above ground (14). The falling-film reactor is a vertical aluminum panel 3.5 m high by 1.0 m wide, in which a liquid flows down the face of the panel in the form of thin, wavy film approximately 3 mm thick for a volume of about 12 liters in the reactor. (The reactor film thickness may not be optimal since a thicker film may provide more absorption of the ultraviolet, though, this was not a variable in the experiment). In the falling-film experiments we circulated 380 liters (100 gallons) of solution through the reactor. The residence time in the falling film was approximately 2.9 seconds per pass. The heliostats have silvered mirrors which reflect about 66% of the solar ultraviolet (300-400nm). We performed experiments at two power levels. In the first experiment with the falling-film, we had approximately 2600 W of UV on the panel or 220 watts of UV per liter of reactor volume. In the second falling-film experiment, the panel was exposed to 5200 watts of UV or 440 watts of UV per liter of reactor volume.

We operated the falling-film at a flowrate of 17 liter/minute (65 gpm). A diagram of the falling-film is shown in Figure 2.

For all three systems, we started with a 30 ppm solution of a model compound, salicylic acid, ($C_7H_6O_3$), chosen for its ease of analysis, and relative complex molecular structure. We used neutral (pH=7)deionized or tap water. Although salicylic acid is not a contaminant, its reaction rate with the UV/TiO_2 process is similar to actual contaminants (3). We varied parameters such as oxidant concentration (100-450 ppm H_2O_2), temperature, and catalyst loading to determine the effect on the destruction rates of the model compound. For the solar systems, the catalyst, titanium dioxide (Degussa-P25 grade, anatase, with an average particle size of 30 nm), was suspended with a mixer in holding tanks and pumped through the reactors.

We analyzed the water samples by UV absorption using a Perkin Elmer UV/VIS spectrophotometer after filtering out the titanium dioxide. The uncertainty in the measurements of the salicylic acid concentration was ± 0.5 ppm and the detection limit was approximately 1 ppm. No measurements of the products of reaction were made in this early stage of the project due to the lack of analytical equipment.

Results and Discussion.

Photocatalytic destruction of the model compound, salicylic acid was observed in all three experimental configurations.

Bench Scale Experiments. The catalyst loading for the bench scale experiments varied from 0 to 0.1%, the hydrogen peroxide (oxidant) concentration ranged from 0 to 300 ppm, and the temperature was varied between 23°C (ambient) and 65° C. The results of these experiments are presented in Figures 3 and 4.

The catalyst loading affected the reaction rate significantly. We found little change in concentration when either the catalyst or the UV light was absent. Slight evaporation caused the concentration to increase in a few runs. With UV and a catalyst loading of 0.1% the initial reaction rate, determined from the initial slope, was 3 times the rate with a 0.02% catalyst loading and UV. With the addition of hydrogen peroxide, reaction rates increased by about an order of magnitude. For example, the combination of UV, 0.1% catalyst and 100 ppm hydrogen peroxide (stoichiometric amount of oxidant) yielded an initial reaction rate that was about 8 times greater than with UV and 0.1% catalyst alone. With 300 ppm H_2O_2, the reaction rate was 10 times greater

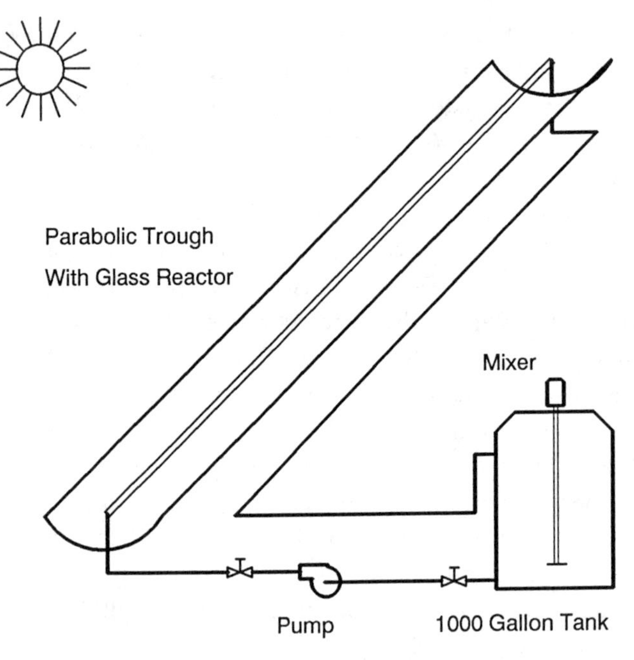

Figure 1. Schematic of parabolic trough system used for solar detoxification studies.

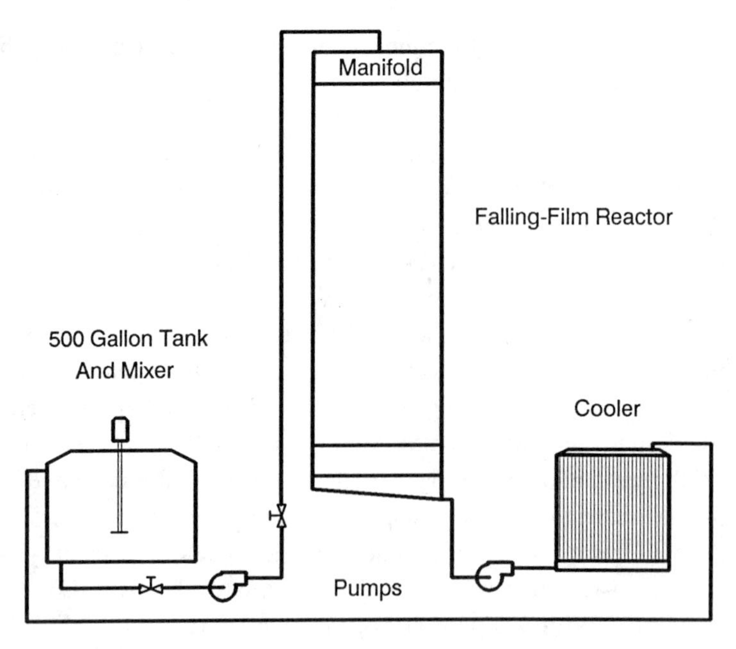

Figure 2. Schematic of falling-film reactor used with heliostats.

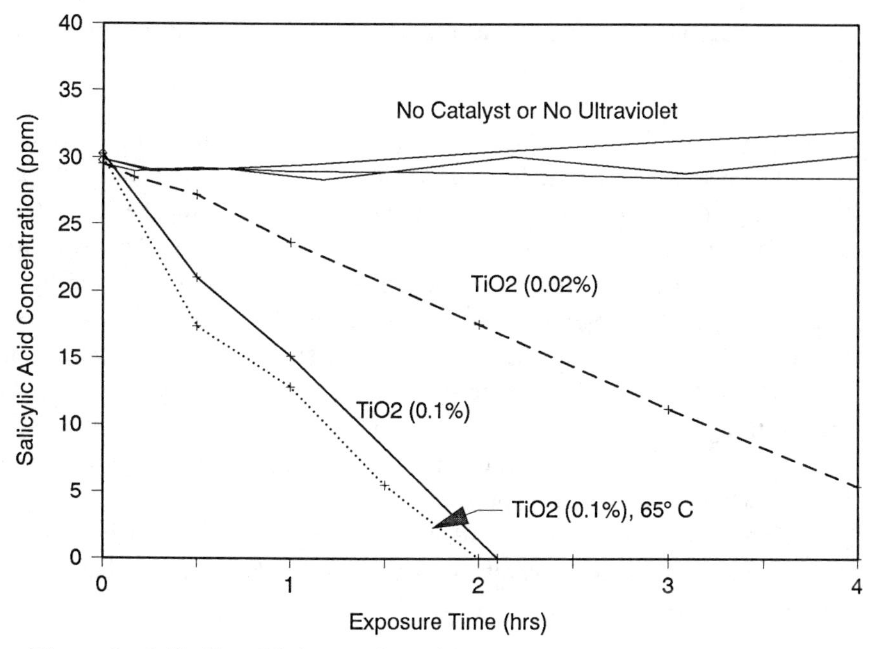

Figure 3. Salicylic acid destruction with bench scale UV light and titanium dioxide.

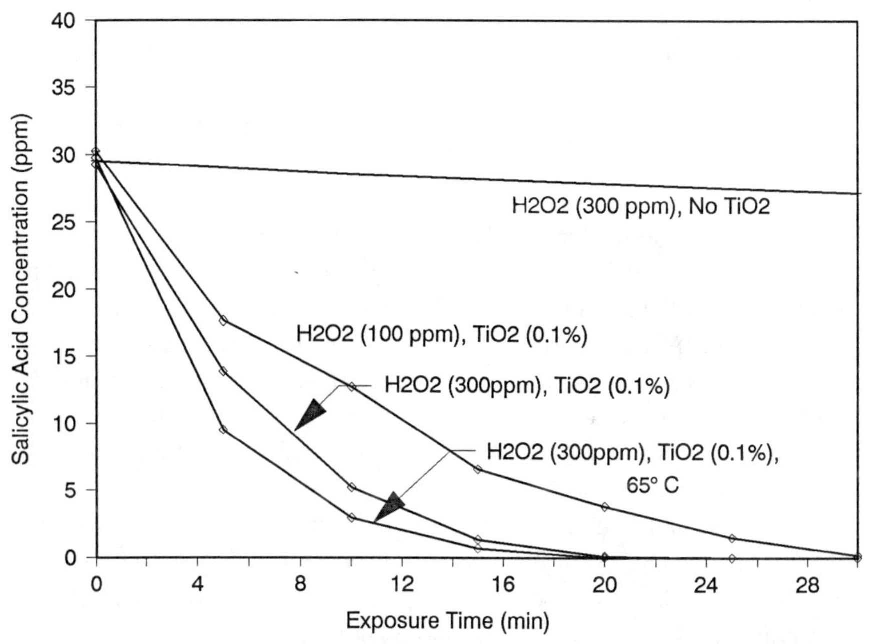

Figure 4. Salicylic acid destruction with hydrogen peroxide, titanium dioxide and bench scale UV light.

than without H_2O_2. Heating the solution to 65°C increased the initial rate sighly more.

Glass-Tube/Trough System. For the trough system we started with 1100 liters of a 30 ppm solution of salicylic acid with 0.1% titanium dioxide and 300 ppm hydrogen peroxide. The destruction results for this test and for the falling film are shown in Figure 5, scaled to represent the *exposure* time. The exposure time is the amount of time the solution is actually exposed to the light, calculated by multiplying the *testing* or *run* time by the ratio of the volume of the reactor to the total volume (11 gallons/300 gallons). In other words, for every 30 minutes of testing time the solution was exposed for about 1.1 minutes for the trough system. As can be seen from Figure 5, the concentration was reduced below 1 ppm, our detection limit, in about 2 minutes of exposure. The initial reaction rate was 32 ± 4 ppm/min.

The uncertainty of the reaction rates is determined by partial derivatives, where the reaction rate, R, is a function of C_1-concentration at $t=t_1$, C_2 - concentration at $t=t_2$, and $\Delta t = t_2 - t_1$:

$$R = \frac{C_1 - C_2}{\Delta t}$$

The uncertainty is:

$$\omega_R = \left[\left(\frac{\partial R}{\partial C_1}\omega_{C_1}\right)^2 + \left(\frac{\partial R}{\partial C_2}\omega_{C_2}\right)^2 + \left(\frac{\partial R}{\partial \Delta t}\omega_{\Delta t}\right)^2\right]^{1/2}$$

where ω_R, ω_{C1}, ω_{C2}, and $\omega_{\Delta t}$ are the measurements uncertainties of the reaction rate, C_1, C_2, and Δt, respectively.

Falling-Film/Heliostat System. The setup for the two falling film experiments was the same except for the UV intensity: 380 liters of 30 ppm solution of salicylic acid with 0.3% titanium dioxide and 450 ppm hydrogen peroxide. For the first falling-film experiment, the solar UV input was approximately 220 watts of UV per liter of reactor. In an exposure time (run time x 12 liters /380 liters) of about 25 seconds, the concentration was reduced to below 1 ppm as can be seen in Figure 5. The initial reaction rate for this test was about 110 ± 15 ppm/min.

The solar UV input for second falling film experiments was about twice the first experiment. At this UV intensity the concentration was reduced below 1 ppm in approximately 12 seconds of exposure. The initial reaction rate at this higher intensity was approximately 310 ± 30 ppm/min.

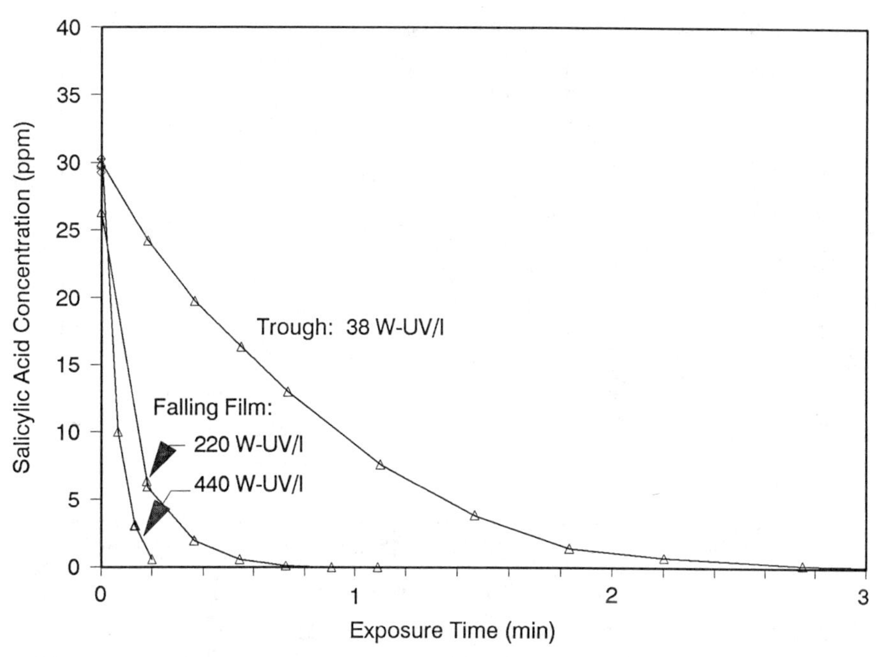

Figure 5. Destruction results with glass-tube/trough and falling-film/heliostat systems.

Thermal Characterization. The temperatures of the glass-tube / trough and falling-film /heliostat systems were not controlled, though a cooling system can easily be installed. In this photocatalytic process only the UV (300-400 nm) portion, which constitutes 3-5% of the solar spectrum is used and the rest of the spectrum is either reflected, scattered or absorbed as heat. In the glass tube/trough system, we determined from the temperature rise across the reactor that about 32% of the effective incident solar power was absorbed, although a much higher percentage of the UV is probably absorbed since titanium dioxide has a strong absorption band in the UV. The temperature of the solution in the tank rose from about 7°C to 53°C. The thermal characterization of the falling-film reactor was completed previously and is described in another paper (16). Between 30 and 50% of the incident power was absorbed on the panel. Again, a higher percentage of UV was likely absorbed.

Performance in Partly Cloudy Weather. The performance of the trough system was evaluated in both clear and partly cloudy weather. The direct normal insolation (DNI) averaged 1032 W/m^2 over the *testing* time for the clear day and 484 W/m^2 on the partly cloudy day. The DNIs for these days are plotted in Figure 6. The results of the change in concentration is shown in Figure 7.

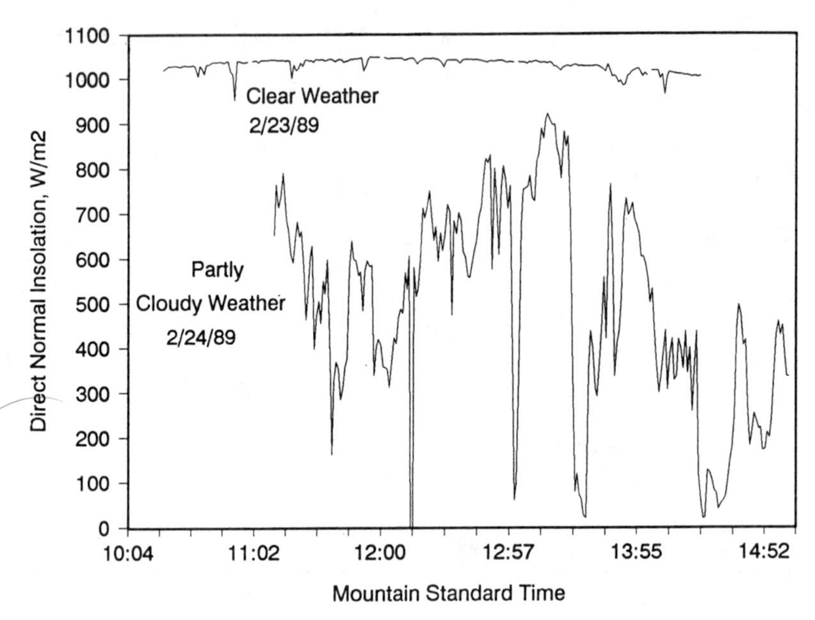

Figure 6. Direct normal insolation during testing periods on clear and partly cloudy days.

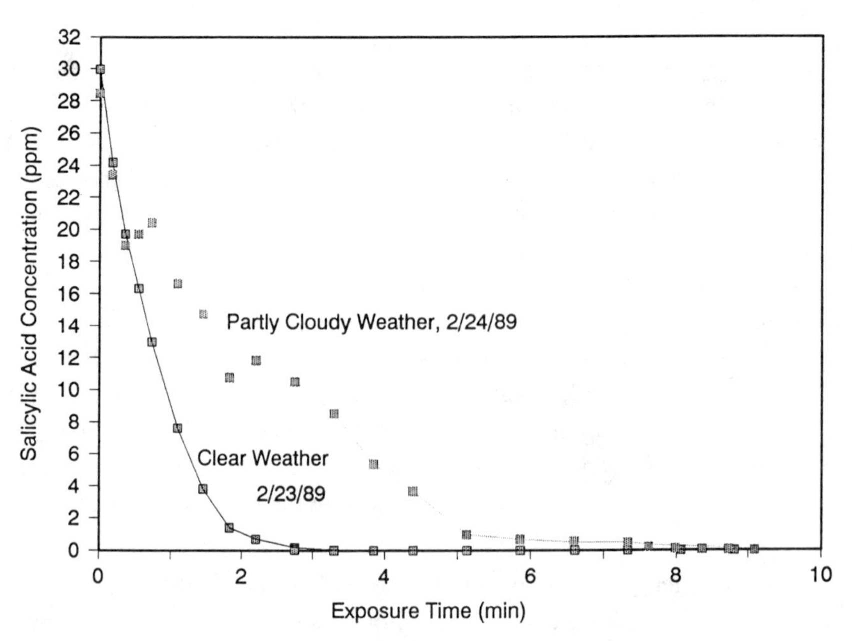

Figure 7. Comparison of destruction of salicylic acid during clear and cloudy weather. Note the time scale is for exposure time.

Under cloudy conditions the concentration of salicylic acid was reduced below 1 ppm in about 5 minutes of *exposure* time, while in clear weather conditions only 2 minutes of exposure were required to obtain the same destruction. This result is as expected from the ratio of the insolations. The concentration appeared to increase during low insolation on partly cloudy weather, but this is due to measurement uncertainties and due to variations in homogeneity of the solution in the tank.

Effect of UV intensity on Reaction Rate. The performance of these three systems can be examined by comparing their initial reaction rates to the specific UV power (watts-UV per liter of reactor) as presented in Figure 8. The reaction rate appears linear with UV power, though the uncertainty in the data is large and more data are needed to establish a relation.

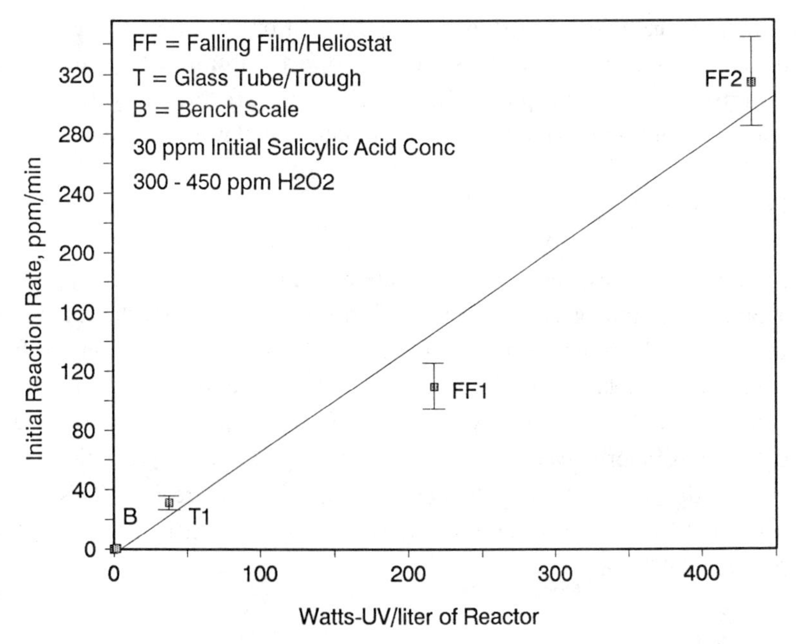

Figure 8. Initial reaction rates for the bench scale, glass-tube/trough and falling-film/heliostat systems.

Conclusions.

We have successfully shown the feasibility of destroying an organic compound, salicylic acid, in water using concentrated sunlight with glass-tube /trough and falling-film / heliostat systems. In the bench scale experiments, we found that the combination of an oxidant and the catalyst under illumination increased the reaction rate by an order of magnitude over that without the oxidant. Destruction of the model compound was observed in less than 15 seconds of exposure with the falling-film reactor. Destruction of salicylic acid with the glass/tube trough system was observed in clear and partly cloudy weather and found to be proportional to the average direct normal insolation. The reaction rates appear to be linear with UV power, though more data are needed to better quantify the relationship. No analysis of the products were conducted in this early stage of the project due to lack of analytical equipment.

We are optimizing the process for each reactor system by varying the catalyst loading and method of support (suspended or fixed), the oxidant type and concentration, and by enhancing the catalyst activity. We plan to evaluate the effect of other parameters such as catalyst deactivation, scavengers, turbidity, and catalyst recovery or immobilization on the process effectiveness and economics. With each system we hope to obtain a high degree of destruction in a single pass. In addition, we are planning tests with trichloroethylene and textile dyes and eventually actual waste samples.

Acknowledgments.

We would like to express our appreciation to Dr. Craig Tyner for providing assistance in planning the experiments and for the data from the falling-film system and to the following people for their efforts and cooperation put forth in assembling the hardware, installing the instrumentation, and operating the systems: Earl Rush, Chauncey Matthews, Larry Yellowhorse, Vern Dudley, and Walter Einhorn. We would also like to express our appreciation to Dr. Daniel J. Alpert for providing the helpful suggestions and review. This work is supported by the U.S. Department of Energy under contract DE-AC04-76DP00789.

Literature Cited.

1. Pye, V.I.; Patrick, R.; Quarles, J.; Groundwater Contamination in the United States, Univ. of Pennsylvania Press, 1983.
2. DOE Solar Thermal Technology Program, Sandia National Laboratories and Solar Energy Research Institute, Quarterly Progress Report, First Quarter Fiscal 1989.

3. Matthews, R. W., Sol. Ener., 1987, no. 6, pp. 405-13.
4. Ollis, D. F., Envir. Sci. Tech., 1985, 19, no. 6, pp. 480-84.
5. Matthews, R. W., Wat. Res., 1986, 20, no. 5, pp. 569-78.
6. Ahmed, S.; Ollis, D., Sol. Ener., 1984, 32, no. 5, pp. 597-601.
7. Hsiao, C.; Lee, C.; Ollis, D., J. Cat., 1983, 82, pp. 418-23.
8. Matthews, R. W., J. Cat., 1986, 97, pp. 565-68.
9. Al-Ekabi, Hussain; Serpone, Nick, J. Phys. Chem., 1988, 92, no. 20, pp. 5726-31.
10. Yamagata, S.; Nakabayashi, S.; Sangier, K.; Fujishima, A., Chem. Soc. Jap., 1988, 61, no. 10, pp. 3429-34.
11. Borup, M. B.; Middlebrooks, E. J., Wat. Sci. Tech., 1987, 19, pp. 381-90.
12. Tyner, C. E.; Pacheco, J. E.; Haslund, C. A.; Holmes, J. T., Proc. Haz. Mat. Man. Conf., Rosemont, Illinois, March, 1989.
13. Barbeni, M.; Pramauro, E.; Pelizzetti, E.; Chemosphere, 1986, 15, no. 9, pp 1913-16.
14. Maxwell, C.; Holmes, J. T., Central Receiver Test Facility Experimental Manual, Sandia National Laboratories, Albuquerque, NM and Livermore, CA, SAND86-1492, January 1987.
15. Cameron, C. P.; Dudley, V. E., The BDM Corporation Modular Industrial Solar Retrofit Qualification Test Results, Sandia National Laboratories, Albuquerque, NM, SAND-85-2317, April 1987.
16. Holmes, J. T.; Haslund, C. A., Proc. Dep. Ener. Mod. Conf., Oak Ridge, TN, October, 1988.

RECEIVED December 29, 1989

Chapter 4

Supported Catalysts in Hazardous Waste Treatment

Destruction of Inorganic Pollutants in Wastewater with Hydrogen Peroxide

Randy A. Bull and Joseph T. McManamon

FMC Corporation, Peroxygen Chemical Division, P.O. Box 8, Princeton, NJ 08543

Inorganic pollutants in wastewater can be oxidatively destroyed with hydrogen peroxide using a supported metal catalyst system. Laboratory work has demonstrated the effectiveness of these heterogeneous catalysts in oxidizing reduced sulfur compounds (e.g., sulfide, thiosulfate) more rapidly and with better efficiency than conventional homogeneously catalyzed processes. These catalysts are also effective for oxidizing and detoxifying mercaptans and cyanide in wastewater. Metals and supports studied include transition metals on silica, alumina, and various natural and synthetic zeolites. The choice of metal and support varies with the pollutant. A significant advantage of heterogeneous catalysts is that they are not subject to the pH limitations often found with soluble catalysts, working equally well in acid and alkaline conditions. Further, they can be used in treating continuously flowing wastewater streams without the need for continuous addition of metal salts. This eliminates the possibility of secondary pollution problems due to the added metal and reduces cost. The potential for using this catalyst concept for treating a wider variety of toxic pollutants will be discussed.

Inorganic pollutants such as sulfide, thiosulfate, cyanide, etc. are commonly found in a variety of wastewater streams. Because of their toxicity and effect on the environment they have been the subject of much process

0097–6156/90/0422–0052$06.00/0

development for their removal from wastewater. These processes use various techniques, including precipitation, complexation, and oxidation. Oxidation processes, unlike precipitation or complexation reactions, result in the destruction of the pollutant rather than its conversion to another form which may require secondary disposal or treatment.

Hydrogen peroxide is one oxidant that has seen successful commercial application in treating these pollutants. However, like all wastewater treatment technologies there are aspects of the treatment process that can be improved by the introduction of new technology and concepts. For example, one problem associated with the use of peroxide for treatment of pollutants like sulfide and cyanide has been the need for catalysts or pH adjustment to obtain effective oxidation. Treating large volumes of water makes pH adjustment a difficult and often costly step. The use of a soluble catalyst requires the addition of a metal to the waste stream. Most metal catalysts for H_2O_2 are transition metals and some, for example, copper, may be considered toxic. This can cause secondary treatment problems, requiring removal of the catalyst from the treated stream, and can also be costly because catalyst must continuously be added. In addition, the added metals could produce unwanted sludge that would require disposal.

An alternative to soluble metal catalysts is the use of solid or supported catalysts. Up to now little has been reported on the use of such materials for waste treatment (1-3). Our work has demonstrated that metal catalysts supported on a solid are effective for destroying inorganic pollutants in wastewater and do not suffer the limitations often found with soluble catalysts.

Experimental

Catalyst Preparation. Commonly available catalyst supports were used. Mordenite was washed with 0.02M NaOH to obtain the sodium form. Zeolites 13X and LZY52 were purchased as the sodium form. The zeolites were used as either extrudate (1/4" x 1/16") or as 14 x 30 mesh size obtained by grinding in a mortar.

Metal ion exchanges were carried out by agitating 50 g of the zeolite support in 100 mL of a 10 to 20 mM solution of metal ion at 50°C for 15 minutes to 1 hour. The solution was decanted and the solid washed well with water and dried at 110°C. This procedure resulted in catalysts containing about 0.2% w/w of the metal.

In general experiments were carried out by either batch or continuous procedures. Specific parameters are noted in the tables.

Batch Experiments. A solution of the pollutant of the desired concentration in deionized water was prepared and the pH adjusted to the desired value with NaOH or H_2SO_4. The H_2O_2 was added, followed by pH readjustment if necessary and the catalyst added to the solution to commence the reaction. Aliquots were taken periodically for analysis.

Continuous Experiments. A jacketed glass column was filled with about 20 g of the catalyst. A solution of the pollutant was prepared, pH adjusted with NaOH or H_2SO_4 and this solution pumped at a constant rate upflow through column bed. Temperature was regulated by room temperature water circulation through the column jacket. Aliquots were taken from the column effluent for analysis.

Analysis. Analyses of initial and residual pollutant were carried out using titrimetric, colorimetric or potentiometric procedures. Titrimetric procedures for thiosulfate, mercaptans, and cyanide are standard procedures(4). Colorimetric analyses based on those methods were facilitated by using LaMotte-Pomeroy or Chemetrics kits for sulfide and Chemetrics kits for thiosulfate. Potentiometric methods for sulfide and cyanide are found in Corning or Orion specific ion electrode manuals. Occasionally analyses by one method were verified by a second method.

Detailed Sulfide Experiments. Ni/Y and Fe/Y catalysts were evaluated in more detailed experiments at conditions similar to those of a geothermal power plant. The reaction parameters studied were reaction time (residence time in the column), H_2O_2/sulfide molar ratio, and temperature.

Sulfide feed solutions of ~90 mg/L were prepared from a stock sulfide solution of ~0.7% S^{2-} using deionized water that had been previously purged for at least one hour with nitrogen to remove oxygen. The pH was adjusted to 8.9 with 1 N H_2SO_4 and the exact sulfide concentration determined potentiometrically. This solution was pumped upflow through the catalyst column, which contained 7 g of supported catalyst. A 2.5% H_2O_2 solution was added to the sulfide stream through a "T" just prior to entering the column at a rate that resulted in the desired stoichiometry. This minimized the contact time between sulfide and peroxide before contacting the catalyst. Runs at 24°C were actually run at ambient temperature, which varied from 23-25°C. Temperature in the 38°C run was maintained by circulating hot water through the column jacket.

Effluent was analyzed for sulfide by collection directly into a solution of a sulfide antioxidant buffer (2M NaOH, 0.2M ascorbic acid and 0.2M Na_2 EDTA), which quenched any further reaction between residual sulfide and peroxide. Peroxide was analyzed colorimetrically by a

Chemetrics test kit based on the Fe-SCN colorimetric method.

Control Experiments. Control experiments were run to determine sulfide oxidation in the absence of any catalyst and in the presence of a sodium exchanged zeolite. Uncatalyzed runs were carried out in flasks at ambient temperature, sampling at 30, 60, and 120 seconds. Continuous flow runs were carried out identically to catalyzed experiments.

Results

Screening Experiments. To demonstrate the concept of heterogeneous catalysis in this application, three key inorganic pollutants, thiosulfate, sulfide, and cyanide, were chosen for study.

Thiosulfate Oxidation. A comparison of thiosulfate oxidation catalyzed by a supported metal catalyst system compared to soluble catalysts in batch experiments is shown in Table I. In all cases except nickel the heterogeneous catalyst resulted in better thiosulfate removal than the soluble catalyst. Particularly noticeable is the beneficial effect of copper supported on zeolite Y, which was studied at two pH levels. At both pH 6.4 and 9.1, 96 and 97%, respectively, of the thiosulfate was removed after fifteen minutes at 20°C. Compared to less than 90% removal using the soluble copper catalyst. An iron catalyst supported on Y zeolite also showed a higher thiosulfate removal than soluble iron.

Although it is difficult to compare homogeneous and heterogeneous catalytic reactions directly because of the difficulty of quantifying active sites in the latter, results of this work show thiosulfate oxidation with the solid catalysts to take place as or more rapidly than with soluble catalysts. The real value of a heterogeneous catalyst system, however, is in treating continuously flowing waste streams, where they would eliminate the need for a catalyst addition step to the treatment process.

Table II presents continuous treatment results for several supported metal catalyst systems evaluated in a column apparatus. These results show the effectiveness of treating continuous flow thiosulfate waste streams. At flow rates which result in five minute contact times of the solution with the catalyst, most yielded more than 70% removal of thiosulfate at ~20°C.

The metals studied are typical H_2O_2 catalysts and all show activity when supported on a variety of zeolites ranging from the natural zeolite mordenite to the synthetic zeolites X and Y. The best catalyst systems were iron on mordenite and copper on either zeolite X or Y, where 87, 82, and 80%, respectively, of the thiosul-

Table I. Comparison of Supported Catalysts to Soluble Catalysts

Catalyst (amount)[b]		Residual Thiosulfate, ppm 4 Mins	Residual Thiosulfate, ppm 15 Mins	% Removal 15 Mins	Residual, ppm 30 Mins	% Removal 30 Mins
Cu^{2+}	10 ppm	62	37	85	32	87
C/Y	5g	62	9	96	7	97
Cu^{2+}[a]	10 ppm	75	31	88	27	89
Cu/X[a]	5g	62	7	97	6	98
Cu/Al_2O_3	5g	62	30	88	15	94
Cu/SiO_2	5g	75	56	78	10	96
Ni^{2+}	10 ppm	150	100	60	90	64
Ni/Y	5g	150	150	40	125	50
Fe^{2+}	10 ppm	100	75	70	30	88
Fe/Y	5g	150	25	90	20	92

Conditions: 20–22°C
pH 6.4
Initial thiosulfate = 250 ppm
Solution volume = 100 mL
$H_2O_2:S_2O_3^{2-}$ = 0.6:1

a) pH = 9.1
b) Omission of support indicates soluble system

Table II. Thiosulfate Destruction with Supported Metal Catalysts

Catalyst	Thiosulfate In, ppm	Thiosulfate[a] Out, ppm	% Removal
Cu/mordenite	203	60	70
Fe/mordenite	203	26	87
Cu/X	225	40	82
Cu/Y	250	50	80
Fe/X	250	125	50
Fe/Y	250	56	78
Ni/X	250	85	66
Ni/Y	250	50	80
Na/Y	225	100	55

Conditions: Ambient temperature = 19–21°C
Reaction time = 5 min @ 2 mL/min flow rate
pH = 8.9
$H_2O_2:S_2O_3^{2-}$ = 0.6:1

a) Sample taken after 30 minutes of operation.

fate was removed compared to 55% when a non-catalytic sodium exchanged Y zeolite was used. These data also show stability of activity. The analyses reported are for column effluent after thirty minutes in operation. Data acquired at 15 and 120 minutes show similar removal efficiencies, demonstrating the utility for treating continuous streams.

Sulfide Oxidation. Sulfide removal can also be effectively carried out using supported metal catalysts. For sulfide, pH is particularly important to the rate of oxidation and the amount of peroxide consumed. At the low stoichiometric mole ratio of 1:1, the supported catalyst systems demonstrate an advantage in both. Figure 1 shows the relative insensitivity of the supported catalyst system to pH compared to an uncatalyzed system. Supported catalysts effected 93 to 99% sulfide removal after 15 minutes at pH 7.7 and ~20^oC, compared to 89% in the absence of catalyst. At neutral or alkaline pH the uncatalyzed oxidation of sulfide with H_2O_2 can rapidly take place, so the catalyzed system shows only a slight improvement. However, at acid pH, where the uncatalyzed reaction is slow, the catalyzed system shows marked improvement over the uncatalyzed reaction. This demonstrates that supported catalysts are not subject to pH limitations often observed in sulfide oxidation.

To illustrate the utility of supported metal catalysts for treatment of continuous flow streams, several experiments were run in columns packed with the catalyst. Results are presented in Table III. At reaction times of two minutes (the time the solution was in contact with the catalyst) at ambient temperature (20^oC) a catalyst of nickel on zeolite Y resulted in 100% removal of sulfide. Iron on zeolite Y caused more than 96% removal. In comparison a non-catalytic sodium on zeolite Y resulted in only 80% sulfide destruction. A further advantage of more efficient H_2O_2 consumption was also found with the metal catalysts compared to the Na Y control experiment. Similar results at 30 and 60 minutes show the utility of these catalysts for continuous use.

In the experiments using supported catalysts, residual peroxide still remained after the 98-100% removal of sulfide. In the non catalytic case little or no peroxide, as well as measurable sulfide, remained in the effluent. This suggests that the supported catalyst system is effecting a mechanism of sulfide destruction that is more efficient than in an uncatalyzed system.

The unexpected advantage of lower H_2O_2 utilization for catalyzed sulfide oxidation was observed at pH 7.7, which may relate to the mechanism of oxidation. Uncatalyzed sulfide oxidation is proposed to take place by one of two mechanisms, depending upon pH:

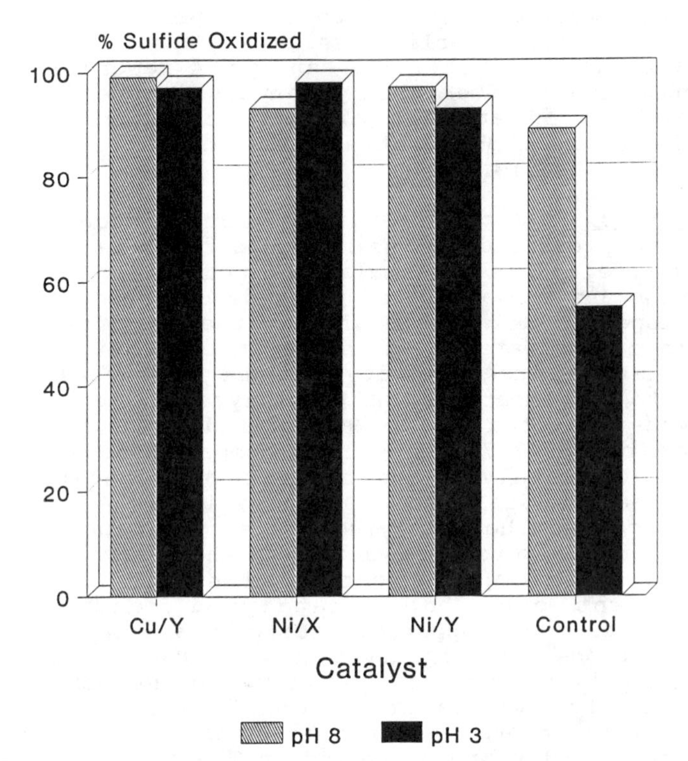

Figure 1 Sulfide Oxidation - Batch Experiments

Table III. Sulfide Removal—Continuous Flow Column

Catalyst	Sulfide In, ppm	Sulfide Out, ppm ½ Hour		(Residual H_2O_2)[a] 1 Hour	
Ni/Y	180	0	(20)	0	(15)
Fe/Y	150	2.5	(50)	5	(25)
Na/Y	125	25	(2.5)	25	(0)

Conditions: pH = 7.4
Ambient temperature = 19–21%
Reaction time = 2 minutes @ 6 mL/min flow rate
H_2O_2:S^{2-} = 1

a) Valves in parentheses are residual H_2O_2 in effluent (ppm)

$$H_2S + H_2O_2 \xrightarrow{H+} S^o + 2H_2O \qquad (1)$$

$$S^{2-} + 4H_2O_2 \xrightarrow{OH-} SO_4^{-2} + 4H_2O \qquad (2)$$

At pH 7.5-7.7, mechanism (2) will begin to influence H_2O_2 stoichiometry. This uncatalyzed sulfide oxidation mechanism consumes four moles of H_2O_2 per mole of sulfide. The experiments in Table III were carried out at 1:1 ratio. The data for catalyzed reactions in Table IV contain higher H_2O_2 residual compared to the uncatalyzed reaction at pH 7.5, while still effectively removing the sulfide. This suggests that the catalyst is causing the oxidation to proceed by equation (1) where only one mole of H_2O_2 is consumed per mole of sulfide.

Detailed Sulfide Experiments. Two catalyst systems were chosen for more detailed study, Ni/Y and Fe/Y. Conditions were chosen to approximate those present at a geothermal steam power plant. These are summarized in Table IV, as are the ranges of time and peroxide stoichiometry studied.

Results for the Ni/Y system at 24°C are presented in Figures 2 and 3. Figure II shows sulfide oxidation as a function of H_2O_2 stoichiometry at various reaction times. As expected, higher peroxide levels effected more complete oxidation even at short times. At longer reaction times, however, the 2:1 and 1:1 stoichiometries perform as well as the 4:1. When compared to the control experiments at 4:1 stoichiometry shown in Figure 3, the benefit of the supported catalyst is evident. Although not quantified, all of the Ni/Y experiments produced a slightly hazy effluent, indicating that some colloidal sulfur was formed. Whether or not this would cause a problem will have to be discussed with potential users of this technology.

Figure 4 illustrates the effect of temperature on sulfide oxidation with Ni/Y catalyst. As seen with H_2O_2 stoichiometry and reaction time (Figure 2), differences are more pronounced at shorter reaction times. More extreme conditions minimize differences brought on by other variables.

Preliminary experiments involving Fe/Y catalyst (Figure 5) yielded results similar to those with Ni/Y, but H_2O_2/S^{2-} ratio appeared to play a more important role. At 1:1 only 70-75% of the sulfide was oxidized even at 60 seconds retention time. A 2:1 ratio markedly raised this to 97%. In general the Fe/Y catalyst system produced no observable colloidal sulfur on other precipitates. No loss of iron from the Fe/Y catalyst was detected by atomic absorption analysis. However, a more detailed study is required to fully define useful catalyst life.

Table IV. Conditions for Sulfide Experiments

Sulfide Concentration	~90 ppm
H_2O_2/Sulfide mole ratio	1:1, 2:1, 4:1
Temperature	24°C (75°F) 38°C (100°F)
Reaction Time	15 sec, 30 sec, 60 sec
pH 8.9	

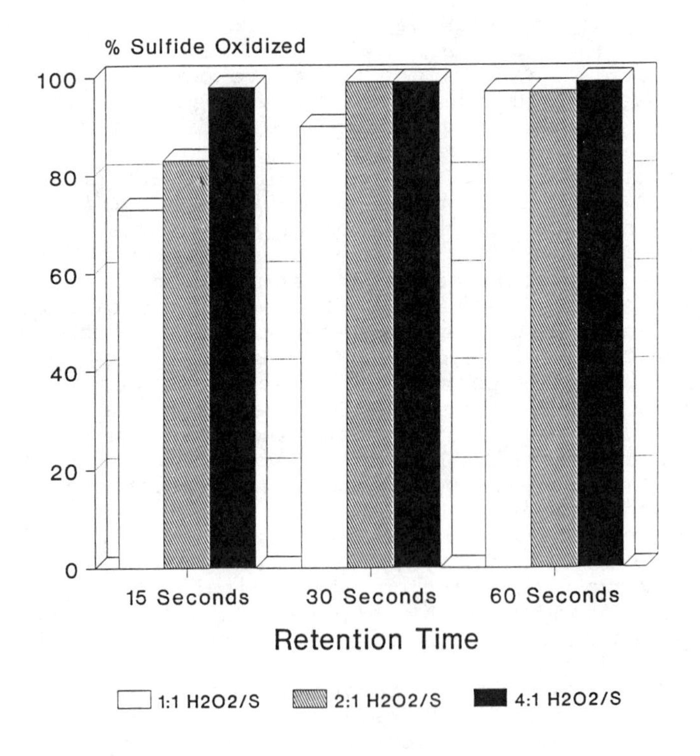

Figure 2 Sulfide Oxidation - Effect of Retention Time and H_2O_2/ Sulfide Ratios

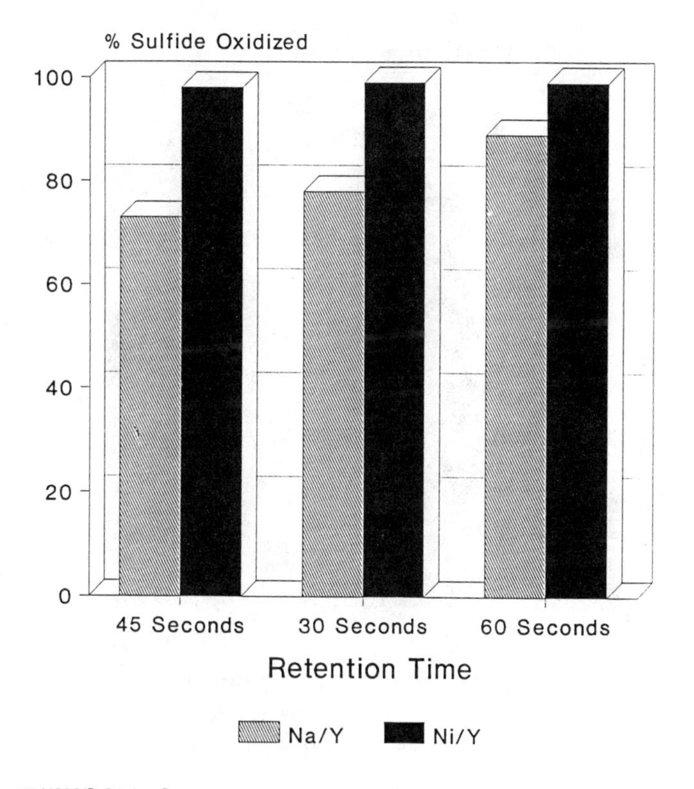

4:1 H202/S, 24 deg C

Figure 3 Control Experiments - Ni/Y and Na/Y

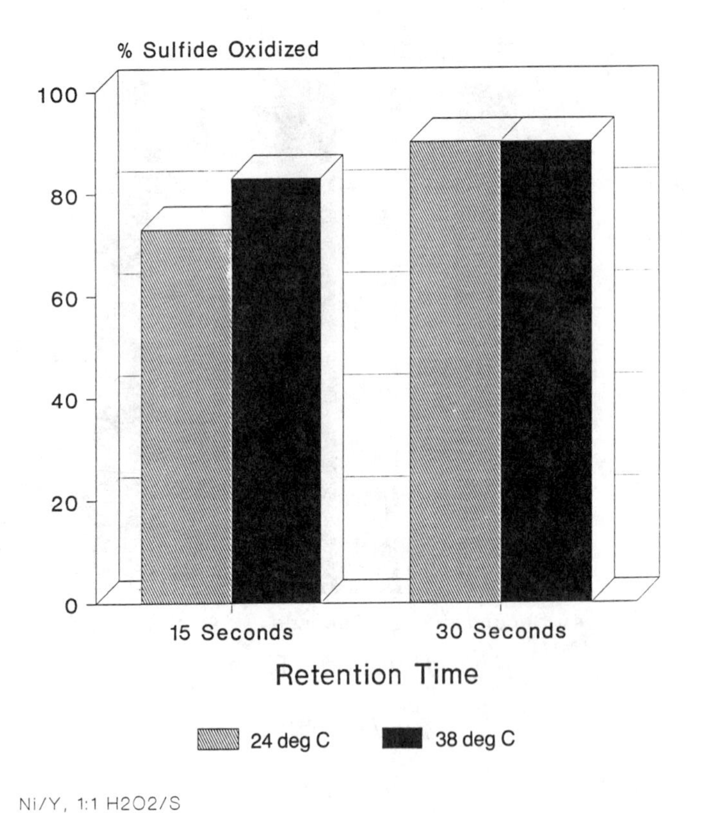

Figure 4 Sulfide Oxidation - Effect of Temperature

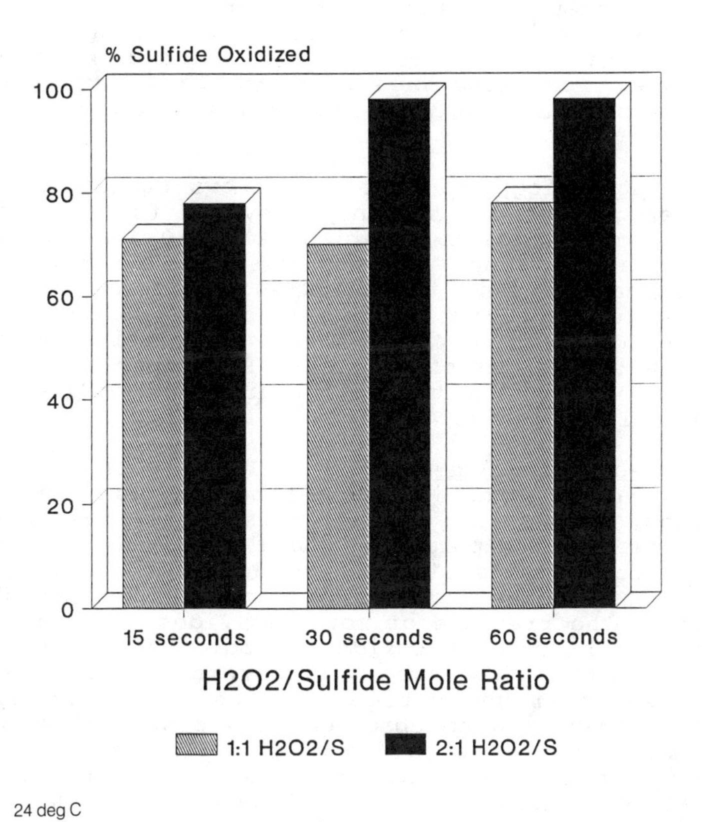

Figure 5. Sulfide Oxidation—Fe/Y Catalyst

An atomic absorption examination of the Ni/Y catalyst after this series of experiments revealed the loss of about 15% of that initially present. However, the catalyst showed no measurable loss of sulfide removal activity. Metal loss and catalyst life will be the subject of additional study.

Mercaptan Oxidation. Table V presents data for the oxidation of mercaptan sulfur with a supported metal catalyst system. In a batch experiment both nickel/Y and copper/X showed catalytic activity at two pH levels, 4 and 7.6, in the destruction of ethyl mercaptan. Uncatalyzed reactions between mercaptans and H_2O_2 are generally slow as shown by the control example in Table V.

Cyanide Detoxification. Table VI shows results of using the heterogeneous metal catalyst system for the removal of CN^- from an aqueous solution with H_2O_2. Silver/Y and copper/Y catalysts show 95% and 74% cyanide removal after one hour at ambient temperatures. Existing processes using soluble silver (5) and other soluble metal catalysts (6) require 40 to 50^oC for effective detoxification at similar reaction times. By comparison the use of only H_2O_2 or a non-catalytic sodium/Y showed less than 20% removal of cyanide.

Catalyst Poisoning/Regeneration. It is well known that sulfide readily forms insoluble precipitates with metals like iron, copper, and nickel. Thus it is reasonable to expect that sulfide poisoning would be a problem with the present catalyst system. Indeed in the absence of peroxide the supported catalyst rapidly turned black, indicating the presence of metal sulfides. However, upon addition of hydrogen peroxide the color returned to the original state and active sulfide removal was observed. The poisoning/regeneration process could be cycled numerous times with no observed loss of activity and no measurable loss of metal from the support. A similar sulfiding of the catalyst took place when insufficient peroxide was added to the feed water and, again, was reversible with added H_2O_2. This could potentially be utilized as an indicator for peroxide control in a commercial process where sulfide levels could vary with time, thus providing a mechanism for most efficient use of oxidant.

Summary

An advantage of this process is that there are no restrictive pH limitations on the effectiveness of the catalyst. Practical pH ranges are from 4 to 12 and depend upon the pollutant. For example, at low pH thiosulfate will disproportionate to yield sulfur and other sulfur

Table V. Mercaptan Removal Using H_2O_2 and Heterogeneous Catalyst System

Catalyst /Support	pH	Residual Mercaptan Sulfur, ppm	% Removal
Cu/X	4.3	27	75
Cu/X	7.6	27	75
Ni/Y	4.2	13	88
Ni/Y	7.7	13	88
None[a]	9.2	75	30

Conditions: $[H_2O_2]$ = 350 ppm
Mercaptan Sulfur = 107 ppm as ethyl mercaptan
Temperature = 50 °C
Reaction Time—1 hour

a) Two hours reaction time

Table VI. Cyanide Detoxification Using Heterogeneous Catalyst System

Catalyst	Residual Cyanide, ppm	% Removal
Cu/Y	50	74
Cu/X	75	62
Ag/Y	10	95
Ag/X	15	92
Ni/Y	150	21
Na/Y	150	21
None	175	7

Conditions: $[H_2O_2]$ = 245 ppm
$[CN^-]$ = 190–200 ppm
pH = 10.5
Ambient temperature = 20–22°C
Reaction time = 1 hour

oxide species. Cyanide below pH 6-8 will result in the formation of toxic HCN gas. Thus, suitable pH conditions will be determined by the pollutant. No pH values have been found which deactivate or poison catalyst activity. At pH more than 11 there is a chance of increased H_2O_2 consumption due to decomposition.

The waste treatment process described in this paper provides these advantages:

- efficient catalysis of H_2O_2 for the destruction of oxidizable inorganic pollutants from wastewater.

- a heterogeneous catalyst support which shows little sensitivity to pH.

- an efficient process for treating continuous flow waste streams with H_2O_2 without the need for continuously adding a soluble catalyst

Future work will involve characterization of the catalyst, catalyst life, and optimization of conditions and chemical usage.

LITERATURE CITED

1. Urban, U.S. Patent 4,358,427 - 1981

2. Brown, U.S. Patent 3,572,836 - 1972

3. Haberman, U.S. Patent 4,696,749 - 1987

4. Standard Methods for the Examination of Water and Wastewater, 16th Ed.; Greenberg, A.E.; Trussell, P.R.; Clesceri, L.S. Eds.; American Public Health Association: Washington, D.C. - 1987

5. Flacher, J.; Knorre, H.; Pohl, G. V.S. Patent 3,970,554 - 1976

6. Mathre, O.B. U.S. Patent 3,617,567 - 1971

RECEIVED November 10, 1989

Chapter 5

Destruction of Mixtures of Pollutants by UV-Catalyzed Oxidation with Hydrogen Peroxide

D. W. Sundstrom, B. A. Weir, and K. A. Redig

Department of Chemical Engineering, University of Connecticut, Storrs, CT 06269

Hazardous organic compounds present in many water supplies and industrial wastes must often be reduced to very low concentration levels. This project investigated the destruction of mixtures of organic compounds by ultraviolet catalyzed oxidation with hydrogen peroxide as the oxidizing agent. Benzene and trichloroethylene were used as the components because they are common priority pollutants and differ greatly in structure. The combination of ultraviolet light and hydrogen peroxide was effective in decomposing both components individually and in mixtures. The rate of reaction of trichloroethylene was much lower in a mixture with benzene than as a single component. The strong interaction between components demonstrates the need to study actual mixtures instead of attempting to predict mixture behavior from pure component data.

Many water supplies and industrial wastes contain hazardous organic chemicals that must be removed prior to use. These compounds are often present at low concentration levels and require high degrees of removal. Packed bed aeration and activated carbon adsorption are currently the most widely applied technologies for removing residual organic compounds. These methods alone are nondestructive and simply transfer the compounds between phases.

In recent years, advanced oxidation processes involving combinations of ozone, hydrogen peroxide and ultraviolet light have received widespread attention (1-6). The photochemical action of UV light on ozone or hydrogen peroxide produces hydroxyl radicals and other reactive species that attack the organic molecules. The combination of UV light and an oxidant can give fast reaction rates and high degrees of removal. Under suitable operating conditions, the final products are mainly carbon dioxide and water.

Although ozone is usually a stronger oxidizing agent than hydrogen peroxide, it has several process disadvantages. Ozone is

0097–6156/90/0422–0067$06.00/0

an unstable gas that must be generated on site and transferred into the liquid phase. The ozone containing bubbles can also strip volatile components into the air. Hydrogen peroxide solutions are readily stored for mixing with the contaminated water according to process demand. Several companies have recently started marketing the UV/peroxide technology for removing toxic organics from water supplies.

Sundstrom et al. (7-9) have studied the destruction of individual aliphatic and aromatic compounds by UV light catalyzed oxidation with hydrogen peroxide. The results demonstrated that the combination of UV light and peroxide is effective in destroying a wide variety of hazardous compounds present in water at low levels. For the chlorinated aliphatics, the reacted chlorine was converted quantitatively to chloride ion, indicating that the chlorinated structures were effectively destroyed. In the case of the aromatics, many intermediates were formed which could be eliminated by extending the treatment time.

Most studies on UV catalyzed oxidation processes have emphasized pure compounds or complex natural mixtures, such as humic matter. In practice, mixtures of known chemical compounds are often found in contaminated water. A knowledge of the interactions between reacting components is necessary for the design of oxidative treatment systems. The purpose of this research was to investigate the destruction of mixtures of benzene and trichloroethylene by the UV/peroxide process.

Experimental Methods

All experiments were conducted in a batch photochemical reactor with an inside diameter of 7.5 cm, a length of 30 cm and a total volume of about 1.5 liter (7). The lower two-thirds of the reactor was jacketed to permit temperature control by a circulating liquid. An ultraviolet lamp entered through a central opening at the top, and other openings were used to add liquids, withdraw samples and measure temperature. The UV source was a low pressure mercury vapor lamp from Ace Glass with an outside diameter of 0.9 cm and a length of 25 cm. The output of the lamp was about 2 watts at the 254 nm resonance line. The contents of the reactor were agitated continuously by a magnetic stirrer.

The reactor was initially charged with 1 liter of distilled water containing a phosphate buffer and dissolved organic compounds. The hydrogen peroxide was added to the reactor and the lamp was turned on. Samples were analyzed for benzene and trichloroethylene by a Perkin-Elmer gas chromatograph equipped with flame ionization detectors. Hydrogen peroxide concentrations in the samples were determined by a glucose oxidase-peroxidase method (7).

Results

Choice of Compounds.

Benzene and trichloroethylene (TCE) were selected as the organic compounds for this study since they are common priority pollutants found in contaminated water and differ greatly in structure and properties. For example, the absorption coefficient for benzene is 16 times larger than that for trichloroethylene. These compounds were previously investigated separately and found to have similar rates of reaction (7,8).

The concentrations of benzene and trichloroethylene were chosen to provide the same range of organic carbon concentration for each component. The benzene concentrations varied from 0.05 to 0.2 mM (3.9 to 15.6 ppm) and TCE concentrations from 0.15 to 0.6 mM (19.7 to 78.8 ppm). The hydrogen peroxide concentrations were selected to give 0.75 to 3 moles of peroxide per mole of organic carbon in solution.

Single Components.

The effects of pollutant and hydrogen peroxide concentrations on the rates of destruction of the pure compounds are illustrated in Figures 1 and 2. For both components, the rates of reaction increased with increasing hydrogen peroxide concentration and decreased with increasing pollutant concentration. For a given peroxide and organic carbon concentration, the reaction rates were similar, with trichloroethylene having a slightly faster rate than benzene.

The semi-log plots of fraction of pollutant remaining versus time are nearly linear, suggesting first order reactions. The kinetics are not true first order as evidenced by the decrease in reaction rate with increasing pollutant concentration. However, pseudo first order rate constants were calculated for each run to facilitate comparison of reaction rates between runs.

The reaction mechanisms for the oxidative destruction of the pollutants are complex and involve hydroxyl radicals as the major reactive species. The dependence of rate constants on concentration probably results from the role of hydroxyl radicals in the chemical reactions. The rate of generation of hydroxyl radicals depends mainly upon the UV light intensity and its absorption by hydrogen peroxide in the solution. As pollutant concentration is decreased at constant peroxide concentration, the number of hydroxyl radicals available to react with each pollutant molecule is increased.

Mixtures of Components.

The reaction rates of benzene and trichloroethylene in mixtures increased with increasing hydrogen peroxide concentration, as illustrated in Figures 3 and 4 for a mixture containing 0.1 mM benzene and 0.3 mM trichloroethylene (each 0.6 mM in organic carbon). Both components in the mixture exhibited nearly linear behavior on semi-log plots of pollutant concentration versus time. Pseudo first order rate constants were obtained from the slopes and plotted versus hydrogen peroxide concentration in Figure 5. As peroxide concentration was increased from 0.9 to 3.6 mM, the rate constants increased from 0.0185 to 0.046 min^{-1} for benzene and from 0.0167 to 0.0313 min^{-1} for trichloroethylene. The rate increases probably resulted from a larger supply of hydroxyl radicals at higher peroxide concentrations.

The reaction rates for two binary mixtures containing the same

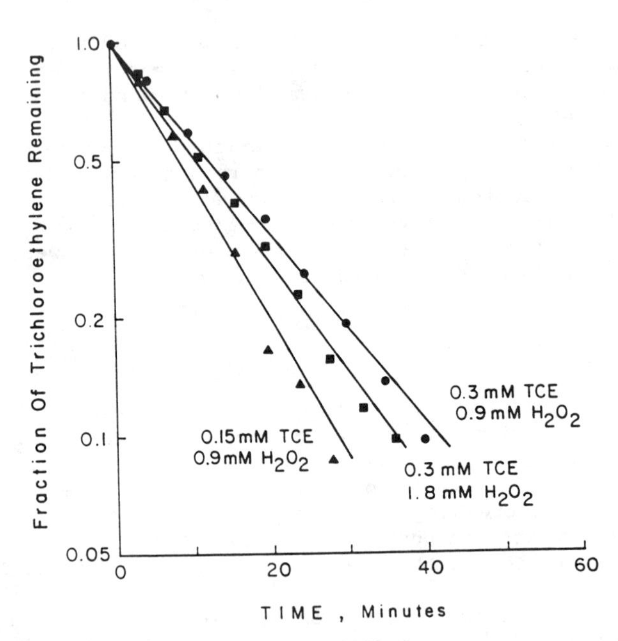

Figure 1. Effect of initial trichloroethylene and hydrogen peroxide concentrations on the decomposition of trichloroethylene at 25°C and pH 6.8.

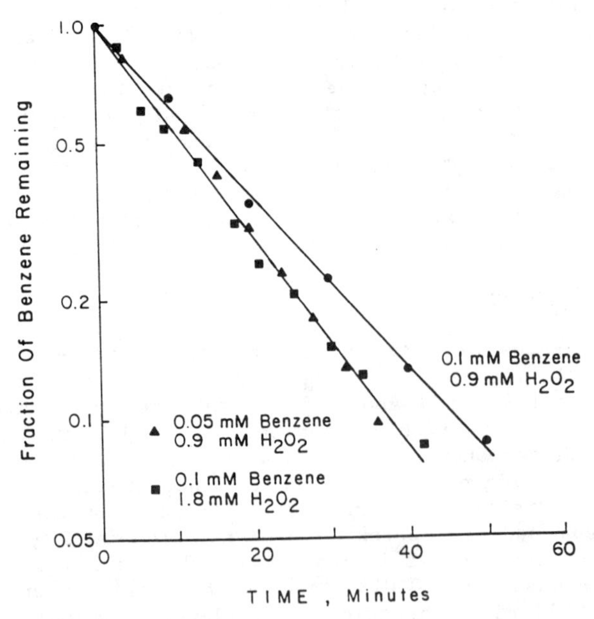

Figure 2. Effect of initial benzene and hydrogen peroxide concentrations on the decomposition of benzene at 25°C and pH 6.8.

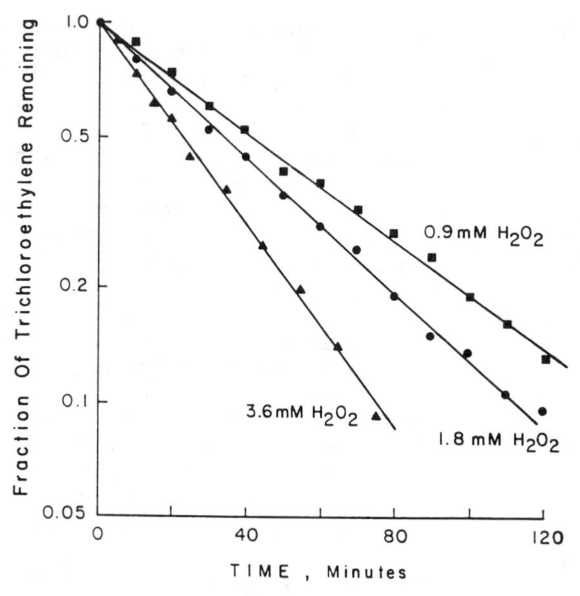

Figure 3. Effect of initial hydrogen peroxide concentration on the rate of destruction of trichloroethylene in a mixture initially containing 0.1 mM benzene and 0.3 mM trichloroethylene at 25°C and pH 6.8.

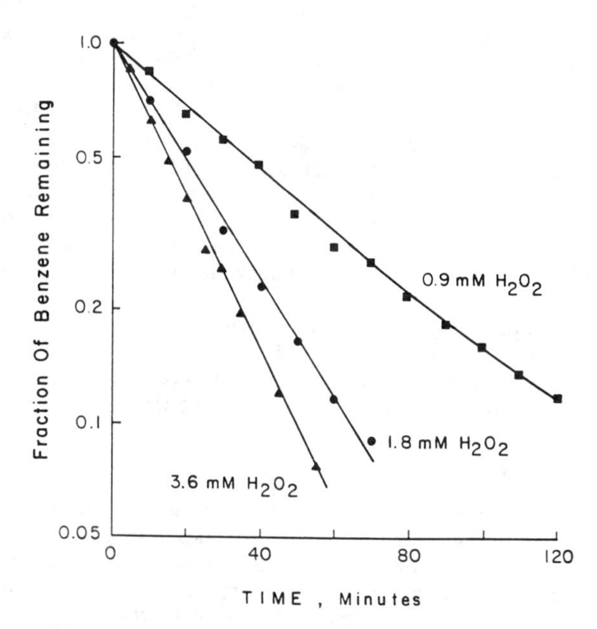

Figure 4. Effect of initial hydrogen peroxide concentration on the rate of destruction of benzene in a mixture initially containing 0.1 mM benzene and 0.3 mM trichloroethylene at 25°C and pH 6.8.

total organic carbon and peroxide concentrations are shown in Figure 6. One mixture initially contained 0.05 mM benzene and 0.3 mM TCE and the other mixture contained 0.1 mM benzene and 0.15 mM TCE. The rate of benzene disappearance was nearly the same in both mixtures even though the initial concentrations differed by a factor of 2. In contrast to the single component results, the reaction rate for trichloroethylene increased as its concentration in the mixture was increased. Since an increase in trichloroethylene concentration was associated with a decrease in benzene, these results suggest strong interactions between the reacting components.

The effects of adding the second component to solutions containing either 0.1 mM benzene or 0.3 mM trichloroethylene are shown in Figures 7 and 8. In both cases the rate constant for benzene or trichloroethylene decreased significantly as the other component was added. The change in reaction rates was more pronounced for trichloroethylene where the rate constants decreased by a factor of 3 when 0.1 mM benzene was added to 0.3 mM trichloroethylene. A doubling of the hydrogen peroxide concentration moderated the decrease in the benzene rate constant but had little effect on the decrease in the trichloroethylene rate constant.

In Figures 7 and 8, organic carbon concentrations increased as the second component was added to the solutions. Thus, part of the decreases in rate constants could be attributed to higher pollutant concentrations. A fairer method of comparison would be to relate rate constants for pure components and mixtures at the same total organic carbon concentration. These comparisons are made in Figures 9 and 10 for solutions containing 1.5 moles of hydrogen peroxide per mole of organic carbon. In both cases, the rate constants for benzene alone were similar in magnitude to those for benzene in a mixture. The rate constants for trichloroethylene in the mixtures, however, were about one-half of their values as pure components.

These results show that the aromatic component, benzene, has a strong adverse effect on the rate of destruction of the aliphatic component, trichloroethylene. This effect may be due not only to the benzene itself, but also to the reaction intermediates formed as benzene is oxidized. Hydroxyl radical attack of benzene yields several aromatic intermediates, including the hydroxycyclohexadienyl radical (10), phenol (8), and the three dihydroxybenzene isomers (8). The presence of benzene and its aromatic oxidation products may inhibit the destruction of trichloroethylene through competition for available UV photons and hydroxyl radicals.

Conclusions

The strong interaction between components demonstrates the need to study actual mixtures instead of attempting to predict their behavior from pure component data. As shown for trichloroethylene in this study, the rate of destruction of a compound may be substantially lower as a component in a mixture than as a pure component. The volume of a reactor system may be seriously underestimated if the design ignores potential interactions between reacting components.

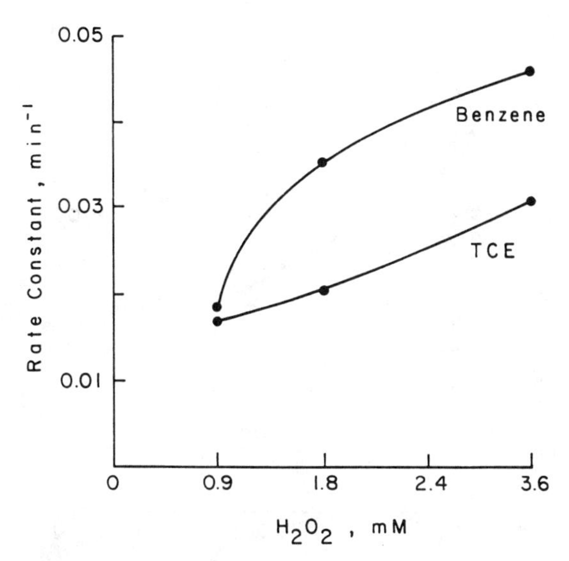

Figure 5. Effect of hydrogen peroxide concentration on apparent first order rate constants for a mixture initially containing 0.1 mM benzene and 0.3 mM trichloroethylene at 25°C and pH 6.8.

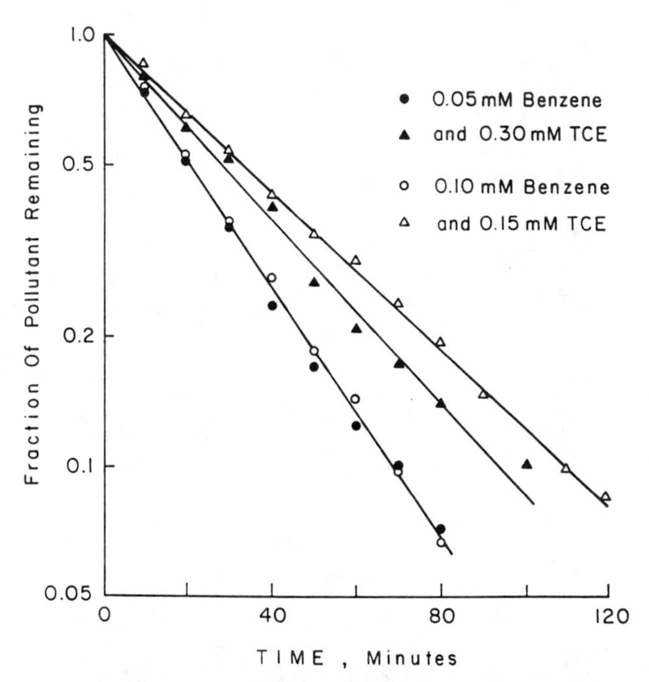

Figure 6. Effect of mixture composition on the rate of destruction of benzene and trichloroethylene at 25°C, pH 6.8 and 0.9 mM initial hydrogen peroxide concentration.

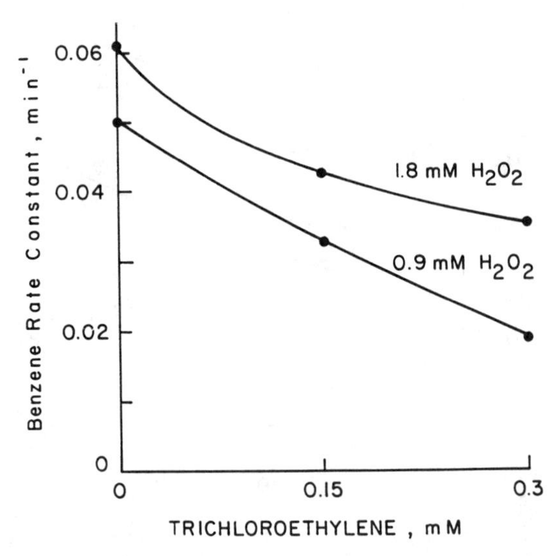

Figure 7. Effect of initial trichloroethylene and hydrogen peroxide cencentrations on apparent first order rate constants for benzene in mixtures containing 0.1 mM benzene at 25°C and pH 6.8.

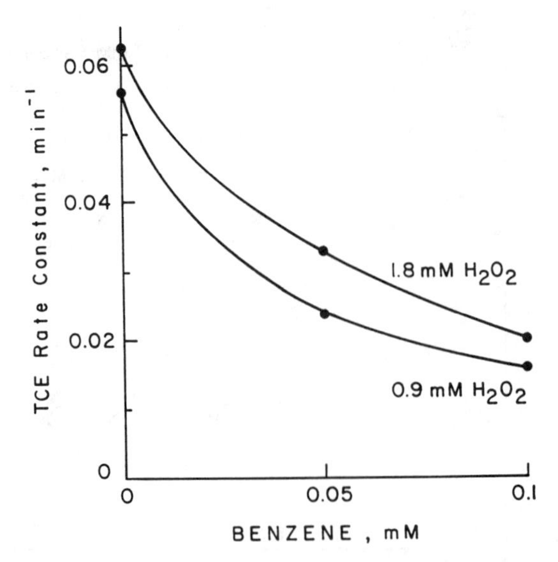

Figure 8. Effect of initial benzene and hydrogen peroxide concentrations on apparent first order rate constants for trichloroethylene in mixtures containing 0.3 mM trichloroethylene at 25°C and pH 6.8.

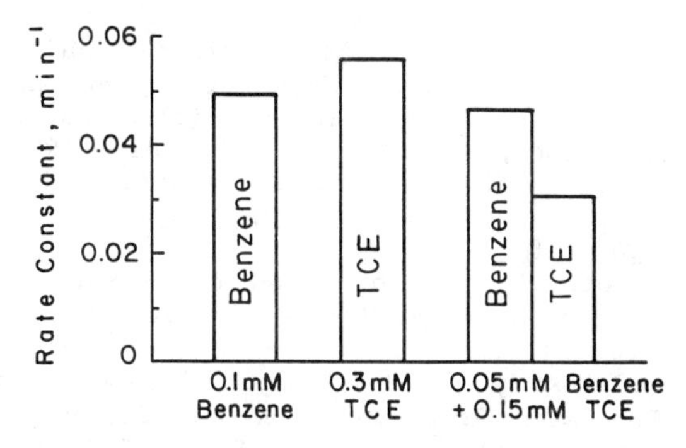

Figure 9. Comparison of apparent first order rate constants for 0.1 mM benzene alone, 0.3 mM trichloroethylene alone, and a mixture with 0.05 mM benzene and 0.15 mM trichloroethylene, all at 25°C, pH 6.8 and 0.9 mM hydrogen peroxide.

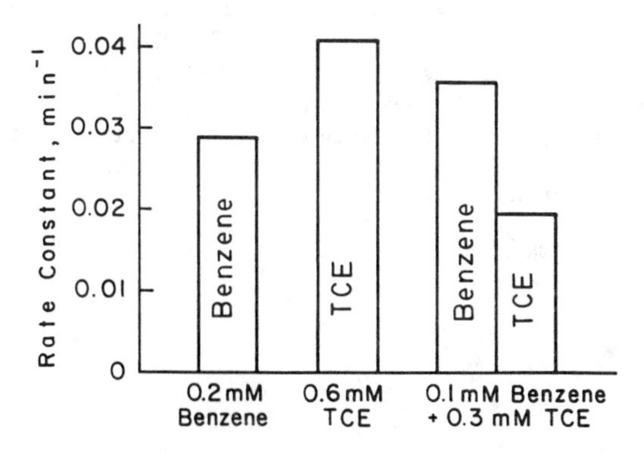

Figure 10. Comparison of apparent first order rate constants for 0.2 mM benzene alone, 0.6 mM trichloroethylene alone, and a mixture with 0.1 mM benzene and 0.3 mM trichloroethylene, all at 25°C, pH 6.8 and 1.8 mM hydrogen peroxide.

Literature Cited

1. Zeff, J.D. "UV-OX Process for the Effective Removal of Organics in Wastewaters"; *AIChE Symp. Ser*, **1977**, 73, (167), 206.
2. Prengle, H.W.; and C.E. Mauk. "New Technology: Ozone/UV Chemical Oxidation Wastewater Process for Metal Complexes, Organic Species and Disinfection"; *AIChE Symp. Ser.*, **1978**, 74, (178), 228.
3. Peyton, G.R.; F.Y. Huang; J.L. Burleson; and W.H. Glaze. "Destruction of Pollutants in Water with Ozone in Combination with UV Radiation. 1. General Principles and Oxidation of Tetrachloroethylene"; *Environ. Sci. Technol.*, **1982**, 16, 448.
4. Koubek, E. "Photochemically Induced Oxidation of Refractory Organics with Hydrogen Peroxide"; *Ind. Eng. Chem. Proc. Des. Dev.*, **1975**, 14, 348.
5. Ho, P.C. "Photooxidation of 2,4-dinitrotoluene in Aqueous Solution in the Presence of Hydrogen Peroxide"; *Environ. Sci. Technol.*, **1986**, 20, 260.
6. Glaze, W.H.; J-W Kang; and D.H. Chapin. "The Chemistry of Water Treatment Processes Involving Ozone, Hydrogen Peroxide and Ultraviolet Radiation"; *Ozone Sci. Eng.*, **1987**, 9, 335.
7. Sundstrom, D.W.; H.E. Klei; T.A. Nalette; D.J. Reidy; and B.A. Weir. "Destruction of Halogenated Aliphatics by Ultraviolet Catalyzed Oxidation with Hydrogen Peroxide"; *Hazard. Waste Hazard. Mat.*, **1986**, 3, 101.
8. Weir, B.A.; D.W. Sundstrom; and H.E. Klei. "Destruction of Benzene by Ultraviolet Light-Catalyzed Oxidation with Hydrogen Peroxide"; *Hazard. Waste Hazard. Mat.*, **1987**, 4, 165.
9. Sundstrom, D.W.; B.A. Weir; and H.E. Klei. "Destruction of Aromatic Pollutants by UV Light Catalyzed Oxidation with Hydrogen Peroxide"; *Environmental Progress*, **1989**, 8, 6.
10. Dixon, W.T.; and R.O.C. Norman. "Electron Spin Resonance Studies of Oxidation. Part IV. Some Benzenoid Compounds; *J. Chem. Soc.*, **1964**, 1964, 4857.

RECEIVED November 10, 1989

Chapter 6

UV Destruction of Phenolic Compounds under Alkaline Conditions

Harry M. Castrantas and Robert D. Gibilisco

FMC Corporation, Peroxygen Chemical Division, P.O. Box 8, Princeton, NJ 08543

This study provides data showing the range of conditions under which phenols and substituted phenols can be oxidatively destroyed using UV-H_2O_2. UV-H_2O_2 was found to be considerably more effective in destroying phenol and substituted phenols than UV. Significant reaction rates were seen at lamp outputs of 60,000 and 240,000 uW-sec/cm^2 over a phenol concentration range of 10-1000 mg/L, a pH range 4-10 and mole ratios of H_2O_2/Phenol of 2.8/1 to 8.3/1. By being effective under alkaline as well as acid conditions, UV-H_2O_2 avoids the acid limitations of Fenton's reagent (Fe + H_2O_2) which is a well know chemistry used for destroying phenolic compounds.

Uncatalyzed hydrogen peroxide is used commercially to destroy pollutants such as inorganic sulfides, thiosulfate, hypochlorite and formaldehyde. Without activation however, hydrogen peroxide will not destroy the more difficult to oxidize pollutants such as phenols, chlorinated hydrocarbons and alcohols. Hydrogen peroxide can be activated to form hydroxyl radicals (OH·) which can destroy these organics and others. This increased ability to destroy organics is related to the substantially higher oxidation potential of the hydroxyl radical (2.80V) compared to hydrogen peroxide(1.78V).

One method of generating hydroxyl radicals is by adding a soluble iron salt to an acid solution of hydrogen peroxide (Fenton's reagent) (1). The process is used commercially to destroy organics such as phenols, but is

0097–6156/90/0422–0077$06.75/0

limited to reactions carried out under acid conditions (2-4).

$$Fe^{2+} + H_2O_2 \longrightarrow Fe^{3+} + OH^- + OH\cdot \quad (1)$$

$$Fe^{3+} + H_2O_2 \longrightarrow Fe^{2+} + HOO\cdot + H^+ \quad (2)$$

Hydrogen peroxide can also be activated by UV, O_3 and UV-O_3 to generate active species such as hydroxyl radicals (5-7). Each system has it's utility depending upon the organics being destroyed.

UV-H_2O_2, the subject of this study, has the following advantages over the iron-H_2O_2-acid system: (1) the reaction is not limited to an acid pH range and (2) an iron catalyst is not required.

The UV light activates the hydrogen peroxide converting it to hydroxyl radicals (8).

$$H_2O_2 \xrightarrow{UV} 2\ HO\cdot \quad (1)$$

$$HO\cdot + H_2O_2 \longrightarrow HOO\cdot + H_2O \quad (2)$$

$$HOO\cdot + H_2O_2 \longrightarrow HO\cdot + H_2O + O_2 \quad (3)$$

$$HOO\cdot + HOO\cdot \longrightarrow H_2O_2 + O_2 \quad (4)$$

The literature has several references on the use of UV-H_2O_2 to destroy otherwise difficult to oxidize compounds, e.g., alcohols, organic acids, chlorinated hydrocarbons, aromatics and phenol (9,10). UV light alone will attack some molecules by bond cleavage and free radical generation, but usually at a much slower rate than with UV-H_2O_2 systems.

This study was undertaken to determine the effect of variables such as: UV light intensity, phenol concentration, ratio of H_2O_2/phenol, pH and effect of substituted phenols on reaction rates. Phenols were chosen for the UV-H_2O_2 study since (1) phenols are commonly found as pollutants in wastewater and groundwater; (2) many waters containing phenolics are alkaline.

Experimental

Equipment and Operation. A continuous flow ultraviolet unit (Ultraviolet Technology Inc., Model 40 L) in which the lamps were suspended in a tubular arrangement outside and parallel to a teflon reactor tube was used for this study (Figure 1). The unit normally holds four 1.5 cm by 84 cm low pressure UV lamps (# 10002) but the unit was modified to hold eight of these lamps. The lamps emit approximately 90% of the ultraviolet energy at 253.7 nm. The lamps were spaced 5.3 cm from each other and were 2.7 cm from the surface of the teflon reactor tube (5 cm.

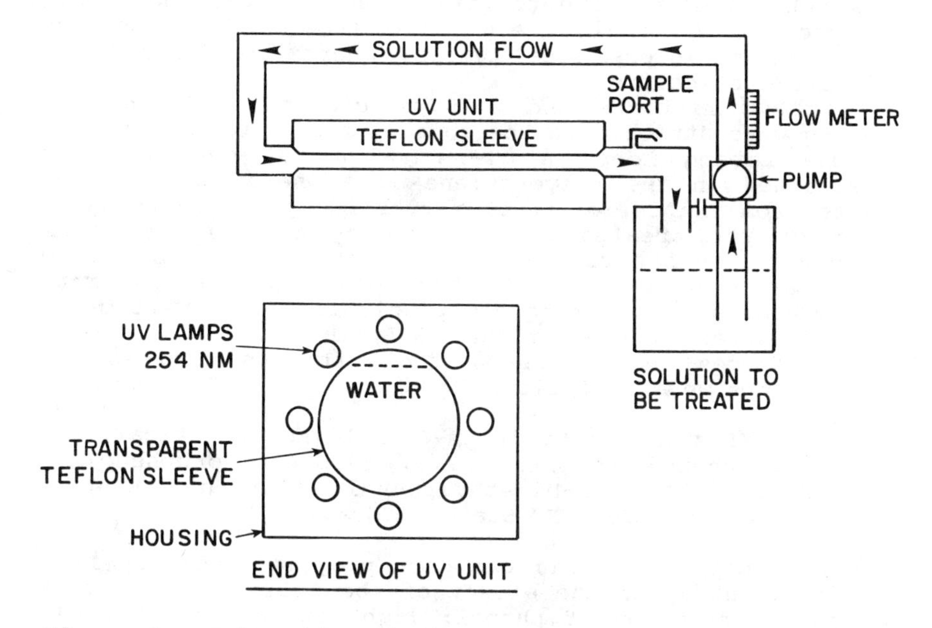

Figure 1. Schematic diagram of UV apparatus used for phenolic compound oxidation.

diameter by 91 cm long). Every other lamp was equipped with an aluminum reflector. Four switches permitted the use of 2, 4, 6 or 8 lamps. Each lamp drew 51.7 watts based on current measurements of 0.45 amperes per lamp at 115 volts. These numbers translated to a calculated energy output at the lamp's surface of 34,592 uW-sec/cm^2 which differed somewhat from the manufacturer's rated output for a new lamp of 30,000 uW-sec/cm^2. Since the phenol study was conducted with new lamps, the latter output was arbitrarily selected for referencing in this study. The actual energy reaching the teflon tube will be significantly less than 30,000 uW-sec/cm^2 existing at the surface of each lamp. The energy reaching the teflon tube was not calculated due to various assumptions that would add to the uncertainty e.g., the effect of reflections from the stainless steel walls, quartz sleeve and the aluminum reflectors.

PVC fittings and 0.95 cm diameter polyethylene tubing were attached to the ends of the UV unit to make a recirculating loop. A glass valve for sampling was positioned in the polyethylene line where the solution exits from the teflon reactor tube but prior to reinjection of the treated solution into an 18 liter glass reservoir. A centrifugal pump (Eastern model D-8 type 304) was used to circulate the solution. The flow rate was measured with a rotameter. Solution flows were maintained between 0.9 L/min and 1.8 L/min.

Sources and grades of the reactants used in the experiments were as follows:

Phenol (Fisher Purified), m-Cresol (MCB Practical), 2-Chlorophenol (Aldrich 99+%), 2,5-Dichlorophenol (Aldrich 98%) 2,5-Dimethylphenol (Aldrich 99+%), Hydrogen Peroxide (FMC Standard Grade)

Experimental Plan. The two part program involved (1) a factorial study on the effect of the following variables on the destruction of phenol: light intensity, pH, mole ratio of H_2O_2/phenol, and phenol concentration. (2) a study of the relative reactivity of substituted phenols with UV and UV-H_2O_2.

Phenol and Substituted Phenols. The phenol compound was weighed (Mettler PE 360 balance) and dissolved in 80 ml of deionized water. This concentrate was added to an 18 liter glass carboy with sufficient deionized water to make up 12 liters. Hydrogen peroxide (when used) was added to the solution in the reservoir and the pH adjusted with aqueous sodium hydroxide or sulfuric acid (Corning Model 255 pH meter). The centrifugal pump was turned on to circulate the solution through the UV unit. During startup, it was occasionally necessary to expel air pockets from the teflon tube by slightly tilting the UV unit upward. When the teflon chamber was free of air pockets, the protective stainless steel cover was placed on the unit and the UV lamps were switched on. Periodically, 80 ml aliquots were removed for immediate temperature readings followed as rapidly as possible by pH, phenol and hydrogen peroxide determinations.

Analytical Methods. Phenols were determined using a test kit (Chemetrics Model P-12) based on the formation of a red color by 4-aminoantipyrine when a phenolic compound was present (11). The color forming reaction is initiated by potassium ferricyanide. An ampoule is stirred in a 25 ml sample of the phenolic solution to dissolve the ferricyanide crystals adhering to it. The ampoule tip is then snapped to draw solution into the vacuum chamber containing 4-aminoantipyrine. The ampoule contents are mixed by inverting several times. The ampoule is then held next to one of two color comparators covering the range 0.1-1.0 mg/L ±4% and 1-12 mg/L ±4% until the closest match is found. For higher concentrations of phenols, serial dilutions are performed. The limit of phenol detection is 0.1 mg/L ± 4%.

H_2O_2 residuals were determined with a test kit (Chemetrics Model HP-10) based on the oxidation of iron by hydrogen peroxide and the formation of the colored ferric thiocyanate complex (12). An ampoule containing ammonium thiocyanate-ferrous solution is introduced into a 25 ml sample of the solution containing H_2O_2. The ampoule tip is then broken allowing the H_2O_2 solution to enter. After inverting several times to insure the contents are mixed, the ampoule is held next to one of two color comparators covering the range 0.1-1.0 mg/L ±4% and 1.0-10.0 mg/L ±4% until the closest match is found. The limit of H_2O_2 detection is 0.1 mg/L ±4%.

Discussion and Results

General. The summary of the experimental conditions and complete rate data for the phenol and substituted phenol runs will be found in Tables I and II. All reactions were conducted at room temperature (21°C-24°C). During reaction, a modest 1°C to 7°C temperature rise was seen. This was attributed to mechanical heating by the centrifugal pump, the heat given off by the lamps and a contribution from the heats of reaction. Based on a flow rate of 0.9 L/min and 1.8 L/min the temperature of the solution inside the teflon tube (1.75 liters) was calculated to be only 0.4°C and 0.2°C respectively higher than at the point of actual measurement at the outlet of the teflon tube.

With the exception of run 7 that used two UV lamps, all other UV runs utilized eight lamps.

A regression equation was developed (Appendix A) using the data from runs 4 and 6-16, Table I. This regression equation was used to generate Figures 5-8. The

Table I-A. Phenol Runs-Summary of Reaction Conditions

RUN	1	2	3	4	5	6	7	8
pH-Initial	4.0	4.0	4.0	9.8	10.0	10.0	7.8	7.8
pH-Final	4.0	3.3	-	9.9	6.1	5.2	6.1	7.5
Temp °C Initial	23	21	22	23	22	23	21	23
Temp °C Final	24	23	26	23	24	25	30	23
Flow Rate-L/min	1.2	1.2	1.1	1.2	1.1	1.2	1.2	1.2
Phenol mg/L-Initial ± 4%	10	10	10	10	10	10	100	100
H_2O_2 mg/L-Initial ± 4%	30	30	0	10	0	30	200	200
H_2O_2/Phenol Mole Ratio	8.3/1	8.3/1	0	2.8/1	0	2.8/1	5.5/1	5.5/1
UV-Lamps-Number Used	0	8	8	8	8	0	2	8
Power Output uW-sec/cm^{2}x10^4	0	24	24	24	24	0	6	24

RUN	9	10	11	12	13	14	15	16
pH-Initial	6.2	6.0	7.8	8.0	9.2	6.0	6.2	9.6
pH-Final	6.8	7.4	3.2	4.3	7.4	-	3.2	4.3
Temp °C Initial	21	23	21	22	22	24	22	23
Temp °C Final	24	23	24	27	31	29	29	2.9
Flow Rate-L/min	1.2	1.3	1.2	1.2	1.6	1.7	1.8	1.0
Phenol mg/L-Initial ± 4%	10	10	10	100	1000	1000	1000	1000
H_2O_2 mg/L-Initial ± 4%	30	30	30	100	1000	3000	1000	3000
H_2O_2/Phenol Mole Ratio	8.3/1	8.3/1	8.3/1	2.8/1	2.8/1	8.3/1	2.8/1	2.8/1
UV-Lamps-Number Used	8	8	8	8	8	8	8	8
Power Output uW-sec/cm^{2}x10^4	24	24	24	24	24	24	24	24

Table I-B. Rate Data-Phenol Runs
(See Table I-A For Reaction Conditions)

Run 1							
Time-Minutes	0	5	15	40			
% Phenol Remaining	100	100	100	100			
% H_2O_2 Remaining	100	100	100	100			
Run 2							
Time-Minutes	0	5	16				
% Phenol Remaining	100	25	6				
% H_2O_2 Remaining	100	83	67				
Run 3							
Time-Minutes	0	4	5	15	40		
% Phenol Remaining	100	100	100	93	77		
Run 4							
Time-Minutes	0	5	15	25	45	75	
% Phenol Remaining	100	72	55	39	28	6	
% H_2O_2 Remaining	100	72	61	61	50	38	
Run 5							
Time-Minutes	0	9	45				
% Phenol Remaining	100	100	100				
Run 6							
Time-Minutes	0	5	15	25	45		
% Phenol Remaining	100	100	100	100	100		
% H_2O_2 Remaining	100	100	100	100	100		
Run 7							
Time-Minutes	0	8	18	48	78	123	153
% Phenol Remaining	100	75	75	50	50	40	40
% H_2O_2 Remaining	100	92	85	69	69	68	68
Run 8							
Time-Minutes	0	5	15	35	75	135	210
% Phenol Remaining	100	75	40	30	25	9	<1
% H_2O_2 Remaining	100	86	82	218	76	70	62

Continued on next page

Table I-B. *Continued*

Run 9								
Time-Minutes	0	5	10	20	40			
% Phenol Remaining	100	47	47	27	7			
% H_2O_2 Remaining	100	100	89	83	83			
Run 10								
Time-Minutes	0	5	10	30	34			
% Phenol Remaining	100	28	17	<1	<1			
% H_2O_2 Remaining	100	83	83	67	67			
Run 11								
Time-Minutes	0	5	10	15	30			
% Phenol Remaining	100	27	20	13	<1			
% H_2O_2 Remaining	100	67	67	67	33			
Run 12								
Time-Minutes	0	5	15	25	37	67	107	
% Phenol Remaining	100	75	50	45	40	16	9	
% H_2O_2 Remaining	100	95	80	45	40	55	55	
Run 13								
Time-Minutes	0	5	15	38	53	68	128	188
% Phenol Remaining	100	100	90	80	55	45	35	25
% H_2O_2 Remaining	100	90	82	73	68	63		
Run 14								
Time-Minutes	0	5	15	45	95	145	235	295
% Phenol Remaining	100	90	90	70	55	45	35	25
% H_2O_2 Remaining	100	987	90	88	86	85	80	76
Run 15								
Time-Minutes	0	10	20	65	125	185	245	305
% Phenol Remaining	100	90	60	55	45	35	25	25
% H_2O_2 Remaining	100	92	91	88	78	74	69	62
Run 16								
Time-Minutes	0	10	33	64	120	180	240	300
% Phenol Remaining	100	90	70	55	55	45	25	25
% H_2O_2 Remaining	100	96	93	90	89	87	84	81

other figures, namely 2-4 and 9-12, were prepared using data from individual runs.

Control Experiments-UV-H_2O_2 vs UV and H_2O_2 Alone for Phenol Destruction. These experiments were conducted to determine the reactivity of phenol with UV-H_2O_2, UV, and H_2O_2 alone under acid and alkaline conditions. At pH 4, 90% of the phenol was destroyed with UV-H_2O_2 at a 8.3/1 mole ratio of H_2O_2/phenol after 15 minutes compared to only a 5% destruction with UV alone and no detectable destruction with H_2O_2 alone (Figure 2). At pH 10, UV or H_2O_2 alone did not result in any detectable destruction of phenol even after 45 minutes. In contrast, UV-H_2O_2 at a 2.8/1 mole ratio of H_2O_2/phenol showed a 72% phenol destruction after 45 minutes (Figure 3). The data showed UV-H_2O_2 is more effective than either UV or hydrogen peroxide under either alkaline or acid conditions.

Effect of UV Intensity on Phenol Destruction with UV-H_2O_2. This phase was carried out to determine if the efficiency of oxidation was directly related to UV intensity. The level of intensity was changed by using two lamps (60,000 uW-sec/cm^2) and eight lamps (240,000 uW-sec/cm^2). After 60 minutes, starting with a phenol concentration of 100 mg/L, 40% and 73% phenol destruction occurred using two lamps (60,000 uW-sec/cm^2) and eight lamps (240,000 uW-sec/cm^2). This shows that a four-fold power increase results in only a 25% increase in phenol destruction (100%).

This data is reinforced in the next phase of the study where the effect of phenol concentration at constant UV intensity was evaluated.

Effect of Phenol Concentration at Constant Light Intensity on the Reaction Rate. A constant light source output of 240,000 uW sec/cm^2 (eight lamps) was used for these experiments (Figure 5). It was significant that the amount of phenol destroyed per unit time was higher in the solutions containing the higher phenol concentrations. For example, after 100 minutes, the rate of phenol destruction was approximately 0.9 mg/min., 2.7 mg/min and 5.2 mg/min at initial phenol concentrations of 100, 400 and 1000 mg/L respectively. Not unexpectedly however, the time to destroy phenol to a certain level e.g., 80% was longer at higher phenol concentrations. To achieve 80% phenol destruction took 58, 189 and 342 minutes respectively at initial phenol concentrations of 100, 400 and 1000 mg/L.

Effect of pH on Phenol Destruction Rate. This factorial study with 1,000 mg/L phenol was conducted over an initial pH range of 6.2-9.8 (Figure 6).

The largest effect of pH was observed during the initial phases of the reaction when the fastest rates

Table II-A. Sustituted Phenol Runs - Summary of Reaction Conditions

Run	17	18	19	20	21
Substituted Phenol	M-Cresol	M-Cresol	2-Chlorophenol	2-Chlorophenol	2,5-Dimethylphenol
pH-Initial	8.0	7.8	7.9	8.0	8.4
pH-Final	7.2	7.5	6.7	7.2	7.6
Temp ^oC-Initial	23	22	22	22	23
Temp ^oC-Final	25	27	26	28	28
Flow Rate-L/min	1.1	1	1	0.9	1.2
Sub Phenol mg/L-Initial $\pm$ 4%	11.5	11.5	13.6	13.6	13
H_2O_2 mg/L-Initial $\pm$ 4%	25	0	20	0	0
H_2O_2/Subphenol Mole Ratio	5.5/1	0/1	5.5/1	0/1	0/1
UV Lamps-Number Used	8	8	8	8	8
Power Output uW-sec/cm^2x10^4	24	24	24	24	24

Run	22	23	24
Substituted Phenol	2,5-Dimethyphenol	2,5-Dichlorophenol	2,5-Dichlorophenol
pH-Initial	8.1	8.5	8.3
pH-Final	7.2	7.6	7.0
Temp ^oC-Initial	24	24	22
Temp ^oC-Final	26	28	26
Flow Rate-L/min	1.1	1	1.2
Sub Phenol mg/L-Initial $\pm$ 4%	13	17.3	17.3
H_2O_2 mg/L-Initial $\pm$ 4%	20	0	20
H_2O_2/Subphenol Mole Ratio	6.2/1	0/1	5.5/1
UV Lamps-Number Used	8	8	8
Power Output uW-sec/cm^2x10^4	24	24	24

Table II-B. Rate Data-Substituted Phenol Runs
(See Table II-A For Experimental Conditions)

Run 17										
Time-Minutes	0	5	10	25	60					
% M-Cresol Remaining	100	31	19	6	<1					
% H_2O_2 Remaining	100	80	60	40	36					
Run 18										
Time-Minutes	0	5	15	30	45	60	90	120	180	240
% M-Cresol Remaining	100	100	88	44	44	44	44	38	19	<1
Run 19										
Time-Minutes	0	5	15	25	40	70				
% 2-Chlorophenol Remaining	100	45	23	9	<1	<1				
% H_2O_2 Remaining	100	100	75	75	75	45				
Run 20										
Time-Minutes	0	5	20	35	60	80	110			
% 2-Chlorophenol Remaining	100	50	30	20	10	<1	<1			
Run 21										
Time-Minutes	0	5	20	30	60	120				
% 2,5 Dichlorophenol Remaining	100	67	17	<1	<1	<1				
Run 22										
Time-Minutes	0	5	20	30	60	80				
% 2,5 Dichlorophenol Remaining	100	40	10	<1						
% H_2O_2 Remaining	100	100	75	75	75	45				
Run 23										
Time-Minutes	0	5	18	33	48	63	93	123	138	210
% 2,5 Dimethylphenol Remaining	100	80	70	60	50	50	50	40	30	<1
Run 24										
Time-Minutes	0	5	30	40						
% 2,5 Dimethylphenol Remaining	100	40	<1	<1						
% H_2O_2 Remaining	100	100	75	75						

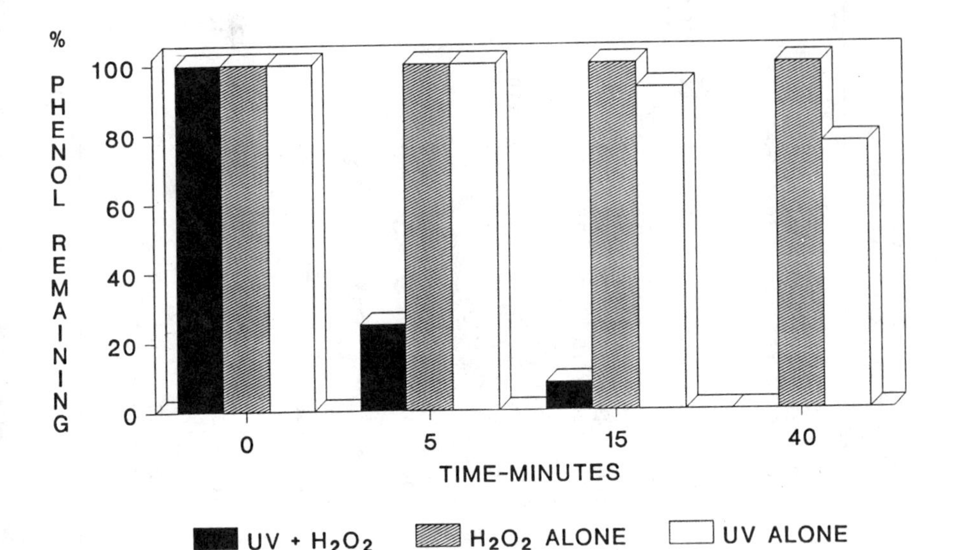

Figure 2. Effect of UV, UV-H_2O_2 and H_2O_2 on phenol destruction. Reaction conditions: 10 mg/L phenol, pH 4, 22°C, 8.3/1 mole ratio H_2O_2/phenol. Data from Table I, runs 1, 2, 3.

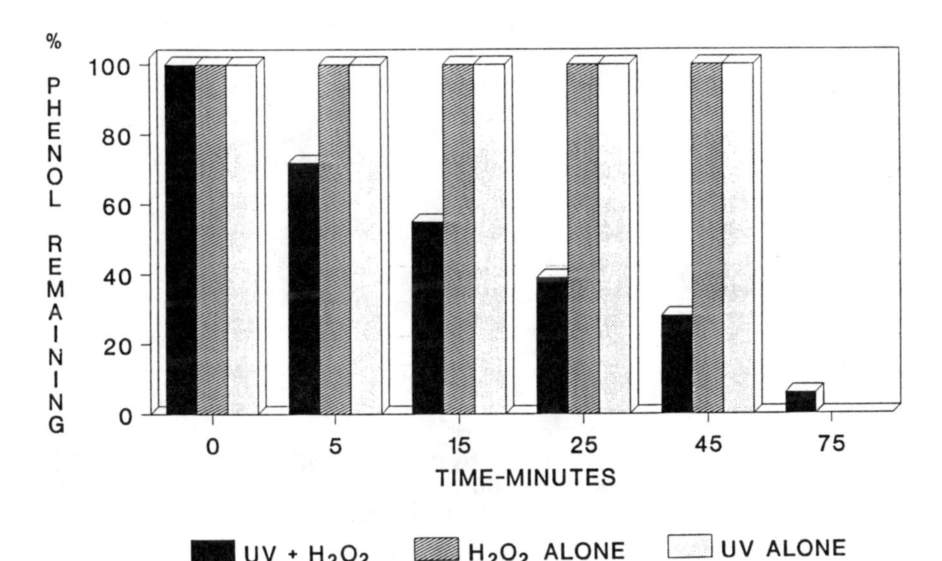

Figure 3. Effect of UV, UV-H_2O_2 and H_2O_2 on phenol destruction. Reaction conditions: 10 mg/L phenol, pH 10, 23°C, 2.8/1 mole ratio H_2O_2/phenol. Data from Table I, runs 4, 5, 6.

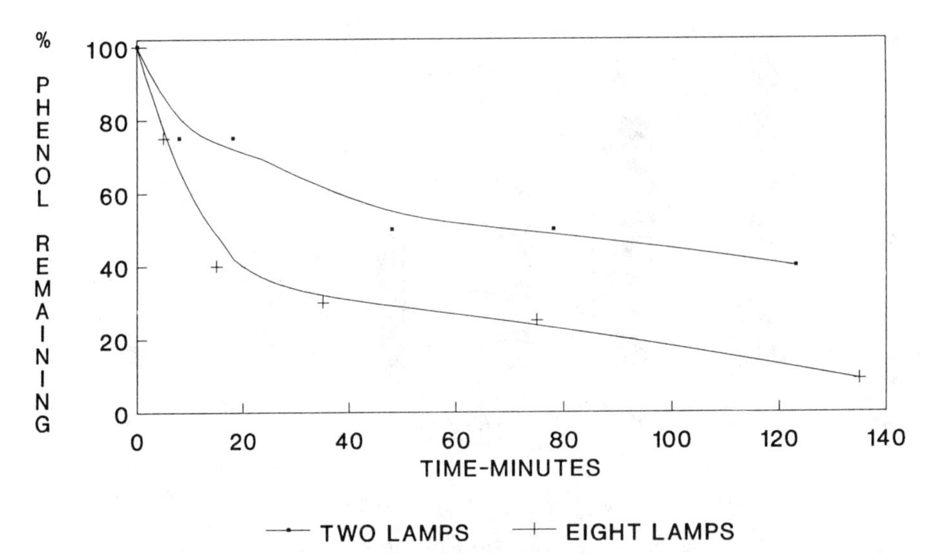

Figure 4. Effect of UV intensity on phenol destruction. Reaction conditions: 100 mg/L phenol, pH 8, 22°C, 5.5/1 mole ratio H_2O_2/phenol. Data from Table I, runs 7, 8.

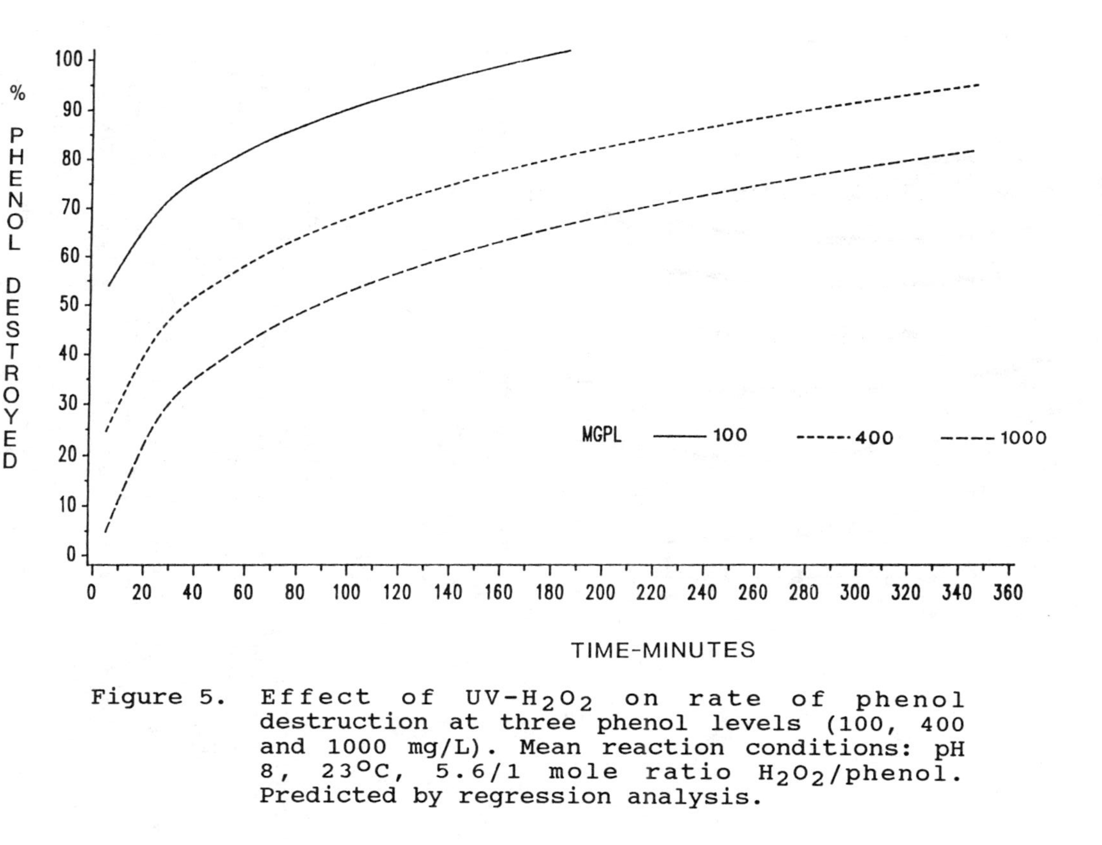

Figure 5. Effect of UV-H_2O_2 on rate of phenol destruction at three phenol levels (100, 400 and 1000 mg/L). Mean reaction conditions: pH 8, 23°C, 5.6/1 mole ratio H_2O_2/phenol. Predicted by regression analysis.

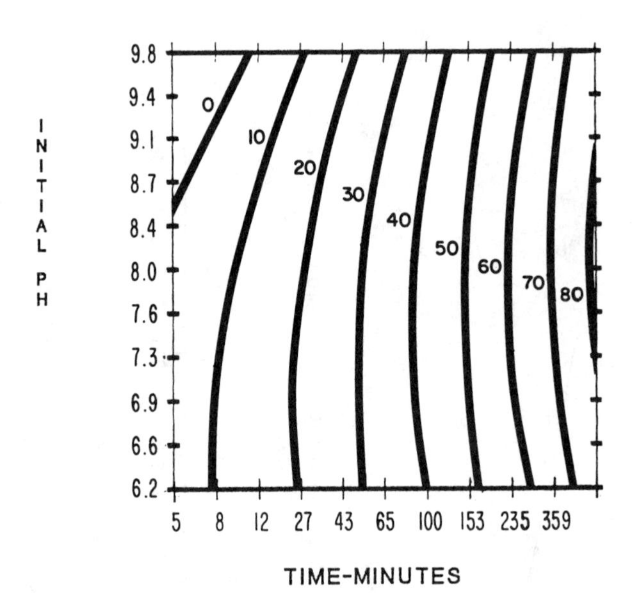

Figure 6. Effect of initial pH on rate of phenol destruction. Mean reaction conditions: 1000 mg/L phenol, 26°C, flow rate 1.7 L/min, 5.5/1 mole ratio H_2O_2/phenol. Predicted by regression analysis.

occurred at pH 6.2 -7.3. The effect of pH on the rate decreased as phenol destruction approached completion, showing a small rate advantage at pH 8.0. The fastest rates of phenol destruction occurred approximately at an initial pH 8.

Effect of H_2O_2/Phenol Mole Ratio on Destruction Rate. The effect of H_2O_2/phenol mole ratio on the rate of phenol destruction depended on the starting phenol concentration.

At low phenol concentration (10 mg/L), higher mole ratios of H_2O_2/phenol yielded the fastest rates of phenol destruction (Figure 7). At high phenol concentration (1000 mg/L) the lower mole ratios of H_2O_2/phenol yielded the fastest reaction rates (Figure 8). The higher concentration of H_2O_2 in the solutions containing the higher phenol concentrations may be increasing UV absorption hence suppressing the free radical oxidation of phenol.

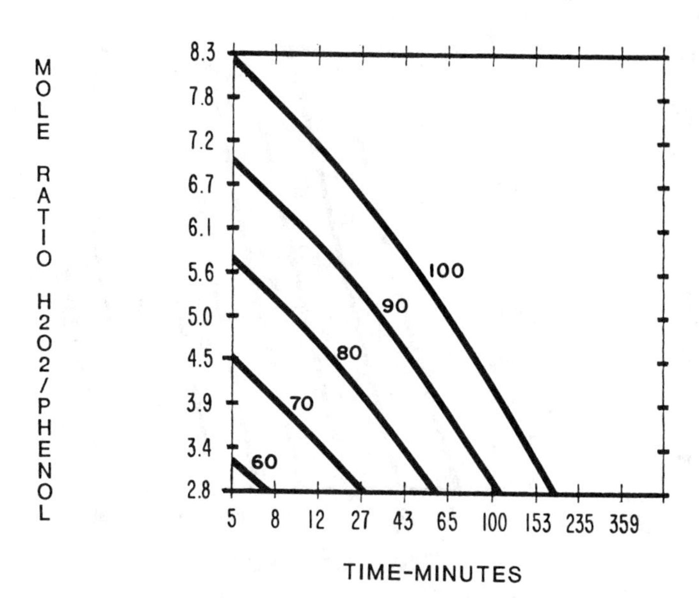

Figure 7. Effect of mole ratio of H_2O_2/phenol on rate of phenol destruction. Mean reaction conditions: 10 mg/L phenol, pH 6.2, 25°C, flow rate 1.5 L/min. Predicted by regression analysis.

Substituted Phenols. The effect of UV-H_2O_2 on substituted phenols was screened at a fixed mole ratio of H_2O_2 /substituted phenol of 5.5/1 and a phenolic concentration of 0.106 mmoles. UV-H_2O_2 significantly increased the rate and degree of destruction of m-cresol, 2-chlorophenol and 2,5-dimethylphenol over UV alone (Figures 9-11). The four substituted phenols were destroyed to 0.1 mg/L or less over a 30 to 60 minute period with UV-H_2O_2. With 2,5-dichlorophenol however, the effect of UV-H_2O_2 was minimal when compared to UV alone (Figure 12). The rate appears to be correlated with the degree of substitution with the higher substituted phenols showing the greater reactivity. With UV-H_2O_2, the reaction rate of 2,5-dimethylphenol and 2,5-dichlorophenol were approximately the same.

The relative reactivity of UV-H_2O_2 with phenol and four substituted phenols at pH 8 is: 2,5-dimethylphenol = 2,5-dichlorophenol > 2-chlorophenol > m-cresol > phenol.

Hydrogen Peroxide Residuals. Examining the UV-H_2O_2 phenol and substituted phenol runs in Tables I and II where the phenolic destruction was over 90%, 33% to 83% of the H_2O_2 remained. The significant amount of H_2O_2 remaining, shows that reductions in H_2O_2 usage could be made while still obtaining satisfactory levels of phenol

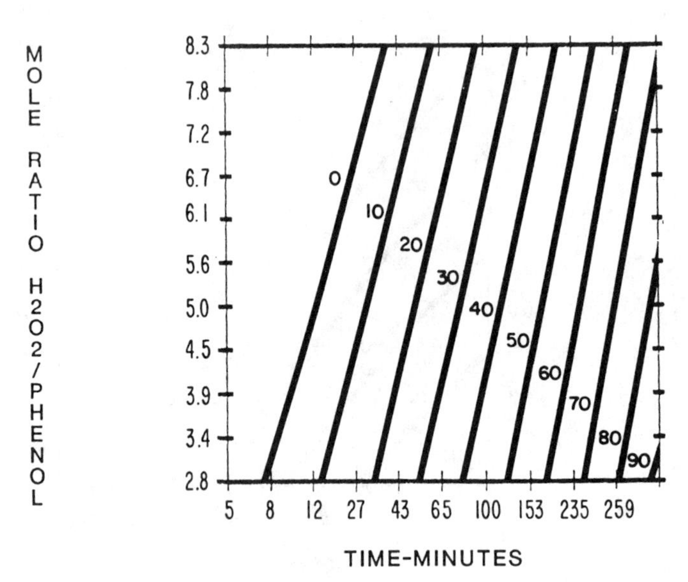

Figure 8. Effect of mole ratio of H_2O_2/phenol on rate of phenol destruction. Mean reaction conditions: 1000 mg/L phenol, pH 6.2, 25°C, flow rate 1.5 L/min. Predicted by regression analysis.

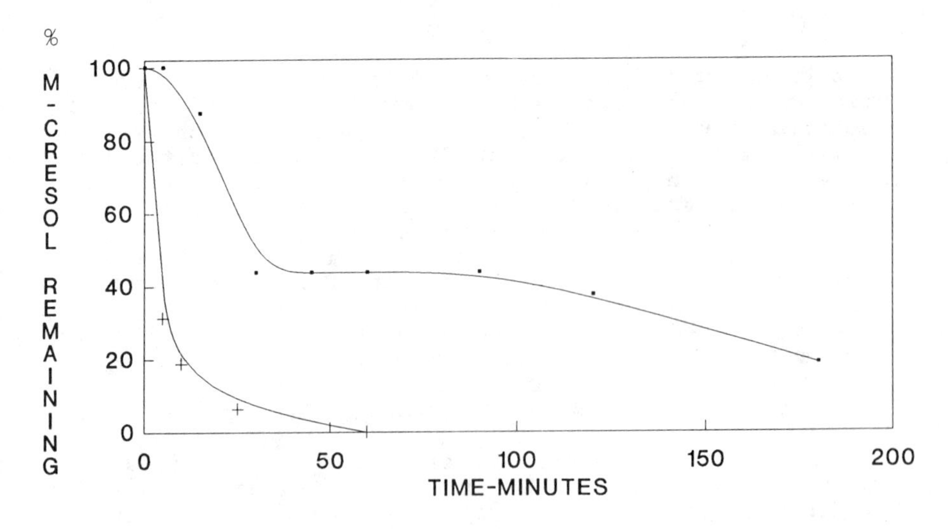

Figure 9. Effect of UV and UV-H_2O_2 on rate of m-cresol destruction. Reaction conditions: 11.5 mg/L m-cresol, pH 8, 22°C. Data from Table II, runs 17, 18.

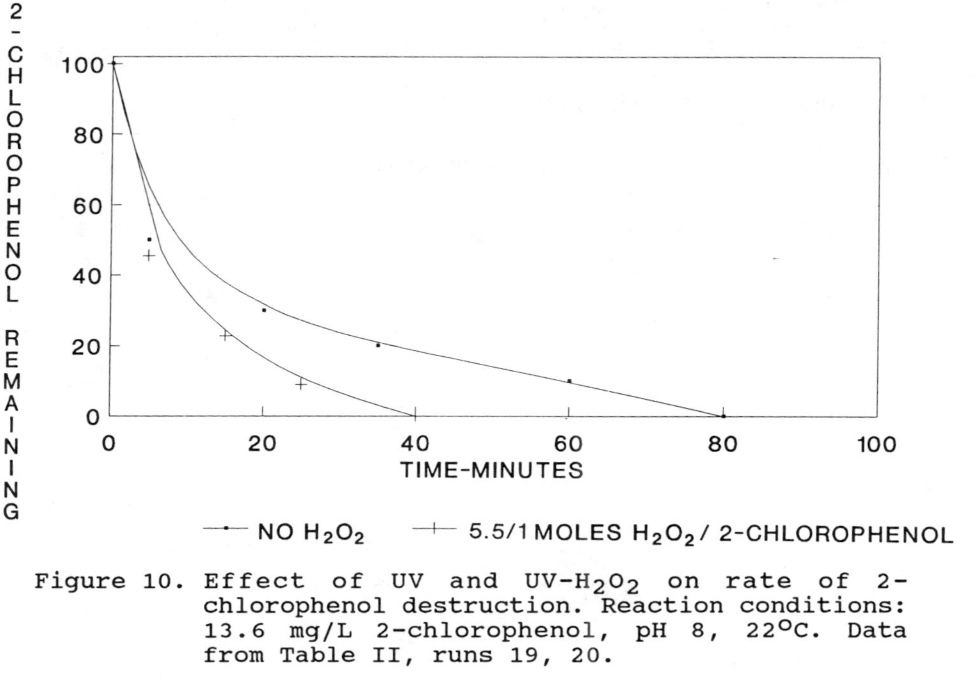

Figure 10. Effect of UV and UV-H_2O_2 on rate of 2-chlorophenol destruction. Reaction conditions: 13.6 mg/L 2-chlorophenol, pH 8, 22°C. Data from Table II, runs 19, 20.

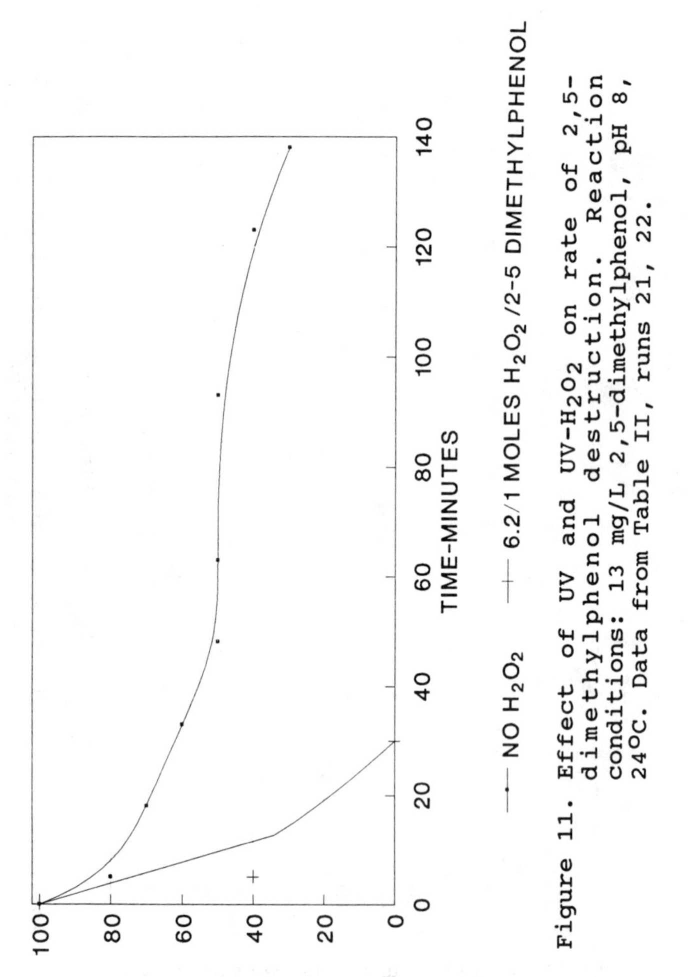

Figure 11. Effect of UV and UV-H_2O_2 on rate of 2,5-dimethylphenol destruction. Reaction conditions: 13 mg/L 2,5-dimethylphenol, pH 8, 24°C. Data from Table II, runs 21, 22.

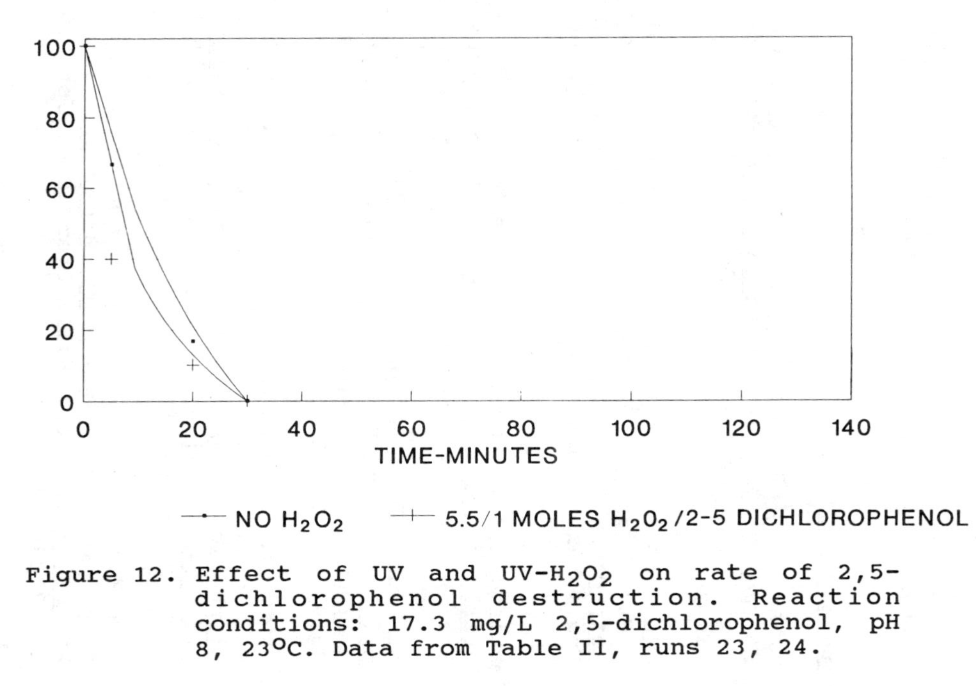

Figure 12. Effect of UV and UV-H_2O_2 on rate of 2,5-dichlorophenol destruction. Reaction conditions: 17.3 mg/L 2,5-dichlorophenol, pH 8, 23°C. Data from Table II, runs 23, 24.

destruction. In some cases, this would result in longer reaction times and increased lamp energy costs. A cost/ benefit analysis would be necessary to determine the optimum economics.

Acknowledgments

The regression equation and the contour plots were prepared by Fred R. Sheldon, Statistician.

Literature Cited

1. Koubek, E.; Photochemically Induced Oxidation of Refractory Organics with Hydrogen Peroxide ; Ind. Eng. Chem. Process Des. Dev. **1975**, 14, No.3, pp. 348-350.

2. Bishop, D.F.; Stern, G.; Fleischman, M.; Marshall L.S.; Hydrogen Peroxide Catalytic Oxidation Of Refractory Organics In Municipal Wastewaters; Ind. Eng. Chem. Process Des. Dev. **1968**, 7, No 1. pp. 110-116.

3. Keating, E.J.; Brown, R.A.; Greenberg, E.S.; Phenolic Problems Solved with Hydrogen Peroxide Oxidation; Ind. Water Eng. Dec.**1978**

4. Eisenhauer, H.R.; Oxidation of Phenolic Wastes Part I Oxidation with Hydrogen Peroxide and a Ferrous Salt Reagent; Water Pollut. Control Fed. Sept **1964**, 36:9, pp.1116-1128.

5. Sundstrom, D.W.; Klei, H.E.; Nalette, T.A.; Reidy D.J.; Weir, B.A.; Destruction of Halogenated Aliphatics by Ultraviolet Catalyzed Oxidation with Hydrogen Peroxide; Hazard. Waste Hazard. Mat. **1986**, 3 No. 1, pp.101-110,

6. Namba K.; Nakayama S.; Hydrogen Peroxide-Catalyzed Ozonation of Refractory Organics. I. Hydroxyl Radical Formation; Bull. Chem. Soc. Jpn. **1982**, 55, pp. 3339-3340.

7. Glaze, W.H.; Kang, J.W.; Chapin, D.H.; The Chemistry of Water Treatment Processes Involving Ozone, Hydrogen Peroxide and Ultraviolet Radiation; Ozone Sci. Tech. **1987**, 9, pp. 335-352.

8. Baxendale, J.H.; Wilson J.A.; The Photolysis of Hydrogen Peroxide at High Light Intensities; Trans. Faraday Soc. **1957**, 53, pp. 344-356.

9. Borup, M.B.; Middlebrooks, E.J.; Photocatalysed Oxidation of Toxic Organics; Wat. Sci. Tech. **1987**, 19, pp. 381-390.

10. Mansour, M.; Photolysis of Aromatic Compounds in Water in the Presence of Hydrogen Peroxide; Bull.Environ. Contam. Toxicol. **1985**, 34 pp. 89-95.

11. Franson, M.A.; Methods for the Examination of Water and Wastewater 16th Ed. American Public Health Association: Washington, DC, **1985**; p 556.

12. Clarke, B.L.; Elving P.J.; Kolthoff, I.M.; Chemical Analysis Vol III Determination of Traces of Metals 3rd Ed. Sandell E.B. Interscience Publishers, Inc.: New York **1959**; p 196.

APPENDIX A

Development of Regression Equation for Destruction of Phenol with UV and H_2O_2

At various H_2O_2/Phenol ratios:

Standard Grade: R Sq.=0.9605 F=104 P>F=0.0001

$$
\begin{aligned}
\%\ Dest = 67.88 &+ 44.71X2 + 7.82X2^2 - 3.83X3 - 5.04X3^2 \\
&+ 11.72X4 - 42.92X6 + 19.07X1X5 \\
&- 17.95X1X6 + 4.50X2X3 = 17.92X2X4 \\
&+ 8.54X2X6
\end{aligned}
$$

Where:

$$
\begin{aligned}
X1 &= (Ratio-2) \\
X2 &= (Ln(Time)-3.748)/2.139 \\
X3 &= (Orig\ pH - 8)/1.8 \\
X4 &= (Flow\ Mean - .39)/.09 \\
X5 &= (Temp\ Mean - 24)/3 \\
X6 &= (Ln(Phenol)-4.46)/2.45
\end{aligned}
$$

Uncertainty limit for this equation at the 5% error level is 12%

RECEIVED January 12, 1990

Chapter 7

Modeling Advanced Oxidation Processes for Water Treatment

Gary R. Peyton

Aquatic Chemistry Section, State Water Survey Division, Illinois Department of Energy and Natural Resources, 2204 Griffith Drive, Champaign, IL 61820

The Advanced Oxidation Processes (AOPs) are water treatment techniques which generate free-radicals in sufficient quantity that they are the primary active species responsible for the degradation of organic contaminants. Although water treatment units based on AOPs have been commercially available since the early seventies, the complex chemistry of such processes has only recently been well enough understood to allow modeling with any generality.

A stoichiometric model for the AOPs such as ozone/UV, ozone/peroxide, etc., has been developed, and we are in the process of verifying its utility for understanding the application of these processes. An eventual goal is the use of the model for prediction of optimum process configurations and parameters, ozone and UV dose requirements, by-products, etc. The model is based on the fact that organic radicals, created by hydroxyl radical attack on organic contaminants, react quickly with oxygen to form peroxyl radicals. These peroxyl radicals can either 1) eliminate superoxide to become stable by-products, or 2) undergo more complicated bimolecular decay to yield hydrogen peroxide and other by-products. The usefulness of the model has already been verified for the methanol/formaldehyde/formic acid system, where superoxide production predominates, and is currently being extended to molecules which do not produce superoxide directly from their peroxyl radicals.

The use of ozone for water treatment is rapidly increasing in the United States. Although ozone is generally considered a powerful oxidizing agent, it is actually quite selective in its reactions. Many drinking water micropollutants of environmental interest react

0097–6156/90/0422–0100$06.00/0

too slowly with ozone for their removal by the direct reaction to be practical. Furthermore, ozonation doses typically used in drinking water treatment are not sufficient to mineralize (convert to carbon dioxide) most micropollutants. As was shown in the early work of Hoigne and Bader (1), the destruction of relatively refractory ($k_{ozone,organic} < \sim 10\ M^{-1}s^{-1}$) compounds during ozonation of water is actually due to the production of hydroxyl radicals which result from ozone decomposition. The development of the Advanced Oxidation Processes (AOPs) is an attempt to take advantage of the rapid, nonselective reactivity of hydroxyl radical.

The AOPs were defined by Glaze (2) as those processes which "...generate hydroxyl radical in sufficient quantity to affect water treatment." The combination of ozone and ultraviolet light (ozone/UV), hydrogen peroxide/UV, and ozone/peroxide are among the most common examples of AOPs. Their greatest value in hazardous waste treatment lies in the ability of hydroxyl radical to eventually oxidize organic contaminants to carbon dioxide. Given the current trends in regulation, the ability to completely eliminate the liability associated with hazardous organic compounds represents a significant advantage over nondestructive treatment methods. Although one of these processes, ozone/UV, has been commercially available since the work of Prengle (3,4) at Houston Research, Inc. in the early seventies, the complex chemistry of the AOPs has only recently been well enough understood to allow detailed modeling of the process. The chemistry of the AOPs has recently been reviewed by Peyton (Peyton, G. R. In Significance and Treatment of Volatile Organic Compounds in Water Supplies; Ram, N.; Cantor, K.; Christman, R.; Eds.; Lewis Publishers, Inc.: Chelsea, Michigan, in press.). The present paper describes a stoichiometric model for the AOPs and presents initial data on its verification for use in understanding and optimizing AOP behavior.

Development and Use of Mechanistic/Kinetic//Mass Balance Models

The mechanistic model has been reported elsewhere (5), and will only be briefly summarized here. It was based on the ozone photolysis studies of Peyton and Glaze (6,7). In those studies, the work of Staehelin and Hoigne (8-11) on base- and peroxide-catalyzed ozone decomposition, and that of Rabani, et al (12), and Ilan, et al (13) on the behavior of peroxyl radicals was adapted to the photolytic ozonation system. Although this version of the model originated during studies of photolytic ozonation, it is generally applicable to ozone/UV, ozone/peroxide, and peroxide/UV treatment. The scheme (5) that includes the reactions which are important in the ozone/UV and ozone/peroxide systems is shown in Figure 1. The extension of this scheme to peroxide/UV treatment requires the addition of the disproportionation reaction of superoxide to form hydrogen peroxide. This reaction is not important in the ozone/UV and ozone/peroxide reaction systems.

When water is treated with ozone, some "utilized" ozone dose, D_u, is taken into solution in the reactor from the feed gas stream. The amount of ozone which is transferred into solution in the reactor

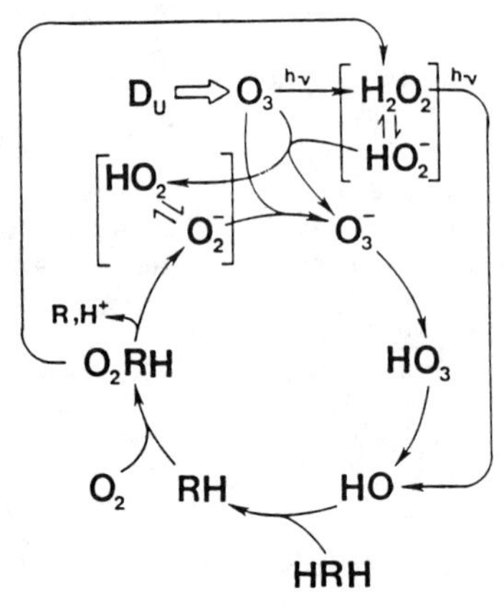

Figure 1. Simplified Reaction Scheme for Advanced Oxidation Processes Using Ozone.

depends on the mass transfer characteristics of the reactor and the concentration of ozone in the feed gas stream and the bulk liquid. This last quantity is usually quite low in these AOPs. Photolysis of aqueous ozone produces hydrogen peroxide (6,7,14). The deprotonated form of peroxide (HO_2^-) can react with ozone (8) to produce ozonide (O_3^-), then hydroxyl radical (OH) (9,11). Hydroxyl radical is a very reactive specie which rapidly attacks organic compounds (typically in $<10^{-6}$ seconds) by abstracting a hydrogen atom or adding to unsaturation, to form an organic carbon-centered radical. The latter reacts quickly with oxygen when present to yield a peroxyl radical. This radical can decompose unimolecularly to produce superoxide (O_2^-, the deprotonated form of the $HO_2\cdot$ radical) or bimolecularly, presumably through a tetroxide structure. The latter pathway produces hydrogen peroxide as a major product, and may in some cases produce some superoxide as well (15). Superoxide reacts quickly with ozone, yielding ozonide and continuing the chain reaction. When ozone is present, this reaction is much faster than superoxide disproportionation to hydrogen peroxide.

The cyclic nature of this reaction scheme suggests that relatively simple but useful mass balance relationships may be derived. Ozone mass balance has been ignored too often in the past literature, presumably on the basis that "autodecomposition" made mass balance impossible. It is now known from the work of Hoigne and Bader (1) that ozone decomposition produces hydroxyl radical, which has very definite and identifiable effects on the chemical system. Therefore, mass balance is now possible in some systems if the appropriate indicator parameters or compounds are followed. In dilute contaminant solutions, the amount of OH lost to reaction with ozone and peroxide must be included by calculation. In natural or potable water, a lack of knowledge of all of the constituents which can react with hydroxyl radical may hinder detailed modeling of the system. For these reasons we chose to do our exploratory work in more concentrated solution, where the calculational burden is eased by the approximation that virtually all OH is consumed by reaction with organic components. After the theory is developed and tested, it can then be extended to more complicated systems, either rigorously, or by the use of surrogate parameters.

The reaction scheme shown in Figure 1 may be used to derive either kinetic or mass balance relationships, depending on whether the rates or cumulative amounts of reactants are considered. The kinetic formulation proves more useful for predicting the instantaneous behavior of a system, while the cumulative model is useful for analyzing laboratory data to determine the course which a reaction has followed. In the following example, we shall use the cumulative model to relate to the variables which are directly observed in the laboratory, i.e., the processing of organic contaminant by the AOP systems. The model compound used for these experiments was methanol (5), since the sequential stable products formaldehyde, formic acid, and carbon dioxide were known and could be easily quantitated. In addition, the hydroxyl radical reaction rate constants of these products are well established. Other compounds have been tested as well, but some assumptions must be made in interpreting that data, since not all of the reaction products could be tracked.

Results of a typical experiment are shown in Figure 2, where the concentration data of the three organic substrates are plotted as the discrete points. The solid lines represent the calculated disappearance curves, using rate constants from the literature. The experimental details are described in the original report (5). The mass balance relationship shown in Equation 1 was derived from the scheme in Figure 1.

$$\frac{R_T}{D_u - A_o - A_h + P_h\,(4/\alpha - 1)} = \frac{1}{(2/\alpha\beta - \gamma - (1 - \gamma)\,\delta)} \quad (1)$$

In this equation, D_u is the cumulative ozone dose which was utilized (removed from the gas stream). Careful measurement and control of the flow and concentration of ozone into and out of the reactor was achieved using a manifold/monitor/flow control system which is described in detail elsewhere (Peyton, G. R.; Fleck, M. J. Submitted to Ozone Sci. Eng. for publication.). Careful measurement of ozone consumption is critical to the use of the mass balance model. R_T is the total amount of organic compound reacted, P_h is the amount of peroxide photolyzed, A_o and A_h are the amount of accumulation of ozone and peroxide, respectively, in the reactor. These variables, which appear on the left-hand side of Equation 1 represent the experimentally measurable quantities. Alpha, beta, gamma and delta are the efficiencies of various steps. The appropriateness of the model may be tested by determining the extent of agreement of the right and left-hand sides of Equation 1, using calculated (e.g., equation 2 below) or reasonable values of the efficiencies. If it is known from the literature, for example, that a particular peroxyl radical decomposes to produce superoxide rather than hydrogen peroxide, it is reasonable to assign a value of unity to γ. Since it is desireable to use the model for predictive purposes, such as process optimization, an eventual goal of model development is to enable the estimation of the value of the parameters γ and δ from the chemical structure of the organic molecule being attacked. For example, it appears that an intermediate peroxyl radical will not be a substantial superoxide producer unless it can eliminate superoxide and hydrogen ion to become a stable organic molecule. Some superoxide may still be produced from the tetroxide, however.

In the kinetic version of the model, the efficiencies are instantaneous values which may be calculated from competition-kinetic terms. The calculation of the instantaneous value of β is shown in Equation 2.

$$\beta = \frac{\sum_i k_i C_{i\,(\text{organics})}}{\sum_j k_j C_{j\,(\text{everything})}} \quad (2)$$

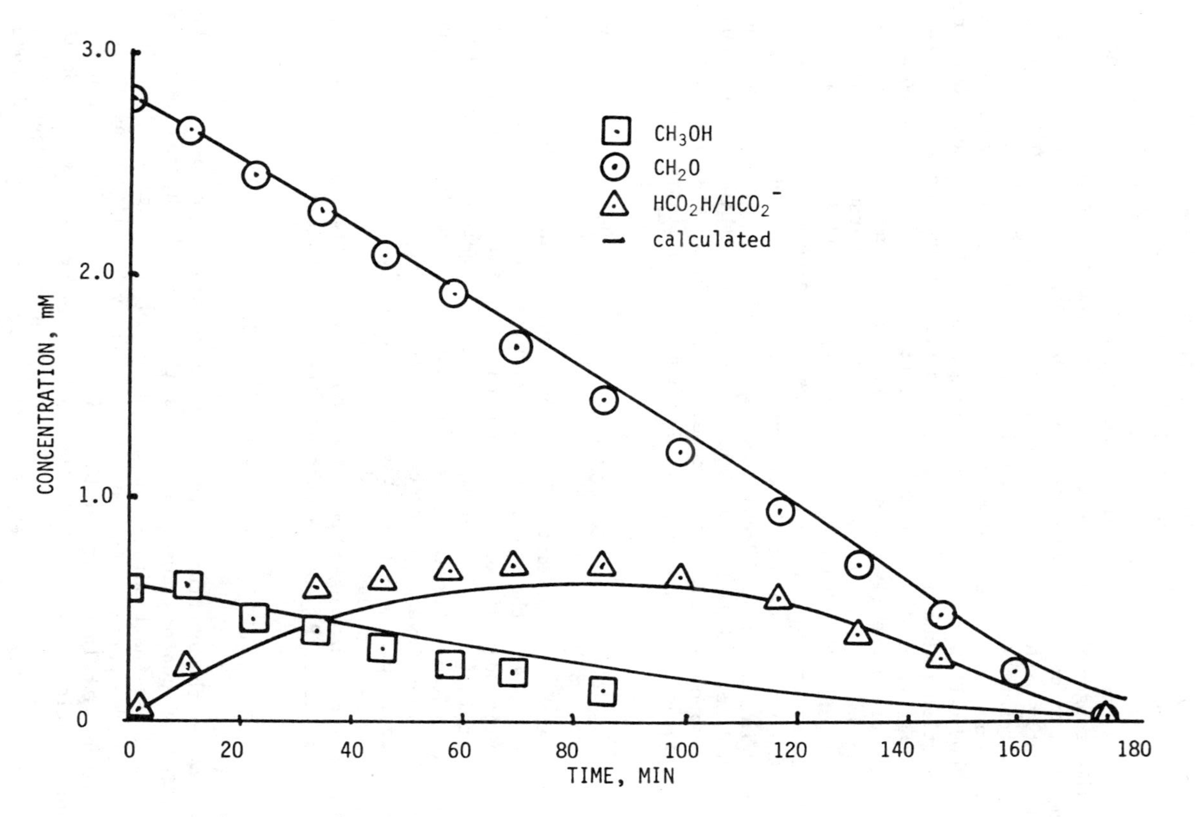

Figure 2. Fit of Calibration Run Data Using Literature Rate Constants.

$$\approx 1 \text{ when organic concentration is high} \tag{3}$$

In this equation, the k's are second-order hydroxyl radical rate constants (L/mol-sec) and the C's are concentrations (mol/L).

In the present (cumulative) example, some efficiencies were evaluated from chemical reasoning and some used as fitting parameters, in order to explore the predictive utility of the model and the range of values which these parameters might take. For example, since the concentration of organic substances is sufficiently high that very little OH could react with anything else, beta was taken to be unity (Equations 2 and 3). Similarly, kinetic calculations were used to demonstrate that bimolecular recombination of $\cdot O_2CH_2OH$ radicals was relatively unimportant. The rate constant for this reaction is known. Therefore, as a first approximation, a value of $\gamma = 1$ was assumed, and no value of δ was needed. Although no side reactions are presently known which would cause the value of α to be less than unity, fixing the values of the other parameters forced α to a value of 0.85 ± 0.03 to obtain the best overall fit. Relaxation of the restriction that $\gamma = 1$ allows the value of α to increase toward unity. Equation 1 was used to analyze laboratory data obtained in nine ozone/UV experiments (5), with the results that the right- and left-hand sides of Equation 1 agreed within a few percent for the above values of the parameters. It is interesting to note that a similar model, using Equation 4 as the initiation step

$$O_3 + H_2O \xrightarrow{hv} \cdot OH + O_2 \tag{4}$$

gave severe disagreement (50-300%) between the right-hand and left-hand sides of the equation which resulted from that model. These results provide additional evidence that the first step in aqueous ozone photolysis is the formation of hydrogen peroxide rather than hydroxyl radical. Despite this and other evidence (6,7,14) to the contrary, many investigators continue to assume that photolysis of aqueous ozone produces available hydroxyl radical directly.

The use of the model described by Equation 1 requires the input of the hydrogen peroxide photolysis rate. The peroxide photolysis rate, P_h, was calculated from Equation 5,

$$P_h = \phi F I_o = \phi\,(1 - 10^{-\epsilon C l'})\, I_o \tag{5}$$

where ϕ is the quantum yield of peroxide disappearance, F is the fraction of photons absorbed, I_o is the flux of photons into the liquid (Einsteins/L·min), ϵ is the extinction coefficient (base 10), and c is the peroxide concentration. The symbol 1' represents the effective path length of the UV radiation, which we have defined as a characteristic parameter of the reactor (5) (Peyton, G. R.; Smith, M. A.; Fleck, M. J.; and B. M. Peyton; manuscript in preparation.). Typical peroxide photolysis data from our laboratory is shown in Figure 3, along with the lines which result from a curve fit of Equation 5. It can be seen by comparison of Equation 5 with Figure 3

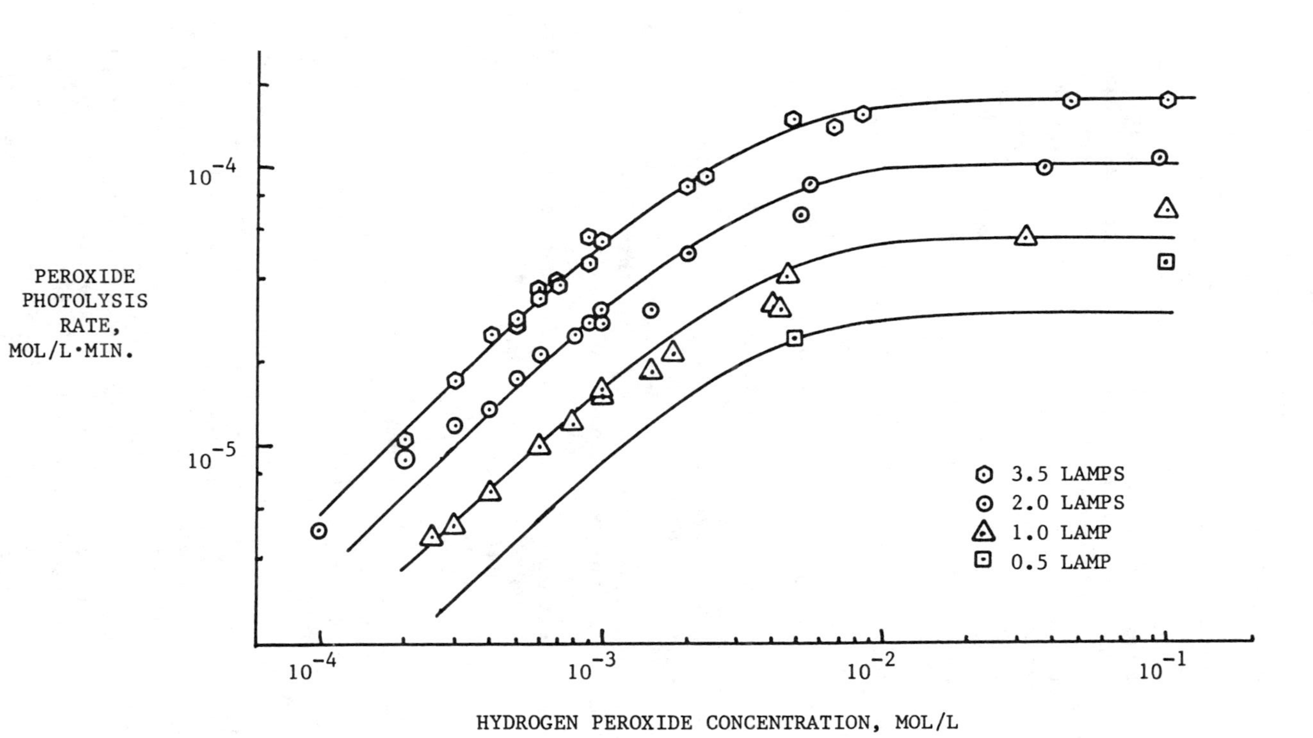

Figure 3. Modeling Hydrogen Peroxide Photolysis Rate in CSTPR.

that in addition to allowing the evaluation of the reactor parameter l′, this method of data analysis also provides an actinometric (chemical) measurement of the UV intensity, I_o.

Predictions of the Model

Although the model is still in the development stage, several consequences have already been either predicted mathematically by the model or have emerged during its use to analyze actual data. The following are observations that relate to the manner in which treatment should be carried out, or to the selection of a particular AOP for treatment of a given stream.

Reaction Phases. When considering AOPs involving ozone, it is convenient to divide the treatment process into three phases, which could be referred to as initiation, destruction, and polishing. These phases differ somewhat from the ones suggested by Lee (16). Initiation involves getting the chain reaction started, either by the Staehelin and Hoigne reaction (Equation 6)

$$O_3 + HO_2^- \longrightarrow OH + O_2^- + O_2 \quad (6)$$

or by ozone photolysis to peroxide followed by peroxide photolysis to hydroxyl radical. In the absence of promoters, initiation may continue well into the destruction phase. This is generally evidenced by an accumulation of hydrogen peroxide.

The destruction phase is that during which the majority of organic removal takes place. Along with that removal occurs the regeneration of superoxide and/or production of peroxide which sustain the reaction. If promoters are abundant, little or no sustaining (additional initiation) of the reaction may be required. This is illustrated diagramatically in Figure 4, where initiation pathways are indicated by the arrows leading to and from hydrogen peroxide, while the promoter pathway is labeled O_2^-. In systems where natural or other complex organic material or reduced transition metals are present, promotion may be caused by electron donators other than superoxide. This has been omitted from Figure 4 in the interest of clarity. The extent of contribution (signified by arrow width) of the various initiator and promoter pathways is a function of solution composition and reaction conditions such as UV intensity, peroxide concentration, etc.

The polishing stage is that in which the concentration of contaminants is reduced to the point that oxidants may be able to successfully compete with contaminants for hydroxyl radical. These reactions were also omitted from Figure 4. When the concentration of organic substrates is low, the majority of hydroxyl radical reacts with ozone, producing superoxide, which converts more ozone to hydroxyl radical. Thus, the system becomes very efficient at ozone destruction but quite inefficient for contaminant removal. During this stage, treatment can be made more effective by reducing the oxidant dose rate, thereby reducing the ozone concentration and thus ozone's ability to compete for hydroxyl radical. For very dilute

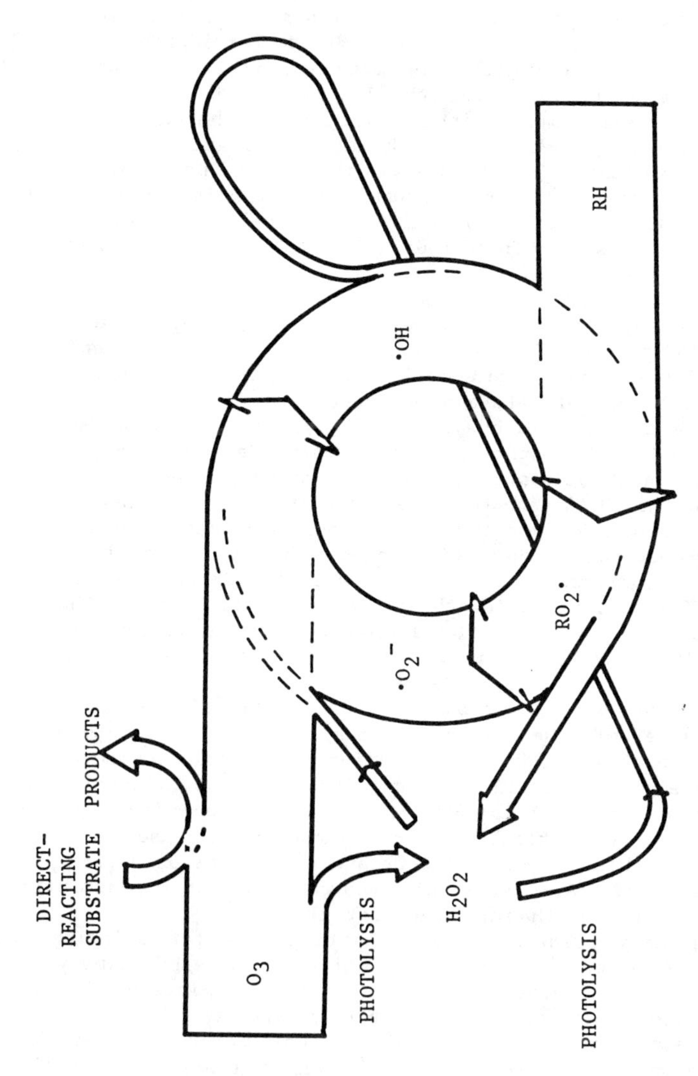

Figure 4. Initiation and Promotion Cycles in the Advanced Oxidation Processes.

contaminants, the entire treatment process could be considered to be in the polishing stage. For more concentrated contaminants, as are frequently encountered in the treatment of hazardous waste, multiple reactors operated at different oxidant dose rates may be economically justifiable.

To illustrate the significance of the balance between these pathways, the destruction of methanol by ozone alone, and by ozone/UV is shown in Figure 5, along with results from an experiment in which the lamps were turned out after a 58-minute initiation period. It is seen that while destruction of methanol by ozone alone is slow, destruction continues after the UV lamps are turned off in the "lights out" experiment, since initiation of the chain reaction has already taken place. Although a significant concentration of peroxide is present (see Figure 5), the rate of methanol destruction can not be accounted for by the Staehelin and Hoigne reaction between ozone and HO_2^- (Equation 6), which was calculated to be a factor of 60 times too slow. This indicates that the chain reaction is sustaining itself.

Ozone/Peroxide Treatment. Since ozone photolysis produces peroxide, it should possible to achieve similar results by replacing UV irradiation with peroxide addition. This is already well established through the work of Nakayama and coworkers (17,18), Duguet, et al, (19,20), and Glaze and coworkers (2,21). Another example is shown in Figure 6, which compares the ozone/peroxide removal of methanol to that by ozone/UV at a similar applied dose rate. The methanol removal rate by ozone/UV is seen to be approximately twice as fast as that by ozone/peroxide. However, if the removal rates are normalized to the utilized dose rates, the efficiencies of removal accomplished by the two AOPs are found to be almost identical. These experiments were run at pH 4-5. Increasing the pH to 6-7 would increase the rate of the Staehelin and Hoigne reaction by two orders of magnitude, presumably bringing the ozone/peroxide reaction rate up to that of ozone/UV.

Figure 6 also provides an additional example of the significance of the different stages in AOP treatment. No consumption of peroxide is seen during the ozone/peroxide treatment. This is because after initiation of the reaction, virtually no more peroxide is required to sustain the chain reaction. Obviously, methanol is a particularly good compound to use in order to demonstrate this effect, since it and all of its oxidation products are good promoters, i.e., their peroxyl radicals produce primarily superoxide, rather than hydrogen peroxide.

In many practical cases, particularly for drinking water, ozone/peroxide may be more economical than ozone/UV, simply because the former is a simpler system to assemble and maintain. However, the above example demonstrates that there are some trade-offs between the two systems. Since the purpose of either UV or peroxide addition during ozonation is to sustain the chain reaction by additional initiation, the system which provides the most efficient initiation is preferable in that sense. Initiation by ozone/peroxide is due to the reaction reported by Staehelin and Hoigne (8), Equation 6.

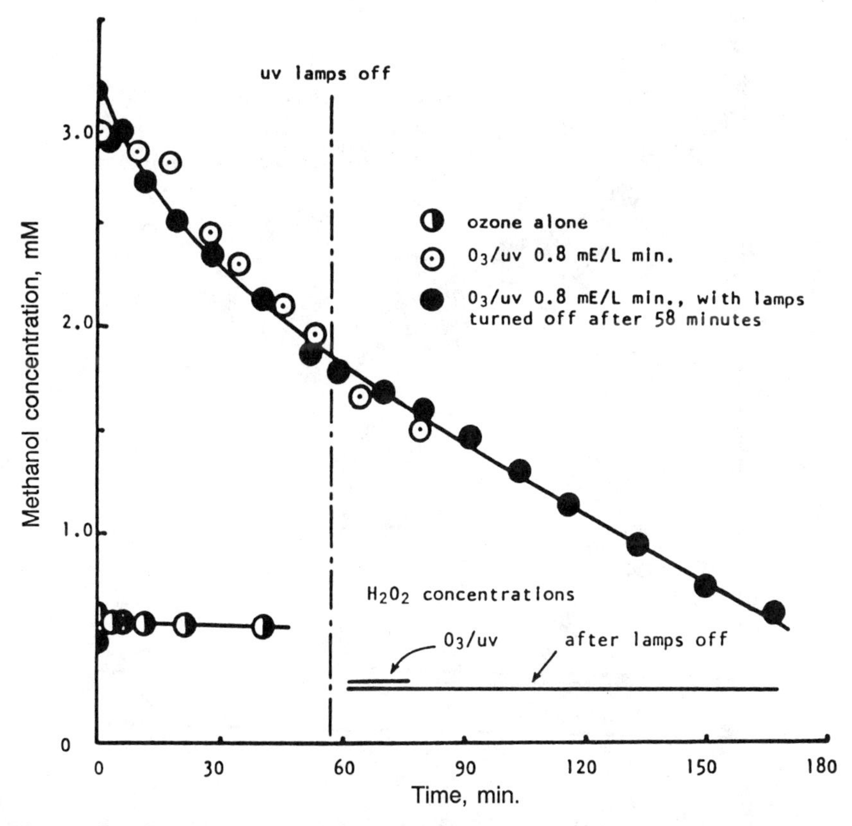

Figure 5. Demonstration of Continuing Free-Radical Reaction After Initiation. Applied Ozone Dose Rate $D_a = 5.26 \times 10^{-5}\ M \cdot min^{-1}$.

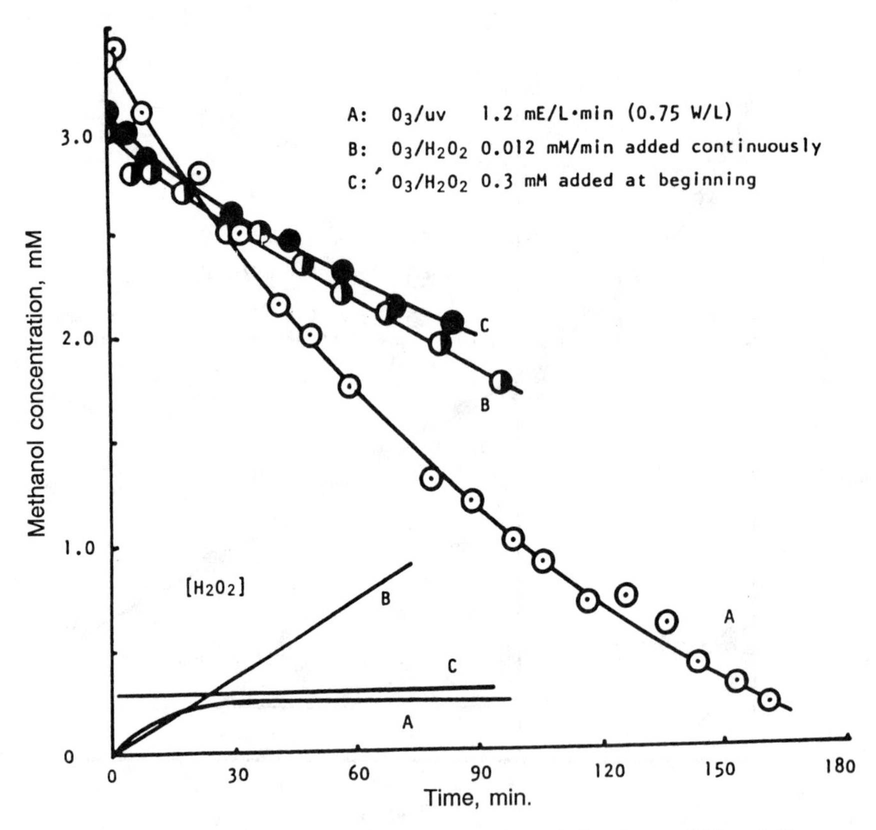

Figure 6. Comparison of Methanol Removal by Ozone/UV and Ozone Peroxide at the Same Applied Dose Rate (5.3×10^{-5} M·min^{-1}).

Initiation in the ozone/UV system may be partly carried out by the same reaction, but as was demonstrated to us by the model, initiation by peroxide photolysis is generally the more important step. This is illustrated in Figure 7 where the relative initiation rate of the two processes (R_p/R_r = rate of photolysis/rate of O_3 - HO_2^- reaction) is plotted versus pH for several different peroxide concentrations. The rate of the Staehelin and Hoigne reaction was calculated from Equation 6, while the rate of photolysis was taken from data obtained in our laboratory (Peyton, G. R.; Smith, M. A.; Fleck, M. J.; and B. M. Peyton; manuscript in preparation.). Below pH 7, initiation by peroxide photolysis is faster than that by the ozone/peroxide reaction.

In laboratory experiments using organic substrates at millimolar concentrations, the pH was found to drop rapidly to 3.5-5.0, due to the production of organic acids. Thus, under these conditions, initiation by peroxide photolysis is much more efficient than by the Staehelin and Hoigne reaction, and ozone/UV treatment may be preferable to ozone/peroxide, despite the fact that the small amount of peroxide required to initiate the reaction must be produced by ozone photolysis. These concentrations are not unreasonable for the types of hazardous waste streams which might be candidates for AOP treatment. On the other hand, for dilute contaminants in natural water or drinking water, the smaller quantities of acids produced by oxidation of the contaminant and the natural organic carbon present may or may not be adequate to overcome the buffering provided by alkalinity. Thus, selection of the appropriate process must be made on a case-by-case basis.

Self-Regulation. The ozone/UV system is "self-regulating," in that UV photolysis of ozone provides an initiator (peroxide) when needed, as at the beginning of treatment. Once the chain production of hydroxyl and superoxide is under way, the rapid reaction of superoxide with ozone keeps the ozone concentration so low that very little photolysis occurs, since the photolysis rate is proportional to the ozone concentration. If at any point in the reaction, the vast majority of the organic molecules in the reactor are not superoxide producers (i.e., if γ in Figure 1 decreases much below unity), then the ozone concentration increases and ozone photolysis again becomes important. This generates peroxide which then sustains the chain by peroxide photolysis as well as by Equation 6. This behavior is mathematically mimicked in the mass balance model.

In the application of the ozone/peroxide system, it is impractical to continually monitor the reaction mixture and adjust the peroxide dose rate. Peroxide may be wasted since the dose rate that is appropriate for initiation may not be needed once the reaction is started. For these reasons, ozone/peroxide may be more suitable for treatment in flow systems with reasonably constant feed water, such as for treatment of municipal water supplies. Ozone/UV treatment may be more suitable for treatment of streams with variable composition, where the self-regulating property may prove to be an economic advantage.

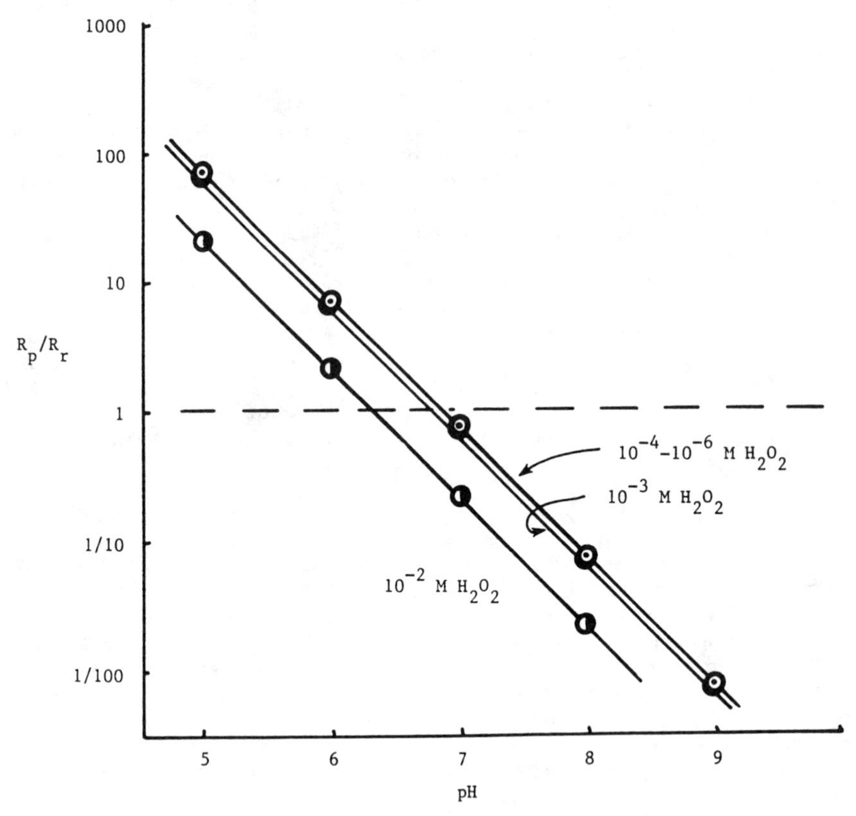

Figure 7. Relative Importance of Chain Initiation by Peroxide Photolysis (R_p) and Ozone/HO_2^- Reaction (R_r) as a Function of pH.

Peroxide/UV. It was seen that hydrogen peroxide was formed and accumulated in the ozone/UV system when the concentration of superoxide producers was low. Under these conditions, and when the pH is less than 7, peroxide photolysis can become a very important hydroxyl-producing pathway. This implies that peroxide/UV may be the treatment system of choice when promoters are absent. Peroxide/UV is simpler in its application than either ozone/UV or ozone/peroxide, in terms of design, operation, maintenance, and understanding. Peroxide/UV treatment is very vulnerable to interference by UV-absorbing solution components, since the extinction coefficient of hydrogen peroxide is very low: 19.6 M^{-1} cm^{-1}, compared to a value of about 3000 for ozone. Additionally, peroxide/UV is more vulnerable to interference by bicarbonate than are ozone/UV and ozone/peroxide, since the carbonate radical anion, formed upon scavenging of hydroxyl by bicarbonate, reacts with and destroys peroxide. In some cases, decarbonation and/or even pretreatment with ozone to decrease UV absorbance may be justified.

Ozone Alone. The original work of Hoigne and Bader (1) showed that hydroxyl radical was responsible for a great deal of organic contaminant removal which occurred during ozonation. Many natural waters contain humic materials which may serve as initiators and/or promoters (10,22). Under these conditions, it may be totally unnecessary to use either UV or peroxide. An example of this is shown in Figure 8, which compares the pilot scale ozone and ozone/UV treatment of trichloroethylene spiked into Savoy, Illinois tap water (Peyton, G. R.; Fleck, M. J. "Field-Scale Evaluation of Aquifer and Wastewater Cleanup Using a Mobile Oxidation Pilot Plant (MOPP)," Research Report RR-037, Hazardous Waste Research and Information Center, Savoy, Illinois, in press.). The two treatments are seen to result in essentially the same removal rate, presumably due to the initiators and promoters present in the water.

Process Selection and Pilot-Scale Testing

The above discussion may help to narrow the choices faced by engineers when confronted by the design of an AOP treatment system for a given task. However, the discussion also emphasizes that presently, the selection and design of a process must be carried out on a case-by-case basis. The above arguments should also help to point out the importance of performing flow experiments at the bench and pilot scale, if the full-sized treatment is to be operated in the flow model.

The conclusions drawn from this work need testing on real-world waste streams, using versions of both processes which are reasonably well optimized. Toward this end, we have assembled the Mobile Oxidation Pilot Plant (MOPP), a pilot scale reactor and ancillary equipment contained in a moving van trailer. The MOPP equipped with computerized data acquisition, research-quality ozone monitor and mass flowmeters for precise mass balance, and an on-board laboratory for near real-time testing. The MOPP will be made available for short-term (3-6 months) testing at field sites, with

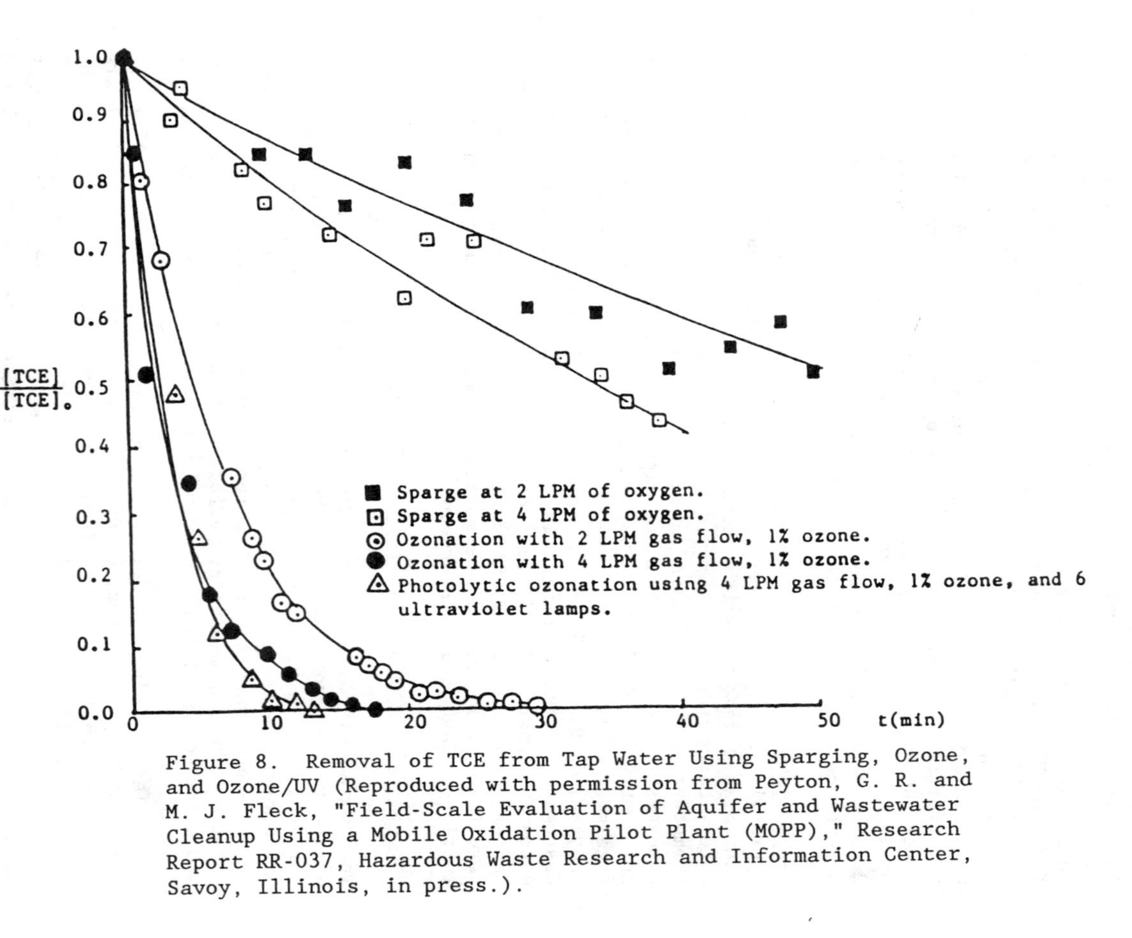

Figure 8. Removal of TCE from Tap Water Using Sparging, Ozone, and Ozone/UV (Reproduced with permission from Peyton, G. R. and M. J. Fleck, "Field-Scale Evaluation of Aquifer and Wastewater Cleanup Using a Mobile Oxidation Pilot Plant (MOPP)," Research Report RR-037, Hazardous Waste Research and Information Center, Savoy, Illinois, in press.).

technical support by the Aquatic Chemistry Section of the Illinois State Water Survey.

Summary and Conclusions

The mechanistic/kinetic/mass balance model presented here, although still in the development and verification stage, shows promise for understanding and interpreting the AOPs. Work is in progress to extend the model to more complicated substrates and streams, as well as to verify the laboratory results at the pilot scale. The following conclusions have resulted from the development and use of the model, so far:

1) Lower oxidant dose rates will be more efficient for the destruction of dilute contaminants, to avoid scavenging of hydroxyl radical by ozone and hydrogen peroxide.

2) Below pH 7, ozone/UV provides more rapid initiation than ozone/peroxide, and can therefore sustain a faster reaction rate when promoters are scarce.

3) The ozone/UV system is self-regulating, and thus can accommodate changes in the promoter content of the reaction mixture more efficiently than can ozone/peroxide. Therefore ozone/UV may be more suitable than ozone/peroxide for streams which vary in composition.

4) In the absence of contaminant photolysis, ozone/peroxide may be more suitable than ozone/UV for treating streams of relatively constant composition, because the former is a simpler system and requires less maintenance.

5) In the absence of promoters, peroxide/UV may be the most appropriate system, since both the ozone/UV and ozone/peroxide system chains will have to be sustained by initiation-type reactions throughout treatment.

6) In the general case, there is no need to simultaneously use both UV and peroxide in combination with ozone. It might be advantageous to use different pairs in successive staged reactors.

7) Selection of an AOP for a particular cleanup task should be made on a case-by-case basis.

Acknowledgments

I would like to acknowledge the efforts of the many coworkers whose results are cited here. I thank Mike Barcelona and the Illinois State Water Survey for allowing me the freedom to pursue this line of investigation, and Pam Beavers for typing the manuscript. I would also like to acknowledge the many contributions of Professor Dr. Jurg Hoigne and his research group to the understanding of aqueous ozone chemistry, upon which the understanding of AOPs is founded.

Literature Cited

1. Hoigne, J.; Bader, T. Water Res. 1976, 10, 377-86.
2. Glaze, W. H.; Kang, J.-W.; Chapin, D. H. Ozone Sci. Eng. 1987, 9, 335-52.

3. Prengle, H. W., Jr.; Hewes, C. G. III; Mauk, C. E. Proc. Second International Symposium for Water and Wastewater Treatment, Montreal, Canada, May, 1975. (Ashton MD: International Ozone Association), 1975, pp 224-252.
4. Prengle, H. W.; Mauk, C. E. Proc. Ozone/Chlorine Dioxide Oxidation Products of Organic Materials, Cincinnati, Ohio, November, 1976, pp 302-320.
5. Peyton, G. R.; Smith, M. A.; Peyton, B. M. Photolytic Ozonation for Protection and Rehabilitation of Ground Water Resources: A Mechanistic Study, University of Illinois, Water Resources Center, Research Report No. 206, 1987.
6. Peyton, G. R.; Glaze, W. H. In ACS Symposium Series, No. 327, Zika, R. G.; Cooper, W. J.; Eds.; 1987, pp 76-88.
7. Peyton, G. R.; Glaze, W. H. Environ. Sci. Technol. 1988, 22, 761-767.
8. Staehelin, J.; Hoigne, J. Environ. Sci. Technol. 1982, 16, 676-681.
9. Staehelin, J.; Hoigne, J. Vom Wasser 1983, 61, 337-348.
10. Staehelin, J.; Hoigne, J. Environ. Sci. Technol. 1985, 19, 1206-1213.
11. Buhler, R. F.; Staehelin, J.; Hoigne, J. J. Phys. Chem. 1984, 88, 2560-2564.
12. Rabani, J.; Klug-Roth, D.; Henglein, A. J. Phys. Chem. 1974, 78, 2089-2093.
13. Ilan, Y.; Rabani, J.; Henglein A. J. Chem. Phys. 1976, 80, 1558-1562.
14. Taube, H. Trans. Farad. Soc. 1957, 53, 656-665.
15. Schuchmann, M. N.; Zegota, H.; von Sonntag, C. Z. Naturforsch. 1985, 40b, 215-221.
16. Lee, M. K. Chem. Water Reuse 1981, 2445-64.
17. Nakayama, S.; Esaki, K.; Namba, K.; Taniguchi, Y.; Tabata, N. Ozone Sci. Eng. 1979, 1, 119-131.
18. Namba, K.; Nakayama, S. Bull. Chem. Soc. Jpn. 1982, 55, 3339-3340.
19. Duguet, J. P.; Brodard, E.; Dussert, B; Mallevialle, J. Ozone Sci. Eng. 1985, 7, 241-257.
20. Duguet, J. P.; Anselme, C.; Mazounie, P.; Mallevialle, J. Water Quality Technology Conference, Baltimore, MD, November, 1987.
21. Glaze, W. H.; Kang, J.-W. J. Amer. Water Works Assn. 1988, 80, 57-64.
22. Peyton, G. R.; Gee, C.-G.; Bandy, J.; Maloney, S. W. In Aquatic Humic Substances: Influence on Fate and Treatment of Pollutants, ACS Advances in Chemistry Series, No. 219; Suffet, I. H.; MacCarthy, P.; Eds.; American Chemical Society, Washington, D.C., 1989; pp 639-661.

RECEIVED January 12, 1990

Chapter 8

Degradation of 2,4-Dichlorophenol in Anaerobic Freshwater Lake Sediments

Juergen Weigel, Xiaoming Zhang, Dava Dalton, and Gert-Wieland Kohring

Department of Microbiology and Center for Biological Resource Recovery, University of Georgia, Athens, GA 30602

In methanogenic sediments the xenobiotic compound 2,4-dichlorophenol is transformed via reductive dechlorination into 4-chlorphenol, which then is sequentially converted via phenol to benzoate (under CO_2 fixation), which via acetate/CO_2/H_2 is degraded finally to CH_4 and CO_2. This proposed sequence is based on determinations of accumulated intermediates and on end products in various enrichment cultures. We conclude that each of the given sequential steps are catalyzed by different anaerobic microorganisms. Transformation of dichlorophenol occurs between 5 and 50°C exhibiting a linear response in the Arrhenius graph between 16 and 30°C. The subsequent transformation steps have similar temperature ranges, with the smallest range for the dechlorination of 4-chlorophenol (17 to 38°C). This dechlorination step apparently is the rate limiting step in the degradation chain of 2,4-dichlorophenol. It also is the most critical one, since accumulated 4-chlorophenol inhibits the other reactions. The pH-optimum for 2,4-dichlorophenol is around 8.5, whereas it is around pH 7.0 for 4-chlorophenol, phenol, and benzoate. Sulfate, nitrate and NaCl (above 2.5%) are inhibitory for the dechlorination. We have obtained a fast growing, 2,4-dichlorophenol dehalogenating, stable enrichment culture. The addition (0.1 % inoculum tested) of this culture to sediments which previously did not exhibit anaerobic dechlorination induced dechlorination of 2,4-dichlorophenol without a noticeable lag period.

It is an important obligation of our society to avoid unnecessary health risks caused through exposures to carcinogenic substances, even at relative low (< 1 ppm) concentrations. The pollution of the environment with chlorinated aromatics is worldwide. Severe contami-

0097–6156/90/0422–0119$06.75/0

nations of groundwater by xenobiotics occur frequently through uncontrolled migration of pollution plumes through the subsurface ground. In other words, we need to control the spread of chemical contaminations in groundwater and drinking water by controlling, directing, and enhancing the degradation of xenobiotics in the subsurface layers before aquifers, the source of our drinking water, become contaminated. A contamination of groundwater with a few ppm 4-chlorophenol makes the water nonpotable. If the right cultures can be obtained, a combined aerobic and anaerobic bioremediation process should be an economically and environmentally safe process to clean up spills and contaminated soil.

To reach such a goal, we need to determine which organisms are required for each individual transformation step involved in the mineralization of such xenobiotics. We need to know what controls and affects the active microbial community in the sediments. But most important, a full understanding is only possible if we understand the enzymatic reactions. Thus, we need to isolate and characterize, first, the key organisms and then, secondly, from these organisms the key enzymes responsible for the dechlorination and total degradation.

Significance of Studying Chlorophenol Transformation and Degradation Under Anaerobic Conditions

Choice of Model Compound. For the below described experiments, we chose chlorophenols as model compounds; specifically we studied 2,4-dichlorophenol, which in anaerobic environments is converted to 4-chlorophenol. This system has several advantages: 2,4-dichlorophenol is a relatively simple compound, it is not too recalcitrant (i.e degradation occurs in weeks), but its first transformation product leads to a more toxic product and a more recalcitrant compound. Thus the degradation occurs in several steps.

Chlorophenols are included in the US-EPA Priority Pollutant Lists (1) since they, especially 4-chlorophenol, are regarded as carcinogenic substances. Generally, chlorinated aromatic compounds have to be regarded as toxic to most organisms, especially higher chlorinated compounds which, due to their lipophilic nature, are increasingly accumulated within a food chain (e.g, 2-9 and lit. cited therein).

Source of Contamination. Various chlorinated aromatic compounds including chlorinated phenols and chlorinated furfurals are produced as chemicals and are then intentionally or by accident released into the environment. They also can be formed unintentionally in various ways. One way is through the chlorination of drinking water which contains aromatic compounds (10]. The source of these compounds can be either from biological origins (e.g., leachate of plant material, degradation products of lignin, etc.) or through industrial pollution (e.g., the effluents from pulping industries or from other chemical industries). A "natural" process during which chlorophenols are formed is a forest fire or generally the combustion of organic matter (11,12). This can lead to an airborne distribution/pollution of remote environments (13). Chlorophenols are used in applications as fungicides in cooling water, paints, and construction materials, as pesticide components, as wood preservatives, and as transformer fluids, to name a few examples (14-17). Large amounts of chlorinated aromatic compounds are produced during the bleaching with chlorine in the paper and pulp industry (18). In one report, chlorophenols have been identified as breakdown products of chlorinated high and low molecular weight lignin (19); however, in another study with a pure culture of *P. chrysosporium* this could not be confirmed (20). Chlorophenols are also formed as "degradation" products from more complex pesticides. Furthermore, Ballschmiter et al. (21) reported the microbial transformation of chlorobenzene (a widespread solvent) to the corresponding chlorophenols.

Biodegradation of Chlorophenols

Aerobic Versus Anaerobic Microbial Degradation of Chlorophenols. Beside chemical (photolysis) degradation (22-24), chlorophenols are slowly degraded by aerobic and anaerobic microorganisms. The aerobic and facultative anaerobic (especially with nitrate as electron acceptor) degradation of halogenated compounds has been studied by several groups. It is a well documented process (e.g. (25-38) to name some representative examples from the last three years). The group of Knackmus in Germany has worked for about 15 years on the mechanisms and on the genetics of the aerobic degradation of haloaromatic compounds. In the US some investigators have worked on developing engineered cultures (e.g, the groups of Chapman and of Focht (39, 40).

However, the aerobic, microaerophilic and facultative anaerobic degradation in surface soil and water can be relatively slow (41; 42), especially with higher substituted chlorinated aromatic compounds. Therefore these compounds may enter the vadose zones, travel through the soil and leak into the aquifer (43-48). Frequently, subsurface zones are anaerobic or rapidly become anaerobic when co-polluted with readily oxidizable substrates. During the recent years it has been shown that chlorinated aromatic compounds can be degraded anaerobically at least as fast as aerobically. Several compounds are degraded even more quickly under anaerobic conditions than under aerobic. The anaerobic degradation occurs via reductive dechlorination (17, 48-52). Thus, anaerobic degradation is most important in subsurface soils and sediments in respect to protecting aquifers through biological degradation of hazardous compounds (48).

Besides using electron acceptors other than oxygen, there are a few more important differences between the aerobic and anaerobic degradation. Frequently pollutants can be totally mineralized through a single aerobic species. It has even been shown that the information for the aerobic degradation of halogenated aromatic compounds can be located on extra chromosomal DNA (plasmids) (53). In contrast to this, the anaerobic degradation of chloroaromatics involves sequentially several organisms which in some steps even depend on a syntropic relationship (45, and lit. cited in there; Wiegel et al. unpubl. results).

Anaerobic Dechlorination of Aromatic Compounds. Much of the work on anaerobic dehalogenation of aromatic compounds has been done with sewage sludge. From the viewpoint of isolating organisms it seems frequently to be a more versatile system to work with than unpolluted soil samples (45; 50-61). The first two steps in the degradation of 2,4-dichlorophenol are a reductive dechlorination. This has been proposed by several researchers (26; 23; 50; 55; 59; 60).

So far, only one anaerobic strain DCB-1, (61) has been isolated that catalyzes the reductive dechlorination of chloroaromatic compounds. It has been isolated from sewage sludge; no organism has been isolated from soil or anaerobic sediments. The isolate, DCB-1, dehalogenates 3-chlorobenzoate. The taxonomic position is still unclear (45). It appears to represent a new type of microorganism with an unusual morphology. Presently it is not known whether this is a unique type of organism or whether it is the first one of a variety of anaerobic organisms able to carry out reductive dehalogenation. It is of high interest to determine whether and in which way this dechlorinater is influenced by the other organisms present in the original enrichment culture and which were required for the total degradation of 3-chlorobenzoate.

Fathepure et al. (54) showed that in sludge hexachlorobenzene was stepwise dechlorinated to 1,3,5-trichlorobenzene which was not further dehalogenated, or to 1,2,4-trichlorobenzene which was dehalogenated to the various dichlorobenzenes. Suflita et al. (48), and Fathepure et al., (54) showed that dehalogenating organisms do not necessarily have to exhibit a high specificity for one specific group of organic compounds. The 3-chlorobenzoate dehalogenating organism, DCB-1, does not dehalogenate chlorophenols, but it reductively dehalogenated 1 ppm tetrachloroethylene. Similar experiences are reported by DeWeerd et al. (55) who studied the relationship between reductive dehalogenation and removal of other aryl substituents by anaerobes. The organisms of our enrichments which dechlorinate 2,4-dichlorophenol, only transforms (as far as we have found) dichloro and higher but no monochlorophenols.

Recently, we have shown that the transformation of 2,4-dichlorophenol in sediments is strongly temperature dependent (56). Dechlorination occurred between 5 and 50°C. The transformation rates increased exponentially between 15 and 30°C, and thus the resulting graph in the Arrhenius plot is a straight line for this temperature range. This means that one can use the Arrhenius function to extrapolate the degradation rates within this temperature range. For example, one can quickly determine the transformation rates at the optimal temperature of 28-30°C and extrapolate from these rates to the slower rates at the lower environmental temperatures e.g., 16°C. This procedure will save time in analyzing degradation potentials in a soil involved in an actual spill. Interestingly, in this temperature range the methane formation showed a similar relationship indicating the possibility that a relationship between methanogenesis and optimal dechlorination rates exists. Interestingly, around 38 - 40°C, no transformation occurred. Two speculations are offered: 1) two different organisms with different T-ranges are involved or 2) in this temperature range the reaction is inhibited by a lack of syntropic interactions between the various required organisms.

The further characterization (see below) of the microbial community which sequentially degrades 2,4-dichlorophenol to methane and CO_2 reveals that the degradation of 2,4-dichlorophenol indeed proceeds via 4-chlorophenol to phenol to benzoate and then via acetate/H_2/CO_2 to methane and CO_2.

Anaerobic Degradation of the Postulated Intermediates Phenol and Benzoate. Knoll and Winter (57) reported that benzoate was formed from phenol when they flushed their sewage sludge fed fermentor with H_2/CO_2. Validly published anaerobes which degrade phenols and/or benzoate in the presence of sulfate to CO_2 and H_2S include the sulfate reducers *Desulfobacterium phenolicum* (phenol, benzoate, cresol, hydroxy-benzoate, etc.) (62) ; *Desulfococcus multivorans*, *Desulfosarcina variabilis* (63); *Desulfotomaculum sapomandens* (benzoate) (64); *Desulfobacterium catecholicum* (65) and *Desulfonema magnum* (66). In contrast to the methanogenic co-cultures, the aromatic compounds can be totally mineralized by a single sulfate reducing organism. The same is true for the facultative anaerobes which degrade aromatic compounds anaerobically and use nitrate as an electron acceptor (47). Under methanogenic conditions, syntropic organisms are more likely to be involved, such as the obligate syntropic *Syntrophus buswelli* (67). This leads to a sequential mineralization. Tiedje et al. (45; 68) postulated that benzoate is degraded to 3 acetates, CO_2 and $3H_2$ or $3H_2S$ depending on which syntropic organism is present. The syntropic partner can be either methanogens (formation of H_2) or sulfate reducers (formation of H_2S). Under methanogenic conditions the H_2 is further converted to CH_4. The products H_2 and acetate are inhibitory for the benzoate degradation, since the reaction is thermodynamically favored toward benzoate (delta $G^{o\prime}$=53 kJ) (69). Schink and Pfennig (70) isolated *Pelobacter acidigallici*, a chemotrophic, non sporeforming organism which is able to utilize neutral phenols. Also the phototrophic *Rhodopseudomonas palustris* (71), is able to photooxidize benzoate under anaerobic/light conditions to benzoyl-S-coenzyme A which is then further degraded to acetyl-S-coenzyme A. The corresponding benzoate-coenzyme A ligase has been characterized recently (72 and lit. cited therein).

Material and Methods

Sampling sites, sampling methods, preparation of samples, media, culture and incubation conditions, as well as the analytical methods have been described before (56) except we used a different wavelength of 275 nm to determine phenol, benzoic and p-hydroxybenzoic acid together with 2,4-dichlorophenol.

The medium for the final enrichment steps for the 2,4-dichlorophenol degrader was made by supplementing filtered lake water with 20 mM potassium / sodium phosphate , pH=8.5, 0.05 % NH_4Cl. 0.25 % NaCl, 0.01% $MgCl_2 \cdot 5\ H_2O$, 0.0025 % $CaCl_2 \cdot 2\ H_2O$, Wolfe's modified trace element and Vitamine solution (56) and between 0.1 and 1%(w/v) yeast extract. The degassed medium was reduced by adding 0.4 g Na_2S and 0.4 g cysteic acid per liter medium.

Results and Discussion

Sequential degradation chain. The first indications for the sequential transformation and degradation of the model compound 2,4-dichlorophenol (Scheme 1) were obtained from accumulated intermediates in respiked cultures and further on in specific enrichment cultures.

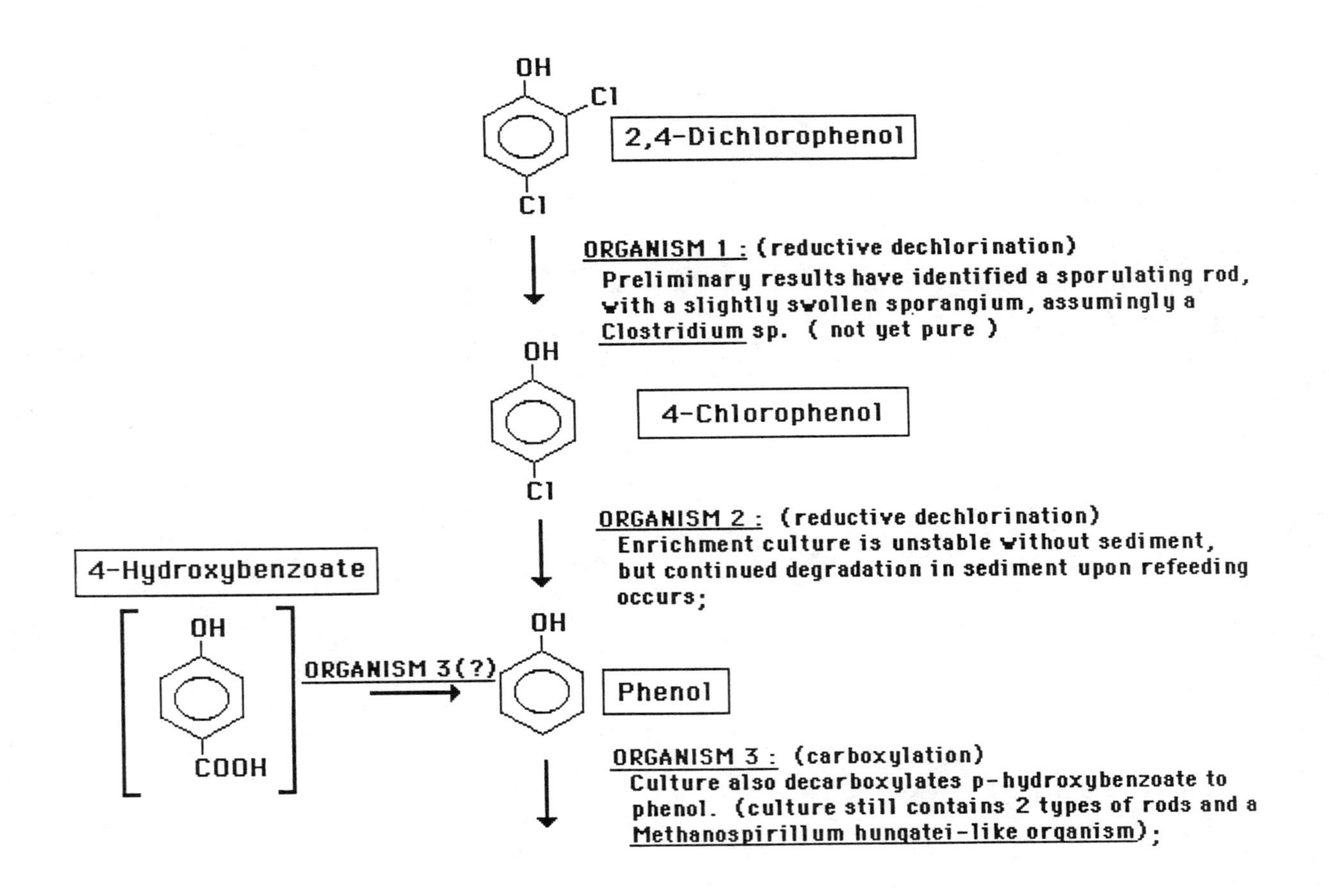
OH
Cl
Cl
2,4-Dichlorophenol
ORGANISM 1 : (reductive dechlorination)
Preliminary results have identified a sporulating rod, with a slightly swollen sporangium, assumingly a Clostridium sp. (not yet pure)
OH
Cl
4-Chlorophenol
ORGANISM 2 : (reductive dechlorination)
Enrichment culture is unstable without sediment, but continued degradation in sediment upon refeeding occurs;
4-Hydroxybenzoate
OH
COOH
ORGANISM 3(?)
OH
Phenol
ORGANISM 3 : (carboxylation)
Culture also decarboxylates p-hydroxybenzoate to phenol. (culture still contains 2 types of rods and a Methanospirillum hungatei-like organism);

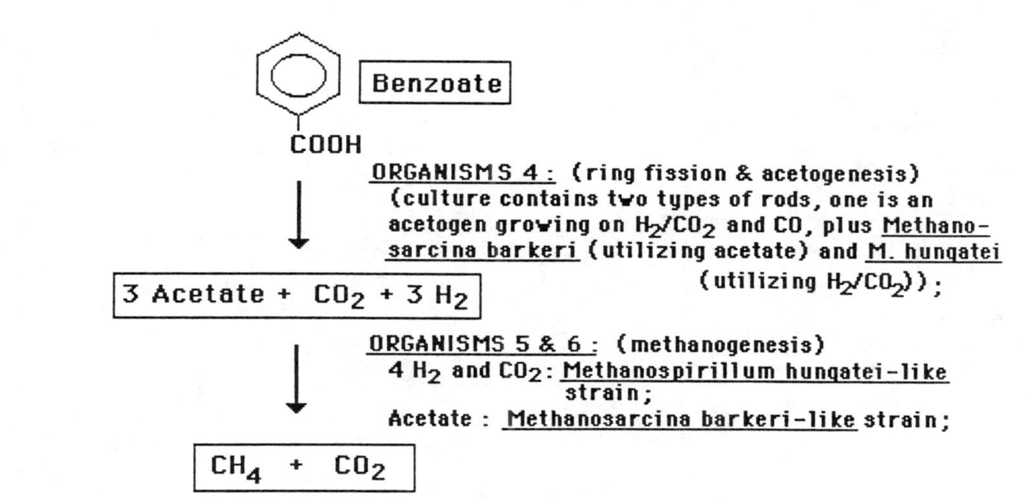

Scheme I. Degradation of 2,4–Dichlorophenol: Enrichment cultures for the various proposed sequential steps.

2,4-dichlorophenol. We respiked sediments, which had been adapted for the transformation of 2,4-dichlorophenol and its subsequent degradation to methane and carbon dioxide, with 2,4-dichlorophenol. It was dechlorinated to 4-chlorophenol with increasing rates up to the third or fourth respike. Dechlorination rates of 50 ppm (310 μmol/l) per day were observed. However, this respiking lead to a high accumulation of the product 4-chlorophenol in stoichiometric amounts as shown in Figure 1. As determined later, long adaptation times for the dehalogenation of 4-chlorophenol (table I) were responsible for the accumulation.

After about 3 weeks of daily respiking, degradation of 4-chlorophenol was evident and the respiking of 2,4-dichlorophenol was stopped.

4-chlorophenol. The produced as well as added 4-chlorophenol was further dechlorinated and phenol accumulated in nearly stoichiometric amounts. In these cultures the next product found was acetate, CO_2 and CH_4. However, when lower concentrations (up to 780 μM) of 2,4 or 4-chlorophenol were used, phenol was not detected as an intermediate presumably due to the fact that phenol degradation rates (table I) were later found to be drastically higher than the 4-chlorophenol transformation rates.

Phenol. When higher phenol concentrations (about 600 uM) were added, benzoate appeared as a product. Again, the later determined benzoate degradation rates were lower than the rates for phenol, thus benzoate had to accumulate when higher concentrations of phenol were converted. However, the higher benzoate degradation rates, when compared to the one for 4-chlorophenol dechlorination, explain why benzoate never accumulated in the sediments. In the phenol degrading enrichment, benzoate was not further degraded. However, p-hydroxybenzoate (but not the ortho- or meta-hydroxybenzoate), when added to the phenol converting enrichment, was immediately transformed in high rates to phenol. Hydroxybenzoate was not detected as an intermediate of 4-chlorophenol or phenol degradation in our sediments or enrichment cultures. We extracted all acids with ether after acidifying the cultures and employing a combined gas chromatography/ mass spectroscopy analysis. Also, in our cultures, added hydroxy-benzoate was not dehydroxylated to benzoate, as observed with a nitrate reducing *Pseudomonas* spec. by Tschech and Fuchs (73). As demonstrated in Figure 2 our cultures decarboxylated added p-hydroxybenzoate to phenol and the accumulated phenol was converted to benzoate via an unknown carboxylation mechanism.

Benzoate. In the early phenol degrading enrichments and in the benzoate degrading enrichment benzoate was degraded to acetate, CO_2 and methane. A benzoate degrading enrichment did not utilize phenol or p-hydroxybenzoate nor 4-chlorophenol. The culture contained two major methanogens (apparently as obligate syntropic partners): *Methanospirillum hungatei* and a *Methanosarcina barkeri*-like organism converting H_2/CO_2, and acetate, respectively, to methane.

The formation of benzoate was seen in three tested freshwater sediments from lakes around Athens, GA., (USA). A similar conversion

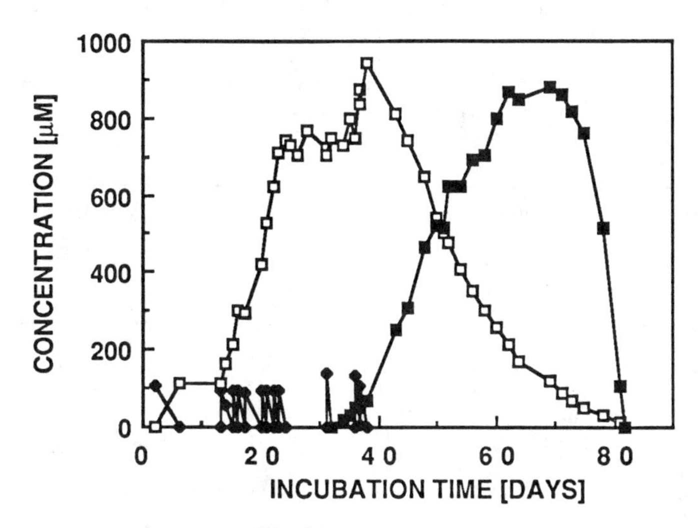

Figure 1: Dechlorination of 2,4-dichlorophenol (-◆-) to 4-chlorophenol in a respiked sample of an anaerobic freshwater lake sediment (nonacclimated), the subsequent accumulation and then dechlorination of 4-chlorophenol (-□-) to phenol (-■-). Sediment samples and methods are described elsewhere (56).

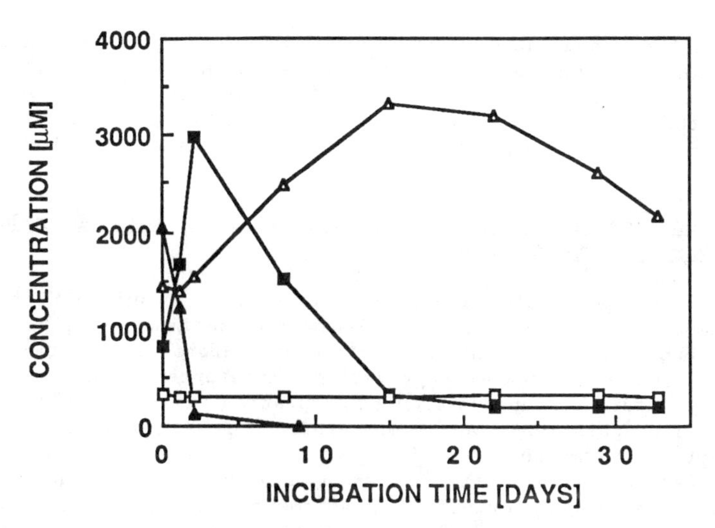

Figure 2: The conversion of p-hydroxybenzoate (-▲-) to phenol (-■-) in a 1:10^{-4} dilution of a phenol-degrading enrichment culture. The intermittent accumulated phenol is converted to benzoate (-△-), but 4-chlorophenol (-□-) is not transformed.

of phenol to benzoate was observed by Knoll and Winter (74) when they purged their sewage sludge fed fermentor with H_2/CO_2.

Table I: Transformation of 2,4-dichlorophenol and of the formed products: Rates and observed adaptation times at 31°C

	Transformation/Degradation Rates* (μM/day)			Adaptation Time (days)		
Compound	Average	Maximal	n**	Average	Miminal	n**
2,4-DCP	245	300	3	7	4	7
4-CP	53	78	4	37	23	5
Phenol	883	2130	5	11	8	16
Benzoate	1180	2080	3	2	0.5	3
p-hydroxy-benzoate***		36000	1	nd	nd	

* degradation rates were obtained in adapted sediments.
** n means the number of the independent samples used for the calculations; 2,4-DCP = 2,4-dichlorophenol; 4-CP = 4-chlorophenol.
*** p-hydroxybenzoate transformation rate was obtained from phenol degrading enrichment culture.

From the accumulated intermediates and the enrichment cultures catalyzing only one or two steps, we constructed the sequential degradation chain depicted in Scheme 1. The lower part of this scheme, the degradation of benzoate, appears to be similar to what has been described by Tiedje for the 3-chlorobenzoate degradation chain (68; 69).

Adaptation Times and Upper Concentration Limits for the Degradation of the Various Intermediates

The different postulated aromatic intermediates required different adaptation times (Table I). In nonadapted sediment samples, 4-chlorophenol required the longest time before degradation was detectable whereas benzoate degradation was detected in less than one day. The differences encountered in adaptation times (table I; Zhang et al. in preparation) and temperature ranges suggest again that different organisms should be involved in the degradation chain.

The long adaptation times are assumed to be the time required for the low number of dechlorinating organisms to grow up before they can exhibit a detectable rate of dechlorination. This has been concluded from enumeration studies (Hale et al., in preparation), from degradation curves of the enrichment culture as depicted in Figure 3 and from curve fittings using mathematical models (J. Struis, unpublished results; Poster # 67 this conference). The obtained transformation/degradation rates and concentration limits

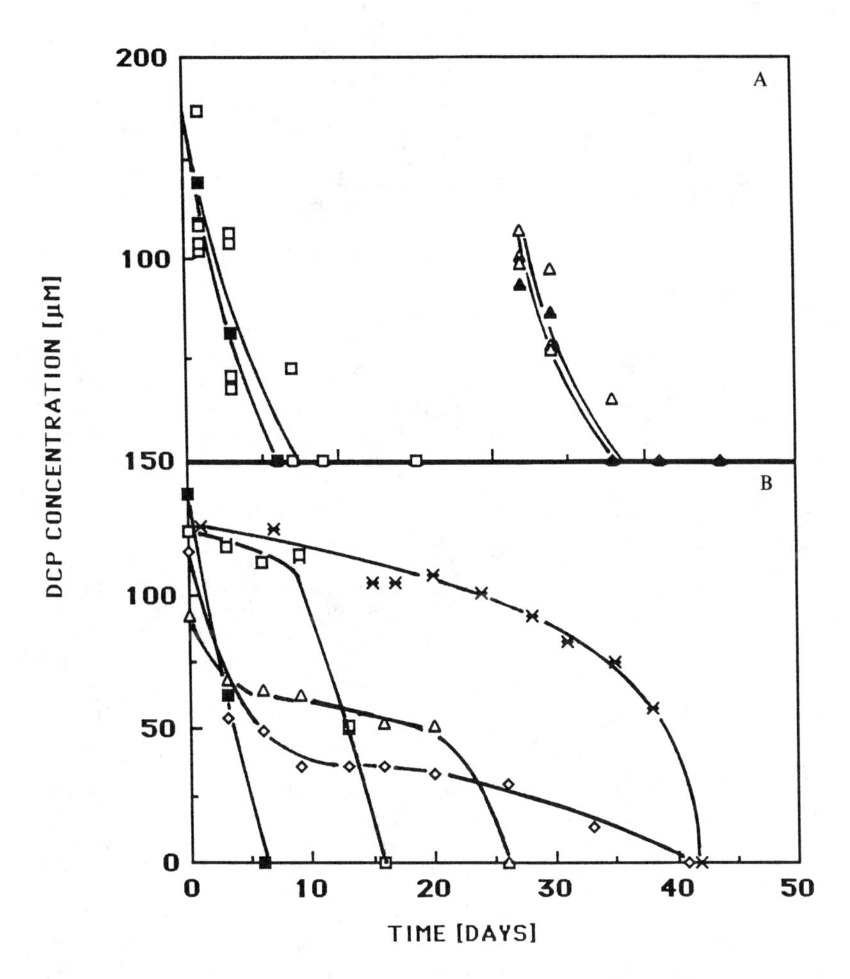

Figure 3: Effect of added carbon sources on the 2,4-dichlorophenol degradation in adapted sediment sample. Closed symbols are the control sediments, i.e., no carbon sources added. The open symbols represent in Figure A: the addition of H_2CO_2, 15 psi; 100 ppm phenol, 100 ppm benzoate; 0.1% cyclohexanone; a mixture of volatile fatty acids (C_3, C_4; and i-C_4) each 0.5% (w/v) and of the corresponding alcohols, respectively. Each of the compounds were tested in 2 parallels; presented data are the average for each compound. The sediment was diluted with 20 mM potassium-sodium phosphate buffer pH 7.0 at a ratio of 1:1 (-■-) and 1:50 (-▲-); in Figure B: addition of 0.5% (w/v) cellobiose (-◇-); fructose plus glucose (-Δ-); pyruvate (-□-); and 0.1% (w/v) adipic acid (-X-). Ratio of sediment to phosphate buffer 1:1.

for a transformation at 31°C were (i.e., the maximal tolerated concentrations in μmole/L of the individual compounds for fresh and for adapted sediments) 1.2 and above 3.1 for 2,4-dichlorophenol, 1.2 and below 3.1 for 4-chlorophenol, 11 and 24 for fresh and 13 and more than 52 with adapted samples for the transformation of phenol and benzoate, respectively. However, the 4-chlorophenol tranformation rates were in most samples significantly lower than for 2,4-dichlorophenol. From these data, it is evident that 4-chlorophenol is the rate limiting step, at least under laboratory conditions.

Temperature and pH Ranges for Degradation of 2,4-Dichlorophenol and Identified Intermediates

The ranges (no figures shown) were as follows: 2,4-dichlorophenol 5-50°C, pH (at least) 5.5-9.0; 4-chlorophenol, 15-40°C; pH 6.0-8.0; phenol 8-50°C, pH 5.5-8.5 and benzoate 4.5-60.5°C, and pH 5.0-8.0. This illustrates, that the 4-chlorophenol transformation has the narrowest boundaries and is therefore a critical step in the degradation chain.

Effect of Added Carbon Sources on the 2,4-Dichlorophenol Degradation in Adapted Sediments

Since the transformations were slow, we added various carbon sources to the sediment samples to obtain enhanced transformation rates and/or to shorten the lag time by providing a suitable carbon and energy sources for the dechlorinating organisms. As can be seen from Figure 4, the addition of the following compounds neither stimulated or inhibited the transformation of 2,4-dichlorophenol: 100 ppm phenol, 15 psi H_2/CO_2, cyclohexanone, and a mixture of 0.5 % each of propionate, butyrate, isobutyrate, propanol, butanol, and isobutanol. The addition of 0.5% pyruvic acid, 0.5% cellobiose and 0.5% each of glucose and fructose caused a delay in the transformation, especially in 1:1 and 1:50 diluted sediment samples. None of the tested compounds led to an increased degradation rate or significant reduction in the transformation rate. However, the adapted sediment samples, i.e., sediment which had consecutively degraded 3 to 4 times 30 ppm chlorophenols, were able to maintain the rates for transforming 30 to 50 ppm 2,4-dichlorophenol, 4-chlorophenol, 2-chlorophenol and 100 ppm phenol over more than 150 days, when the compounds were added every 2-4 days to the sediment. No drastic increase or decrease in the transformation rates were observed after the initial adaptation (Figure not shown). Especially the transformation of 4-chlorophenol was surprising since we were not able to obtain further dechlorination if the samples were diluted 1:2 or higher. We have not yet tested how long an adapted culture will "stay adapted" in the absence of chlorophenols.

Effect of Hydrogen

The dechlorination of 2,4-dichlorophenol in sediment-containing enrichments was stimulated by adding H_2 as gas phase. Our enrichment cultures were kept and grown under a gas atmosphere of 100% H_2 without any inhibition up to 2 atm pressure. Using serum bottles, the optimal

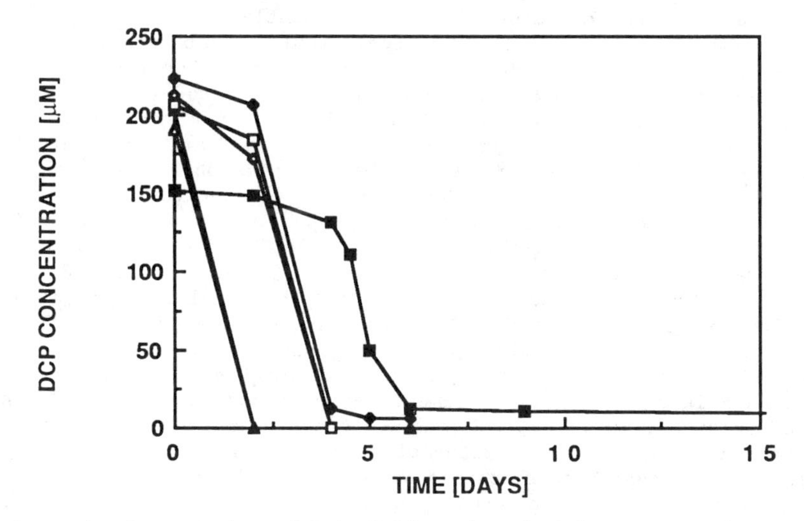

Figure 4: Degradation of 2,4-dichlorophenol with various dilutions (-△- 10^{-1}; -▲- 10^{-2}; -◇- 10^{-3}; -◆- 10^{-4}; -□- 10^{-5}; -■- 10^{-6}) of the final enrichment culture into the mineral medium containing 1% (w/v) yeast extract, pH 8.5.

pressure was around 20 psi. However, the dechlorination of 4-chlorophenol, transformation of phenol to benzoate and the degradation of benzoate were totally inhibited if the gas phase contained as much as 10% H_2 (Figures not shown).

Inhibition of the Various Transformation Steps by the Various Products

The dechlorination of 4-chlorophenol is the slowest reaction in the described degradation chain. Therefore, the influence of the 4-chlorophenol on other reactions was determined. 4-Chlorophenol, up to 2.0 mM, exhibited only a very slight effect on the rate of the 2,4-dichlorophenol dechlorination reaction, but above 4.5 mM total inhibition occurred (Figure not shown). Concentrations between 2 and 4.5 mM leads to nonreproducible values; some of the parallels exhibited an increasing degree of inhibition with increasing concentrations, whereas some did not. Apparently, the effect of the 4-chlorophenol concentrations is strongly influenced by other factors in the sediment. The degradation of benzoate (Figure not shown) was decreased by the addition of more than 0.5 mM, but totally inhibited by the addition of 3 mM 4-chlorophenol. However, 4-chlorophenol had a markedly strong effect on the phenol transformation. Partial inhibition occurred at 200 uM; total inhibition was observed above 1.5 mM 4-chlorophenol as indicated by the arrow in Figure 5A.

Phenol, although more quickly degraded than 4-chlorophenol, decreased the rates of 4-chlorophenol dechlorination when added in concentrations above 2 mM as shown in Figure 5B. Phenol concentrations of 5 mM and above prevented the dechlorination of 4-chlorophenol (2,4-dichlorophenol was not tested). The addition of 100 to 200 μM phenol had a slight activating effect (i.e., decrease in time required for dechlorination) on the 4-chlorophenol transformation reaction. In nonacclimated sediment the presence of phenol reduced the adaptation time (i.e., the time before a measurable degradation can be observed) for 4-chlorophenol dehalogenation. Although, this could indicate that both reactions are catalyzed by one organism, the phenol enrichment could not dehalogenate 4-chlorophenol. Furthermore, the different temperature ranges and adaptation times, do not support this hypothesis.

Enrichment of a 2,4-Dichlorophenol Dechlorinating Culture

Sediments adapted for 2,4-dichlorophenol degradation were used as inoculum for dilution rows. When 1% yeast extract as growth substrate and a pH of 8.5 was chosen, this led to the enrichment of a stable sediment-free culture which dechlorinated 2,4-dichlorophenol but not 2- or 4-chlorophenol (3-chlorophenol not tested). The dechlorinating organism in this enrichment should be an anaerobic sporeformer since a dehalogenating culture withstood several times incubations of 30 min at 85°C. The dechlorinator should not be a sulfate reducer and therefore we assume that it is probably a *Clostridium*. However, so far we were not able to obtain an axenic culture. The enrichment contained still three or four organisms, of which at least one was an acetate producer. Cultures grown in medium containing 1% yeast extract produced 40-50 mM acetate while dechlorinating 150 μM 2,4-dichlorophenol. These cultures contained between 10^7 and 10^8

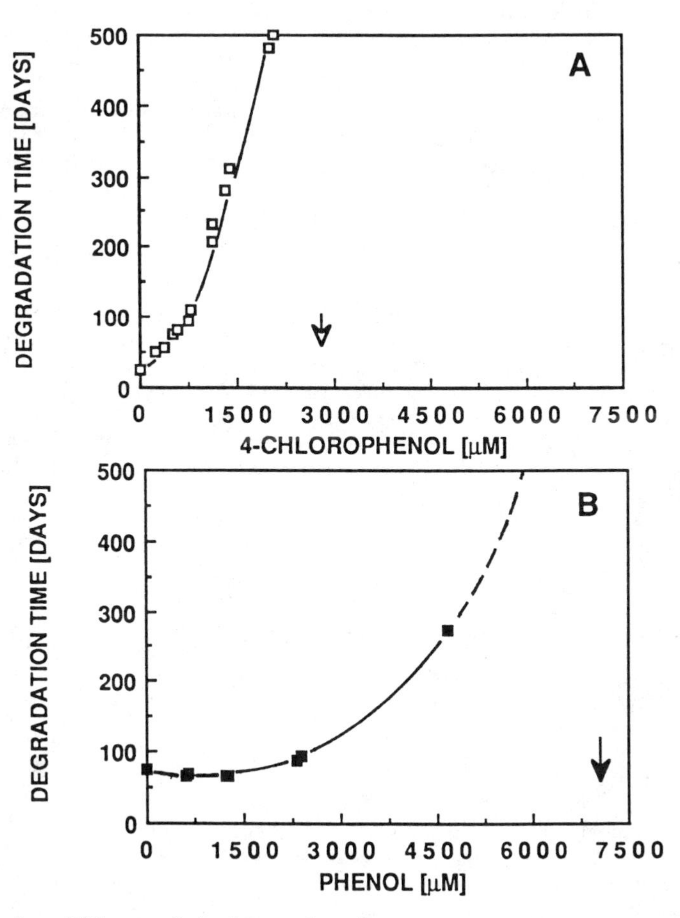

Figure 5: Effect of 4-chlorophenol concentration on the phenol conversion to benzoate in nonacclimated freshwater lake sediments (Figure 5A) and influence of initial phenol concentrations on the dehalogenation of 4-chlorophenol (Figure 5B). (Realize that phenol is significantly faster transformed than 4-chlorophenol so the shown effect is a mixture of an extended lag time and a slow down of the transformation reactions). The arrows indicate concentrations which were not transformed.

dechlorinating cells/ml. However, it is not clear whether the acetate former(s) was also the dechlorinater. Pure strains of a variety of homoacetogenic bacteria could not be induced to dechlorinate 2,4- or 4-chlorophenol. Our enrichment could dechlorinate 1.7 mM 2,4-dichlorophenol totally, but not 3 mM, which was only partly dehalogenated (Figure 6). When cultures which had degraded 150 μM were respiked with 150 μM of 2,4-dichlorophenol, only 30 to 80 μM were dehalogenated. The concomitant respiking with yeast extract increased the further dechlorination only slightly.

The addition of up to 1.5 NaCl, 1 % sodium nitrate, or 0.5 % sodium sulfate to the yeast extract-based medium had no significant effect on the dechlorination. But the addition of 2.5 % nitrate and 3.0% NaCl prevented dechlorination of 2,4-dichlorophenol totally.

Dechlorination Under Sulfate Reducing Conditions

Freshwater sediments from the Athens, GA. area contained between 0.005 and 0.15 mM sulfate. These sulfate concentrations can not support massive growth of sulfate reducers. However, if the sediments were amended with 20 - 25 mM sulfate, sulfate reduction was immediately induced and the dechlorination rates were reduced. In contrast to earlier reports in the literature the dechlorination still occurred, although sulfate reduction took place (75). Dechlorination under sulfate reducing conditions was furthermore demonstrated in the 2,4-dichlorophenol transforming enrichment culture. When sulfate and *Desulfovibrio vulgaris* were added and H_2 in the headgas was exchanged with Argon, dechlorination occurred at a rate reduced by about 60% indicating a competition for <H> between the sulfate reducers and the dechlorinating organisms.

Test of the Enrichment as a Possible Seed Culture

When a marine sediment sample, which previously was unable to dechlorinate 2,4-dichlorophenol, was inoculated with as low as 0.1% (v/v) of the grown up enrichment culture, 180 μM 2,4-dichlorophenol was degraded in less than 6 days. As evident from Figure 7, there was no difference between the 1% and 10% inoculum and only a small increase in the required time for the 0.1% (v/v) inoculation. The sediment had to be diluted 1:1 with 10 mM phosphate buffer, to lower the inhibitory concentration of NaCl found in marine samples. The next experiment should be done as field trial experiment at a contaminated side.

Conclusions

Our results clearly demonstrate that anaerobic dechlorination of chlorinated phenol occurs at noticeable rates in sediments and is strongly temperature dependent. However, the number of dechlorinating organisms appears to be very low and thus long adaptation times are required. The dechlorination and consequent degradation depend on the interaction of several organisms. The dechlorination of the monomeric compound appears to be the most critical step, because it

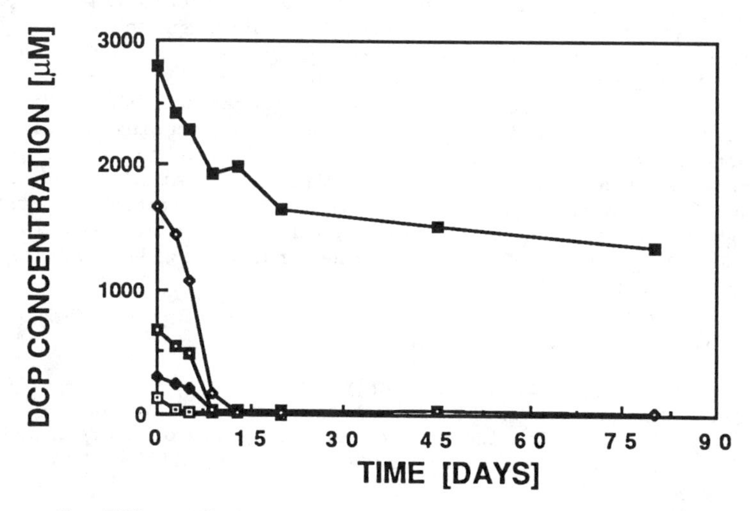

Figure 6: Effect of the initial 2,4-dichlorophenol concentrations on its dehalogenation by the enrichment culture in mineral medium containing 1% yeast extract, pH=8.5.

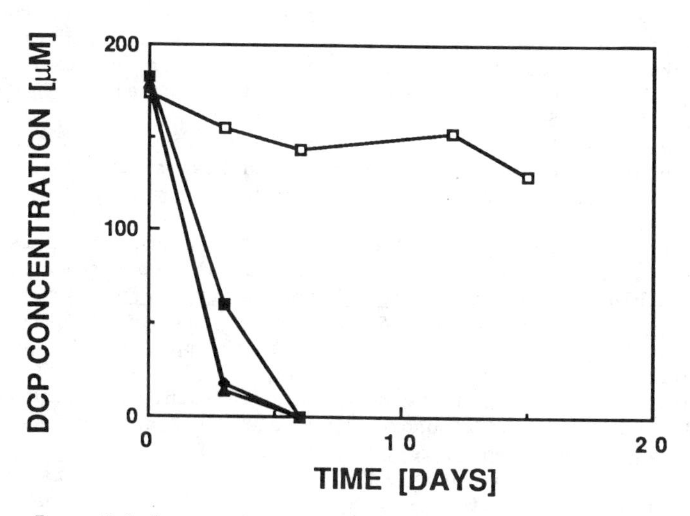

Figure 7: Dehalogenation of 2,4-dichlorophenol in 1:1 with deionized water diluted marine sediments (seagrass bed in a mangrove swamp, Bahamas) after they have been inoculated with the 2,4-dichlorophenol-degrading enrichment. -□- control, no inoculum received; -■- 0.1% (v/v); -◆- 1%; and -▲- 10% inoculum.

is the slowest one, and it occurs over the narrowest pH, temperature, and concentration range.

Using sediment samples which did not exhibit dechlorination, and adding 0.1 % (v/v) of the dechlorinating enrichment culture, the consequent, relatively quickly occurring, successful dechlorination of 2,4-dichlorophenol demonstrates that generally it should be possible to develop successfully seed cultures for bioremediation processes. The dechlorination under sulfate reducing conditions demonstrate that this process is not restricted to methanogenic environments. Under sulfate reducing conditions, it might be necessary to include hydrogen-producing organisms and a suitable substrate. Although, the obtained culture is a good start for degradation of chlorophenols, now more cultures transforming a wider variety of compounds need to be isolated.

Acknowledgments

This work was supported in the early stages through a Cooperative Agreement between the Research Laboratory in Athens of the Environmental Protection Agency and the University of Georgia and in part by a Department of Energy grant (DE-FG09-86ER13614 amendment A003 to J.W.).

Literature Cited

1. Federal Register, 1979, 44:233, Dec. 3 and 1987, 52:131, July 9.
2. Renberg, L. O.; Svanberg, O.; Bentsson, B.-E.; Sundström, G.; Chlorinated guaiacols and catechols bioaccumulation potential in bleaks (Alburnus alburnus, Pisces) and reproductive toxic effects on the harpacticoid Nitocra spinipes (Crustaceae). Chemosphere, 1980, 9:143-150.
3. Renberg, L.; Marell, E.; Sundström, G.; Adolfson-Eric, M.; Levels of chlorophenols in natural waters and fish after an accidental discharge of a wood-impregnating solution. Ambio, 1983, 12:121-123.
4. Lu, P.-Y; Metcalf, R.L.; Cole, L.K.; The environmental fate of ^{14}C-pentachlorophenol in laboratory model ecosystems. In: Pentachlorophenol. Chemistry, Pharmacology and Environmental Toxocology. Environmental Science Research (Rao, K.R.,Ed.). Plenum Press, New York, NY 1978; Vol. 12, pp 53-66.
5. Ehrlich, W.; Mangir,M.; Lochman, E.-R.; The effect of pentachlorophenol and its metabolite tetrachlorohydroquinone on RNA, protein, and ribosome synthesis in Saccharomyces cells; Ecotoxicol. Environ. Safety; 1987, 13:7-12.
6. Pritsos, C.A.; Pointon, M.; Pardini, R.; Interaction of chlorinated phenolics and quinones with the mitochondrial respiration: a comparison of the o- and p-chlorinated quinones and hydroquinones; Bull. Environ. Contam. Toxicol. 1987, 38:847-855.
7. Neilson, A.H.; Allard, A.S.; Reiland, S.; Remberger, M.; Tärnholm, A. Victor, T.; Landner, L.; Tri- and tetrachloroveratrole, metabolites produced by bacterial O-methylation of tri- and tetrachloroguaiacol: an assessment of their bioconcentration

potential and their effects on fish reproduction. Can. J. Fish Aquat. Sci. 1984, 41:1502-1512.

8. Landner, L.; Lindström, K.; Karlsson, M.; Nordin, J. Sörenson, L.; Bioaccumulation in fish of chlorinated phenols from kraft pulp mill bleachery effluents; Bull. Environ. Contam. Toxicol. 1977, 18:663-673.
9. Dougherty, R.C.; Human exposure to pentachlorophenol In: Pentachlorophenol. Chemistry, Pharmacology and Environmental Toxicology. Environmental Science Research (Rao, K.R.,Ed.). Plenum Press, New York, NY 1978; Vol. 12, pp. 352-361.
10. Detrick, R.S; Pentachlorophenol, possible sources of human exposure; Forest Products Journal 1977, 27:13-16.
11. Ahling, B.; Lindskog, A.; Emission of chlorinated organic substances from combustion. In: Chlorinated Dioxins & Related Compounds. Impact on the Environment; Hudsinger, O; Frei, R.W..; Merian, E.; Pocchiari, F. Eds. Pergamon Press Oxford 1982, pp 215-225.
12. Paasivirta, J. Heinola, K.; Humppi, T.; Karjalainen, A.; Knuutinen, J.; Mäntykoski, K.; Paukku, R.; Piilila, T.; Surma-Aho, K.; Tarhaen, J.; Welling, L.; Vihonen, H.; Polychlorinated phenols, guaiacols, and catechols in the environment; Chemosphere 1985, 14:469-491.
13. Salkinoja-Salonen, M.S.; Valo, R.; Apajalahti, J.; Hakulinen, R.; Silaskoski, L.; Jaakola, T.; Biodegradation of chlorophenolic compounds in wastes from wood-processing industry. In: Current Perspectives in Microbial Ecology, Klug, M.J.; Reddy, C.A. Eds.; American Society for Microbiology, Washington DC; 1987, pp. 75-80.
14. Cirelli, D.P.; Patterns of pentachlorophenol usage in the United States of America - an overview; In: Pentachlorophenol. Chemistry, Pharmacology and Environmental Toxicology. Environmental Science Research (Rao, K.R.,Ed.). Plenum Press, New York, NY 1978; Vol. 12, pp. 13-18.
15. Nilsson, C.A.; Norrström, A.; Andersson, K.; Rappe, C.; Impurities in commercial products related pentachlorophenol; In: Pentachlorophenol. Chemistry, Pharmacology and Environmental Toxicology. Environmental Science Research (Rao, K.R.,Ed.). Plenum Press, New York, NY 1978; Vol. 12, pp. 313-324.
16. Rochkind, M.L.; Blackburn, J.W.; Sayler, G.S.; Microbial decomposition of chlorinated aromatic compounds. EPA /600/2-86/090; U.S. Environmental Protection Agency, Cincinnati, OH. 1986.
17. Sahm, K.; Brunner, M.; Schoberth, S.M.; Anaerobic degradation of halogenated aromatic compounds. Microb. Ecol. 1986, 12:147-153.
18. Salkinoja-Salonen, M.S. Apajalahti, J.; Jaakkola, T.; Saarikoski, J. Haulinen, R.; Koistinen, O.; Analysis of toxicity and biodegradability of organochlorine compounds released into the environment in bleaching effluents of kraft pulping. In: Advances in the Identification and Analysis of Organic Pollutants in Water. Keith, L.H.; Ed.; Ann Arbor Science Publishers/The Butterworth Group, Ann Arbor, MI. 1981, Vol. 2, pp. 1131-1164.
19. Neilson, A.H.; Allard, A.-S.; Hynning, P.-A.; Remberger, M.; Landner, L.; Bacterial Methylation of chlorinated phenols and

guaiacols: Formation of veratroles from guaiacols and high-molecular-weight chlorinated lignin. Appl. Environ. Microbiol. 1983, 45:774-783.

20. Pellinen, J.; Joyce, T.W.; Chang, H.-M.; Dechlorination of high-molecular-weight chlorolignin by the white-rot fungus *P. chrysosporium*. Tappi, 1988, 71:191-194.
21. Ballschmiter, K.; Unglert, C.; Heinzmann, P.; Formation of chlorophenols by microbial transformation of chlorobenzenes. Ang. Chem. Int. Ed. Engl. 1977, 16:645.
22. Hwang, H.M.; Hodson, R.E.; Lee, R.F.; Degradation of phenol and chlorophenols by sunlight and microbes in estuarine water; Environ. Sci. Technol. 1986, 20:1002-1007.
23. Boule, P.; Guyon, C.; Lemaire, J.; Photochemistry and environment IV - Photochemical behaviour of monochlorophenols in dilute aqueous solution. Chemosphere 1982, 11:1179-1182
24. Wong, A.S.; Crosby, D.G.; Photodecomposition of pentachlorophenol in Water; J. Agr. Food Chem. 1981, 29:125-130
25. Reineke, W.; Microbial metabolism of halogenated aromatic compounds. In: Microbial Degradation of Organic Compounds. (D.T. Gibson, ed) Marcel Dekker, Inc., New York. 1984, pp. 319-360.
26. Saber, D.L; Crawford, R.L; Isolation and characterization of *Flavobacterium* strains that degrade pentachlorophenol. Appl. Environ. Microbiol. 1985, 50:1512-1518.
27. Steiert, J.G.; Pignatello, J.J.; Crawford, R.L.; Degradation of chlorinated phenols by a pentachlorophenol degrading bacterium. Appl. Environ. Microbiol. 1987, 53:907-910.
28. Van den Tweel, W.J.J.; Kok, J.B. de Bont, J. A. M.; Reductive dechlorination of 2,4-dichlorobenzoate to 4-chlorobenzoate and hydrolytic dehalogenation of 4-chloro-4-bromo, and 4-iodobenzoate by *Alcaligenes denitrificans* NTB-1. Appl. Environ. Microbiol. 1987, 53:810-815.
29. Shimp, R.S.; Pfaender, F.K.; Influence of easily degradable naturally occurring carbon substrates on biodegradation of monosubstituted phenols by aquatic bacteria. Appl. Environ. Microbiol. 1985, 49:394-401.
30. Glecka, G.M.; Maier, W.J.; Kinetic of microbial growth on pentachlorophenol. Appl. Environ. Microbiol. 1985, 49:46-53.
31. Schmidt, E.; Response of a chlorophenol degrading mixed culture to changing loads of phenol, chlorophenol and creosols. Appl. Microbiol. Biotechnol. 1987, 27:94-99.
32. Ruckdeschel, G.; Renner. G.; Effects of pentachlorophenol and some of its known and possible metabolites on fungi. Appl. Environ. Microbiol. 1986, 51:1370-1372.
33. Suflita, J.M.; Miller,G.D.; Microbial metabolism of chlorophenolic compounds in ground water aquifers. Environ. Tax. Chem. 1985, 4:751-758.
34. Larson, P.; Okla, L.; Tranvik, L.; Microbiol degradation of xenobiotic, aromatic pollutants in humic water. Appl. Environ. Microbiol. 1988, 54:1864-1867.
35. Bedard, D.L.; Unterman, R.; Bopp, L.H.; Brennan, M.J.; Haberl, M.L.; Johnson, C.; Rapid assay for screening and characterizing microorganisms for the ability to degrade polychlorinated biphenyls. Appl. Environ. Microbiol. 1986, 51:761-768.

36. Eaton, D.C.; Mineralization of polychlorinated biphenyls by *Phanerochaete chrysosporium*, a lignolytic fungus. Enzyme Microb. Technol. 1985, 7:194-196.
37. Schennen, U.; Braun, K.; Knackmus, H.J.; Anaerobic degradation of 2-fluorobenzoate-degrading, denitrifying bacteria. J. Bacteriol. 1985, 161:321-325.
38. Spain, J.C.; Nishino, S.F.; Degradation of 1.4-dichlorobenzene by a *Pseudomonas* sp. Appl. Environ. Microbiol. 1987, 53:1010-1019.
39. Focht, D.D.; Performance of biodegradative microorganisms in soil: Xenobiotic chemicals as unexploited metabolic niches. Basic Life Sciences 1988, 45:15-30.
40. Chapman, P.J.; Constructing microbial strains for degradation of halogenated aromatic hydrocarbons. Basic Life Sciences 1988, 45:81-96.
41. Ettinger, M.B.; Ruchholft, C.C.; Persistence of chlorophenols in polluted river waters and sewage dilutions. Sewage Ind. Wastes. 1950, 22:1214-1217.
42. Schauerle, W.; Lay, J. P.; Klein, W.; Korte, F.; Long-term fate of organochlorine xenobiotics in aquatic ecosystems. Ecotoxicol. Environ. Safety 1982, 5:560-569.
43. Guthrie, M. A.; Kirsch, E. J.; Wukasch, R. F.; Grady, C.P.L. Jr.; Pentachlorophenol biodegradation II. Water Res. 1984, 18:451-461.
44. Circelli, D.P.; Patterns of pentachlorophenol usage in the United States of America - an overview. In: Pentachlorophenol. Chemistry, Pharmacology and Environmental Toxicology. (K.R. Rao, ed) Plenum Press, New York. 1978, pp. 13-18.
45. Tiedje, J. M.; Boyd, S. A.; Fathepure, B.Z.; Anaerobic degradation of chlorinated aromatic hydrocarbons. Developments in Industrial Microbiology 1987, 27:117-127.
46. Gibson, S.A.; Suflita, J.M.; Extrapolation of biodegradation results to groundwater aquifers: reductive dehalogenation of aromatic compounds. Appl. Environ. Microbiol. 1986, 52:681-688.
47. Berry, D. F.; Francis, A. J.; Bollag, J.M.; Microbial metabolism of homocyclic and heterocyclic aromatic compounds under anaerobic conditions. Microbiol. Rev. 1987, 51:43-59.
48. Suflita, J.M.; Gibson, S.A.; Beeman, R.E.; Anaerobic biotransformations of pollutant chemicals in aquifers. J. Ind. Microbiol. 1988, 3:179-194.
49. Sleat, R.; Robinson, J.P.; The bacteriology of anaerobic degradation of aromatic compounds - a review. J. Appl. Bacteriol. 1984, 57:381-394
50. Suflita, J.M.; Robinson, J.A.; Tiedje, J.M.; Kinetics of microbial dehalogenation of haloaromatic substrates in methanogenic environments. Appl. Environ. Microbiol. 1983, 45:1466-1473.
51. Mikesell, M.D.; Boyd, S.A.; Complete reductive dechlorination and mineralization of pentachlorophenol by anaerobic microorganisms. Appl. Environ. Microbiol. 1985, 52:861-865.
52. Young, L. Y.; Anaerobic degradation of aromatic compounds, In: Microbial Degradation of Organic Compounds. (D.T. Gibson, ed.) Marcel Dekker, New York. 1984, pp. 487-523.

53. Chatterjee, D.K.; Kellogg, S.T.; Hamda, S.; Chakrabarty, A.M.; Plasmid specifying total degradation of 3-chlorobenzoate by a modified ortho-pathway. J. Bacteriol. 1981, 146:639-646.
54. Fathepure, B.Z.; Nengu, J.P.; Boyd, S.A.; Anaerobic bacteria that dechlorinate perchloroethane. Appl. Environ. Microbiol. 1987, 53:2671-2674.
55. DeWeerd, K.A.; Suflita, J.M.; Linkfield, T.; Tiedje, T.M.; Pritchard, P.H.; The relationship between reductive dehalogenation and other aryl substituent removal reactions catalyzed by anaerobes. FEMS Micobiol. Ecol. 1986, 38:331-339.
56. Kohring, G.-W.; Rogers, J.E.; Wiegel, J.; Anaerobic biodegradation of 2,4-dichlorophenol in freshwater lake sediments at different temperatures. Appl. Environ. Microbiol. 1989, 55: 348-353.
57. Knoll, G.; Winter, J.; Anaerobic degradation of phenol in sewage sludge. Benzoate formation from phenol and carbon dioxide in the presence of hydrogen. Appl. Microbiol. Biotechnol. 1987, 25:384-391.
58. Boyd, S.A.; Shelton, D.R.; Anaerobic biodegradation of chlorophenols in fresh and acclimated sludge. Appl. Environ. Microbiol. 1984, 47:272-277.
59. Mikesell, M.D.; Boyd, S.A.; Reductive dechlorination of the herbicides 2,4-D, 2,4,5-T, and pentachlorophenol in anaerobic sewage sludge. J. Environ. Qual. 1985, 14:337-340.
60. Steiert, J.G.; Crawford, R.L.; Microbial degradation of chlorinated phenols. Trends Biotechnol. 1985, 3:300-305.
61. Shelton, D.R.; Tiedje, J.M.; Isolation and partial characterization of bacteria in an anaerobic consortium that mineralizes 3-chlorobenzoic acid. Appl. Environ. Microbiol. 1984, 48:840-848.
62. Bak, F.; Widdel, F.; Anaerobic degradation of phenol and phenol derivatives by *Desulfobacterium phenolicum* sp. nov. Arch. Microbiol. 1986, 146:177-180.
63. Widdel, F.; Pfennig, N.; Dissimilatory sulfate- or sulfur-reducing bacteria. In: Bergey's Manual of Systematic Bacteriology (N.R. Krieg, J.G. Holt, eds). Williams and Williams, Baltimore. 1984, Vol. I. pp. 663-679.
64. Cord-Ruwisch, R.; Garcia, J. L.; Isolation and characterization of an anaerobic benzoate-degrading, spore-forming, sulfate-reducing bacterium, *Desulfotomaculum sapomandens*, sp. nov. FEMS Microbiol. Lett. 1985, 29:325-330.
65. Szewzyk, R.; Pfennig, N.; Complete oxidation of catechol by the strictly anaerobic sulfate-reducing *Desulfobacterium catecholicum* new species. Arch. Microbiol. 1987, 147:163-168.
66. Widdel, F.; Kohring, G.W.; Mayer, F.; Studies on dissimilatory sulfate reducing bacteria that decompose fatty acids. III. Characterization of the filamentous gliding *Desulfonema limicola* gen. nov. sp. nov., and *Desulfonema magnum* sp. nov. Arch. Microbiol. 1983, 134:286-294.
67. Mountfort, D.O.; Bryant, M.P.; Isolation and characterization of an anaerobic syntropic benzoate-degrading bacterium from sewage sludge. Arch. Microbiol. 1982, 133:249-256.

68. Tiedje, J.M.; Stevens, T.O.; The ecology of an anaerobic dechlorinating consortium. Basic Life Science 1988, 45: 3-14.
69. Dolfing, J.; Tiedje, J.M.; Acetate inhibition of methanogenic, syntropic benzoate degradation. Appl. Environ. Microbiol. 1988, 54:1871-1873.
70. Schink, B.; Pfennig, N.; Fermentation of trihydroxybenzenes by *Pelebacter acidgallici* gen. nov. sp. nov., a new strictly anaerobic, non sporeforming bacterium. Arch. Microbiol. 1982, 133:195-201.
71. Harwood, C.S.; Gibson, J.; Anaerobic and aerobic metabolism of diverse aromatic compounds by the photosynthetic bacterium *Rhodopseudomonas palustris*. Appl. Environ. Microbiol. 1988, 54:712-717.
72. Geissler, J.F.G.; Harwood, C.S.; Gibson, J.; Purification and Properties of benzoate-coenzyme A Ligase, a *Rhodopseudomonas palustris* enzyme involved in the anaerobic degradation of benzoate. J. Bacteriol. 1988, 170:1709-1714.
73. Tschech, A.; G. Fuchs, G; Anaerobic degradation of phenol by pure cultures of newly isolated denitrifying pseudomonads. Arch. Microbiol. 1987, 148:213-217.
74. Knoll, G.; Winter, J.; Degradation of phenol via carboxylation to benzoate by a defined, obligate syntropic consortium of anaerobic bacteria. Appl. Microbiol. Biotechnol. 1989, 30:318-324.
75. Kohring, G.-W.; Zhang, X.; Wiegel, J.; Anaerobic dechlorination of 2,4-dichlorophenol in freshwater sediments in the presence of sulfate. Appl. Environ. Microbiol. 1989, 55: (in press).

RECEIVED November 10, 1989

Chapter 9

Determination of Monod Kinetics of Toxic Compounds by Respirometry for Structure–Biodegradability Relationships

Sanjay Desai[1], Rakesh Govind[1], and Henry Tabak[2]

[1]Department of Chemical Engineering, University of Cincinnati, Cincinnati, OH 45221

[2]Risk Reduction Engineering Laboratory, U.S. Environmental Protection Agency, Cincinnati, OH 45268

The key to the evaluation of the fate of toxic organic chemicals in the environment is dependant on evaluating their susceptibility to biodegradation. Biodegradation is one of the most important mechanisms in controlling the concentration of chemicals in an aquatic system because toxic pollutants can be mineralized and rendered harmless. Experiments using an electrolytic respirometer have been conducted to collect oxygen consumption data of toxic compounds from the list of RCRA and RCRA land banned chemicals (phenols and phthalates). The estimation of Monod kinetic parameters were obtained for all the compounds by a graphical method. The first order kinetic constants for the substituted phenols were related to the structure of the compounds by the group contribution method.

It has been estimated that 50,000 organic chemicals are commercially produced in the United States and a large number of new organic chemicals are added to the production list each year (1). The presence of many of these chemicals in the environment could be attributed to inadequate disposal techniques.

Since many of these hazardous chemicals can be detected in wastewater, their fate in wastewater treatment system is of great interest. Of many factors that affect the fate of these compounds, microbial degradation is probably the most important. Biodegradation can eliminate hazardous compounds by biotransforming them into innocuous forms, degrading them

0097–6156/90/0422–0142$06.00/0

by mineralization to carbon dioxide and water. Prediction of the concentration of toxic chemicals in natural and engineered environments and assessing their effects on human and other species due to possible exposure requires information on the kinetics of biodegradation.

Howard et al. (2) have reviewed the different techniques for measuring kinetics of biodegradation. Due to the large number of chemicals, it is imperative that the method selected should not be labor intensive and time consuming. Also, the parameters obtained should be intrinsic; that is, dependant only on the nature of the compound and the degrading microbial community and not on the reactor system used for data collection. If this condition is satisfied, then the parameters obtained can be used for any reactor configuration and can be used in mathematical models to estimate the fate of the toxic chemicals.

It is possible to obtain intrinsic kinetic parameters from batch experiments for a single substrate, provided that the initial conditions are precisely known (3,4). Gaudy et al. (5) have shown that it is possible to use oxygen consumption to determine substrate removal because oxygen utilization during microbial energy production. So, it is possible to obtain intrinsic kinetic parameters from oxygen consumption data from single batch experiments. Grady et al. (6) also demonstrated that the values of the parameters obtained from oxygen uptake are in general agreement with values determined by more traditional methods.

The measurement of oxygen consumption is one of the oldest means of assessing biodegradability, and respirometry is one technique for measuring oxygen consumption. In respirometric methods substrate samples, along with biomass, are kept in contact with a gaseous source of oxygen. Oxygen uptake by microorganisms over a period of time is then measured by change in volume or pressure of the gas phase. An alkali is included in the apparatus to absorb carbon dioxide produced during biodegradation. Samples are usually incubated at constant temperature and are kept away from light, but the procedures vary with different instruments.

Electrolytic respirometry is the most commonly employed method for biodegradation studies. Generally it is used for biochemical oxygen demand (BOD) determination with the exception of studies by Dojlido (7), Oshima et al. (8), Gaudy et al. (9) and Grady et al. (6), who used it for determination of biodegradation kinetics. The data collection is automatic with electrolytic respirometry, so manual collection and analysis of samples is not required to follow the course of substrate degradation. Since the respirometer can be automated, the length of the experiment is less important and thus unacclimated biomass can be used in the experiments.

The objective of this study was to find degradation rates of the Resource Conservation and Recovery Act

(RCRA) and RCRA land banned compounds with the use of mixed cultures obtained from a wastewater treatment plant, so that the rates obtained in the laboratory could be used to predict degradation rates in the environment. Chemicals were obtained from Aldrich Chemical Company (Milwaukee, WI) and were reported 99+% pure. The phenols and phthalates tested were phenol, o-cresol, m-cresol, p-cresol, 2,4-dimethyl phenol, dimethyl phthalate, diethyl phthalate, dipropyl phthalate, dibutyl phthalate and butyl benzyl phthalate.

Experimentation

Electrolytic respirometry studies were conducted using an automated continuous oxygen measuring Voith Sapromat B-12 electrolytic respirometer (Voith-Morden, Milwaukee, WI). This system consists of a temperature controlled water bath which contains the measuring units, a recorder for digital indication, a plotter for direct presentation of the oxygen uptake curves of substrates, and a cooling unit for conditioning and continuous recirculation of water bath contents. The system used had 12 measuring units each connected to the recorder. Each unit, as shown in Figure 1, consisted of a reaction vessel A, with a carbon dioxide absorber (soda lime) mounted in a stopper, an oxygen generator B, and a pressure indicator C. Interconnected by hoses, the vessels form a sealed measuring system so that barometric pressure fluctuations do not adversely affect the results. The magnetic stirrer in the sample to be analyzed provides vigorous agitation, thus ensuring effective gas exchange. Microorganism activity in the sample creates a pressure reduction that is recorded by the pressure indicator which controls both the electrolytic oxygen generation and plotting of the measured values. The CO_2 generated is absorbed by soda lime.

The N_2/O_2 ratio in the gas phase above the sample is maintained constant throughout the experiment by an on/off feedback control loop. As oxygen is depleted, pressure reduction is created in the sample and, as a result, the level of the sulphuric acid in the pressure indicator rises and comes in contact with a platinum electrode. This completes the circuit and triggers the generation of oxygen by electrolysis. The electrolytic cell provides the required amount of oxygen to the reaction vessel by electrolytic dissociation of a $CuSO_4$-H_2SO_4 solution thus alleviating the negative pressure. As the level of the electrolyte in the pressure indicator decreases, the contact with the electrode is broken. This switches off the electrolytic cell. The amount of oxygen supplied to the sample is recorded directly in milligrams per liter by the recorder. The recorder is connected to a microcomputer which records data from the measuring units every 15 minutes.

The nutrient solution is made as per Organization of

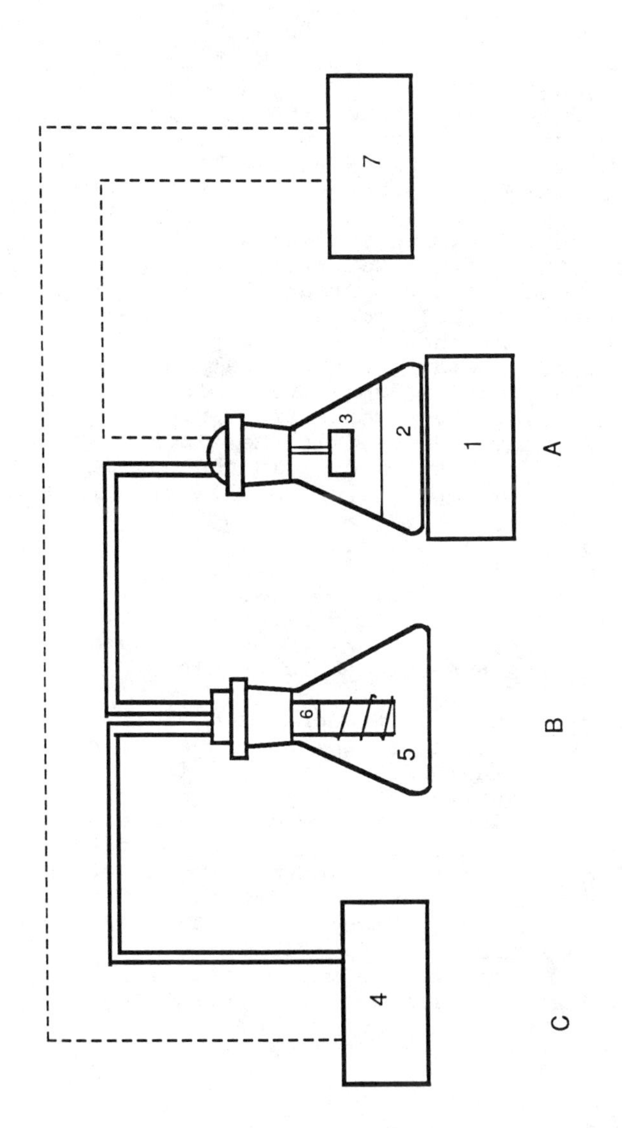

Figure 1. Schematic of a Measuring Unit

Economic Cooperation and Development (OECD) guidelines (10). It contains 10 ml of solution A and 1 ml of each of the solutions B to F per litre of synthetic medium : solution A - KH_2PO_4 8.5 g, K_2HPO_4 21.75 g, $Na_2HPO_4.2H_2O$ 33.4 g, and NH_4Cl 2.5 g; solution B - $MgSO_4.7H_2O$ 22.5 g; solution C - $CaCl_2$ 27.5 g; solution D - $FeCl_3.6H_2O$ 0.25 g; solution E - $MnSO_4.4H_2O$ 39.9 mg, H_3BO_3 57.2 mg, $ZnSO_4.7H_2O$ 42.8 mg, $(NH_4)_6Mo_7O_{24}$ 34.7 mg and $FeCl_3$.EDTA 100 mg; and solution F - Yeast extract 150 mg. The chemicals for each of the solutions A, B, C, D, E and F, are dissolved in 1000 ml of deionized water. The solution D is freshly prepared immediately before the start of an experiment.

The microbial inoculum was an activated sludge sample from The Little Miami wastewater treatment plant in Cincinnati, Ohio, which receives predominantly domestic sewage. The sludge was allowed to settle for about an hour, decanted, and then aerated at room temperature for 24 hours. The dry weight of sludge was determined by drying samples, in duplicates, of 1 ml, 2 ml and 3 ml at 105°C overnight. A concentration of 30 mg/l of sludge as dry matter was used in the experiment. The total volume of the synthetic medium in the 500 ml capacity reactor vessels was brought up to a final volume of 250 ml.

The concentration of the test compounds was 100 mg/l of medium. Aniline was used as the biodegradable reference compound at a concentration of 100 mg/l. The stock solutions were made in distilled water for aniline and phenol with concentration of 5 g/l. Other compounds were added directly to the reaction vessels using microliter syringes.

The typical experimental system consisted of duplicate flasks for the reference substance aniline and the test compounds and a single flask for toxicity control (test compound + aniline at 100 mg/l) and an inoculum control. The contents of the reaction vessels were stirred for an hour to ensure a steady state of the endogenous respiration at the initiation of oxygen uptake measurements. Then the test compound and aniline were added. The reaction vessels were incubated at 25°C in the temperature controlled water bath and stirred continuously throughout the run. The microbiota of the activated sludge used as inoculum were not preacclimated to the test compounds. The incubation period of the experimental run was between 20 to 40 days.

Biokinetic Parameters

Although many models have been proposed for microbial growth and substrate removal, the Monod relationship is among the most popular kinetic expressions used today. The Monod relation, in combination with the linear law for substrate removal, can provide an adequate description of microbial growth behavior. It states that the rate of cell growth is first order with respect to biomass concentration (X) and of mixed order with respect to substrate concentration (S).

$$dX/dt = (S\mu_m X)/(K_s + S) \quad \text{..........(1)}$$

Cell growth is related to substrate removal by the linear law

$$dX/dt = - Y\ (dS/dt) \quad \text{..............(2)}$$

The kinetic parameters of interest are the maximum specific growth rate μ_m, the half saturation constant K_s, and the yield coefficient Y. Here these kinetic parameters were estimated directly from the experimental oxygen uptake curves. The yield coefficient Y and the half saturation constant K_s were determined by the method of Grady et al. (6) and the maximum specific rate constant, μ_m, was determined by the method of Gaudy et al. (10). If the concentration of substrate and biomass are expressed in oxygen equivalents, then the oxygen uptake O_u, at any time in a batch reactor may be calculated from :

$$O_u = (S_o - S) - (X - X_o) \quad \text{.............(3)}$$

where subscript 'o' denotes initial condition.

The yield coefficient Y was estimated from the oxygen consumption at the beginning of the plateau of the oxygen uptake curve by :

$$Y = (1 - O_{up}/S_o) - Y_p \quad \text{................(4)}$$

where Y_p is the product yield and O_{up} is the cumulative oxygen uptake value at the initiation of the plateau.

In this study the product yield was negligible. The oxygen uptake has reached its plateau when the cumulative oxygen uptake value is approximately constant over a long period of time. At this plateau, oxygen uptake is due to endogenous respiration of the microbiota rather than the degradation of the test substrate.

If the assumption that $S \gg K_s$ is made, then the term $S/(K_s + S)$ in Equation 1 approaches one and simplifies the equation. Combining this simplified equation with Equation 2 and integrating and substituting in Equation 3 gives

$$\ln\,[X_o + O_u/(1/Y - 1)] = \ln(X_o) + \mu_m t \quad \text{...(5)}$$

The plot of $\ln[X_o + O_u/(1/Y - 1)]$ versus time will give a straight line with slope μ_m. The oxygen uptake value O_u' associated with one-half of the estimate of μ_m is used in

$$K_s = S_o - O_u'/(1 - Y) \quad \text{................(6)}$$

to get the estimate of K_S. The oxygen uptake curve before the plateau is utilized to get this estimate of K_S. Note that Equation 5 is valid only in the initial period of degradation when $S >> K_S$ and not throughout the degradation period.

Structure-Activity Relationships

Predictive techniques, based on relationship between structure and biodegradability, are important due to the increasing number and type of chemicals entering the environment and the fact that assessing biodegradability may be time consuming and expensive.

In the field of biodegradation there are several studies which have attempted to correlate some physical, chemical or structural property of a chemical with its biodegradation. Based on the type and the location of the substituent groups, Geating (11) developed an algorithm to predict biodegradation. Qualitative relationships for different compounds have been investigated by others, but quantification is required for regulatory purposes.

First order rate constants of biodegradation, or five day BOD values for chemicals, have been correlated using both physical and chemical properties. Paris, et al. (12,13) established a correlation between second order biodegradation rate constants and the van der Waal's radius of substituent groups, for substituted anilines and for a series of para-substituted phenols. Wolfe, et al. (14) correlated second order alkaline hydrolysis rate constants and biodegradation rate constants for selected pesticides and phthalate esters. Several workers have observed a correlation between biodegradability and lipophilicity, specifically octanol/water partition coefficients (log P). Paris, et al. (15) found a good correlation between biodegradation rate constants and log P, for a series of esters of 2,4-dichlorophenoxy acetic acid. Banerjee, et al. (16) obtained a similar relationship for chlorophenols. Vaishnav, et al. (17) correlated biodegradation of 17 alcohols and 11 ketones with octanol-water coefficients using 5-day BOD data. Pitter (18) found a dependence of biodegradation rates on electronic factors, like the Hammett substituent factors, for a series of anilines and phenols. Dearden and Nicholson (19) have correlated 5-day BOD values with modulus differences of atomic charge across a selected bond in a molecule for amines, phenols, aldehydes, carboxylic acids, halogenated hydrocarbons and amino acids. A direct correlation between the biodegradability rate constant and the molecular structure of the chemical has been used by Govind (20) to relate the first order biodegradation rate constant with the first order molecular connectivity index and by Boethling (21) to correlate the biodegradation rate constants with the molecular connectivities for esters, carbamates, ethers, phthalates and acids.

Group Contribution Approach

This approach is widely used in chemical engineering thermodynamics to assess the variety of pure component properties such as liquid densities, heat capacities and critical constants for organic compounds. This method is similar to the Free-Wilson model, which is widely used in pharmacology and medicinal chemistry. Using a group contribution approach, a large number of chemicals of interest can be constituted from perhaps a few hundred functional groups. Property prediction is based on the structure of the compound. According to this method, the molecules of a compound are structurally decomposed into functional groups or their fragments, each having unique contribution towards the compound properties.

The biodegradability rate constant, k, is expressed as a series function of contributions, α_j, of each group of the compound. The first order approximation of this series function representing biodegradation rate constant can be expressed as

$$\ln(k) = \sum_{j=1}^{L} N_j \alpha_j \qquad (7)$$

where N_j is the number of groups of type j in the compound, α_j is the contribution of group of type j and L is the total number of groups in the compound.

Using Equation 7 for each compound, a linear equation is constructed. For a given data set this generates a series of linear equations which are solved for α's using linear regression analysis.

Results and Discussion

Oxygen uptake curves for the test compounds, reference compound aniline, the toxicity control (aniline + test compound) and the endogenous control systems were generated over a period of 20 to 40 days. Within a period of 10 days all the controls, test compounds and aniline revealed the lag phase, biodegradation phase and the plateau region. Figures 2 and 3 illustrate the representative oxygen uptake curves generated for two of the test compounds, phenol and dibutyl phthalate.

These data reveal that all the test compounds, namely phenol, o-cresol, m-cresol, p-cresol, 2,4-dimethyl phenol, dimethyl phthalate, diethyl phthalate, dipropyl phthalate, dibutyl phthalate and butyl benzyl phthalate were biodegradable at the concentration level of 100 mg/l under the experimental conditions. According to OECD (11), the results of the degradation experiment were valid because 60% degradation of control substance aniline was achieved within a period of 28 days. Within a period of 40 days all the compounds were degraded between 80% - 95%.

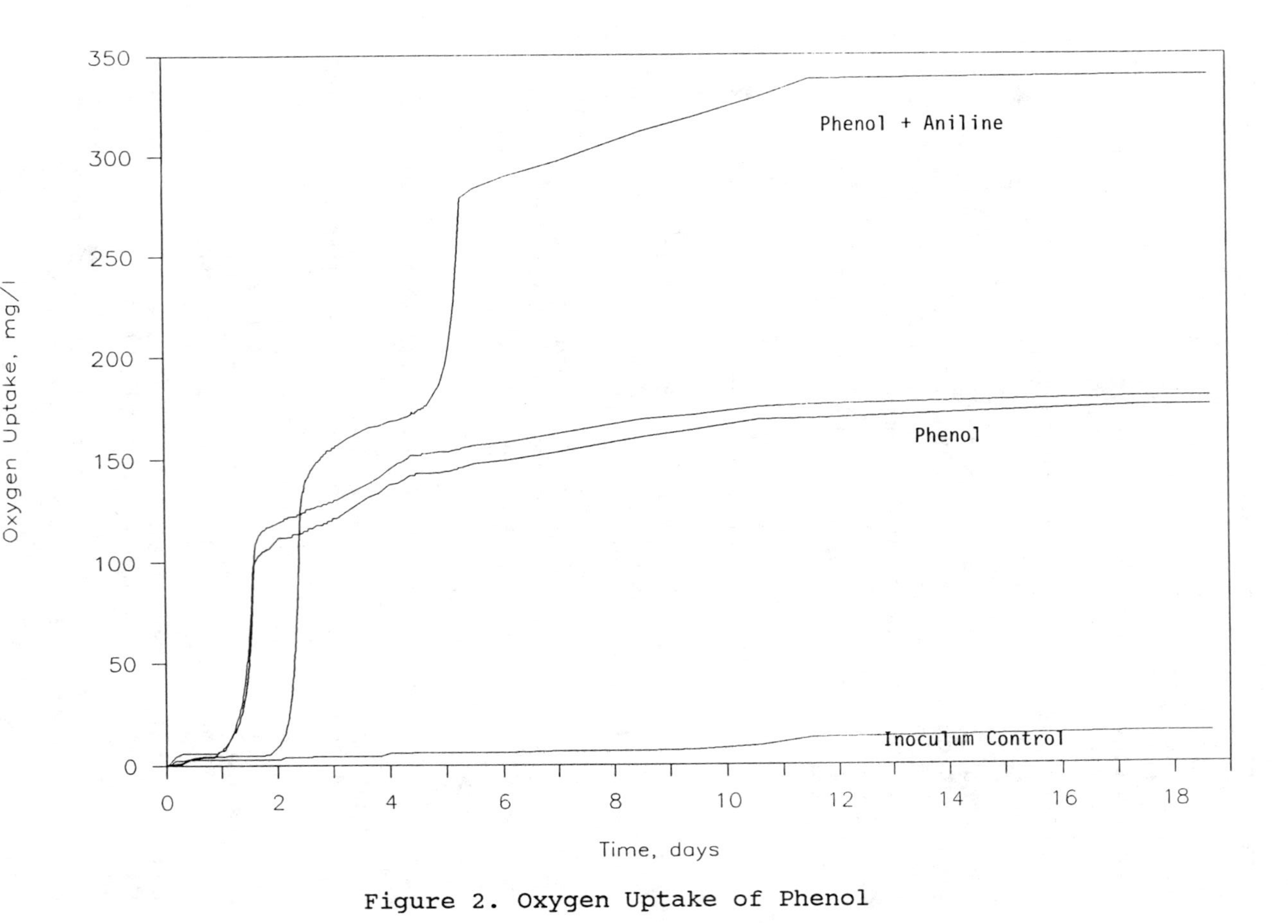

Figure 2. Oxygen Uptake of Phenol

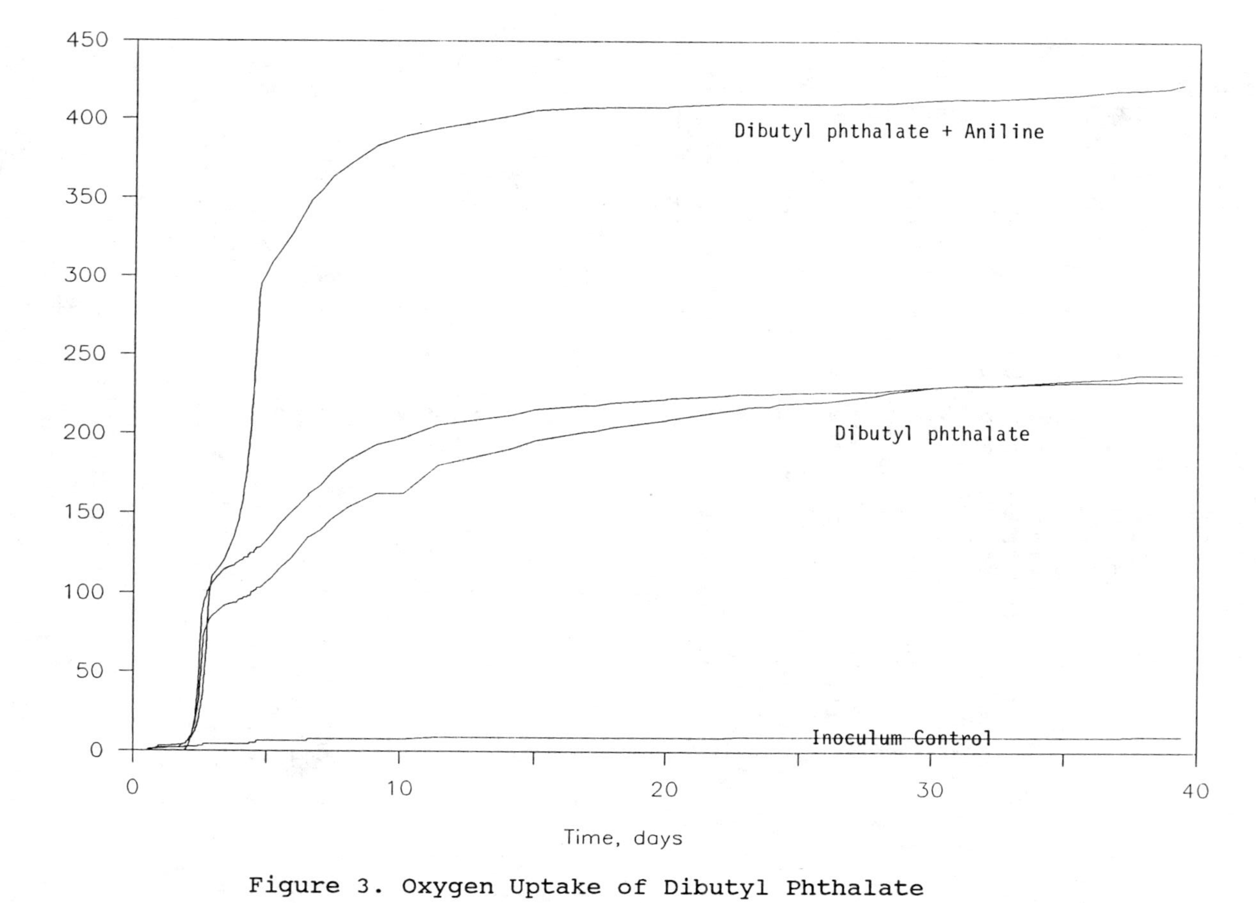

Figure 3. Oxygen Uptake of Dibutyl Phthalate

The toxicity control data (test compound + aniline) revealed that, except butyl benzyl phthalate, none of the test compounds were inhibitory towards the biodegradation of aniline in the sludge. Butyl benzyl phthalate causes a slight inhibitory effect on aniline degradation by increasing the acclimation time. The oxygen uptake curves of all the toxicity controls demonstrated two consecutive degradation phases, one for the test compound and another for positive control compound aniline as shown in Figure 2 and 3, and this behavior is characteristic for cultures in which the test compound is not inhibiting the biodegradation of the control substrate. The difference between the total oxygen uptake of the toxicity control (test compound + aniline), and of the corresponding test compound, was approximately equal to the total oxygen uptake of the control compound aniline.

Estimates of the Monod kinetic parameters, and the maximum specific substrate uptake rate per unit biomass, k_m ($=\mu_m/Y$) for phenols, phthalates and aniline, obtained by the method described before, are given in Table I.

Table I. Estimate of Monod Parameters

Compound	k_m (day^{-1})	Y	μ_m (day^{-1})	K_s (mg/l)
Aniline	16.1	0.38	6.15	6.10
PHENOLS				
Phenol	16.9	0.58	9.82	9.43
o-Cresol	10.0	0.41	4.10	16.41
m-Cresol	17.3	0.46	7.97	17.62
p-Cresol	18.5	0.33	6.11	27.78
2,4-Dimethyl phenol	14.4	0.39	5.62	14.07
PHTHALATES				
Dimethyl phthalate	16.4	0.43	7.07	41.68
Diethyl phthalate	4.5	0.46	3.00	11.67
Dipropyl phthalate	12.0	0.48	5.78	15.81
Dibutyl phthalate	12.0	0.58	6.95	51.38
Butyl benzyl phthalate	12.8	0.61	7.80	36.25

The measured rate constants suggest that, in comparison to phenol, the presence of an o- or p-methyl group renders a phenolic compound less biodegradable. Substituents in the para position have a less pronounced effect than in the ortho position. This also explains the μ_m value of 2,4-dimethyl phenol which is smaller than the one for p-cresol but greater than the one for o-cresol.

The introduction of a m-methyl group seemed to have little effect on the biodegradability of the phenolic compound. Sugatt et al. (22) have reported that the low molecular weight phthalates degrade slightly faster than the high molecular weight ones, but no such trend was observed in this study. Comparison of our results with those reported in the literature (22,23,24,25) was not possible because either the rates were not reported, or they were second order rates with different units. Our values for half saturation constant K_S, are high for some of the compounds. This result might be caused by the linearized graphical approach used for its estimation.

It is necessary to have at least five data points for each unknown variable to get a statistically valid correlation. So it is imperative to ensure that each group for which a contribution is calculated occurs in at least five compounds. The contributions of different groups were calculated from literature data (26). These data were first order constants calculated by the equation :

$$dBOD/dt = k\ BOD \quad \text{.................}(8)$$

The groups and their contribution parameters are given in Table II. The data set used to calculate contribution parameters did not include the cresols, phenol and 2,4-dimethyl phenol. The contribution of the phthalate group was not calculated because either not enough data were available or the only data available were for those phthalates used in this study.

Table II. Group Contribution Parameters

No.	Group		α_j
1	Methyl	CH_3	-3.460
2	Hydroxy	OH	-2.983
3	Aromatic CH	ACH	-0.8340
4	Aromatic C	AC	1.9730

Using the contribution parameters of Table II, first order rate constants were predicted for phenols. The comparison between the predicted values and the values obtained from experimental data, using Equation 8 are given in Table III. The predicted values are within 10% of the experimental values. However, the method described is a first order approximation which assumes that there are no group interactions. Additional studies are needed to verify this assumption.

Conclusion

Based on the oxygen uptake data all the tested phenols and phthalates are biodegradable in activated sludge with our experimental conditions. Similar results have been noted in the literature. All compounds were degraded in excess of 80%. A methyl group in o- or p- position of phenolic compounds decreases the degradation rate in comparison to phenol, while a m-methyl group does not affect the degradation rate. No such trend was observed with phthalates. With the exception of butyl benzyl phthalate, none of the test compounds inhibit aniline degradation. More data are required to determine whether or not unique group contributions exist with respect to its position (o, m, and p), and to include more groups in the structure-activity relationships.

Table III. Comparison of Actual and Predicted ln(k) Values

Compound	Actual ln(k)	Predicted ln(k)	% Error
o-Cresol	-6.0890	-5.8330	4.20
m-Cresol	-5.7706	-5.8330	-1.08
p-Cresol	-5.8659	-5.8330	0.56
2,4-Dimethyl Phenol	-6.2472	-6.4860	-3.82
Phenol	-5.6891	-5.1800	8.95

Nomenclature

k	first order biodegradation constant (hr^{-1})
k_m	maximum specific substrate uptake rate per unit biomass (day^{-1})
K_s	half saturation constant (mg/l)
L	total number of groups
N_j	number of groups of type j
O_u	oxygen uptake (mg/l)
O_{up}	oxygen uptake at the plateau (mg/l)
S	substrate concentration (mg/l)
X	biomass concentration (mg/l)
Y	biomass yield coefficient
Y_p	product yield coefficient
α_j	contribution of j th group
μ_m	maximum specific rate constant (day^{-1})

Literature Cited

1. Blackburn, J. W.; Troxler, W. L.; "Prediction of the fates of organic chemicals in a biological treatment process - An overview"; Environ. Prog. **1984**, 3, 163-65.
2. Howard, P. H.; Banerjee, S.; Rosenberg, A.; "A review and evaluation of available techniques for

determining persistance and routes of degradation of chemical substances in environment"; USEPA Report **560/5-81-011**, 1981.
3. Robinson, J. A.; Tiedje, J. M.; "Nonlinear estimation of Monod growth kinetic parameters from a single substrate depletion curve"; Appl. Environ. Microbiol. **1983**, 45, 1453-58.
4. Simkins, S.; Alexander, M.; "Nonlinear estimation of the parameters of Monod kinetics that best describe mineralization of several substrate concentrations by dissimilar bacterial densities"; Appl. Environ. Microbiol. **1985**, 50, 816-24.
5. Gaudy, A. F., Jr.; Gaudy, E. T.; "Biological concepts for design and operation of the activated sludge process"; USEPA Report **17090 FQJ 09/71**, 1971.
6. Grady, C. P. L., Jr.; Dang, J. S.; Harvey, D, M.; Jobbagy, A.; Wang, X-L.; "Determination of biodegradation kinetics through use of electrolytic respirometry"; Wat. Sci. Tech. **1989**, 21, 957-68.
7. Dojlido, J. R.; "Investigation of biodegradability and toxicity of organic compounds"; USEPA Report **EPA-600/2-79-163**, 1979.
8. Oshima, A.; Tabak, H. H.; Lewis, R. F.; "The evaluation of biological treatability and removability of toxic organic chemicals by respirometry"; In house report, WERL, USEPA, Cincinnati, Ohio, 1984.
9. Gaudy, A. F., Jr.; Rozich, A. F.; Garniewski, S.; Moarn, N. R.; Ekambaram, A.; "Methodology for utilizing respirometric data to assess biodegradation kinetics"; Proc. 42nd Annual Industrial Waste Conference, Lewis Publishers, Chelsea, MI, **1988**, 35-44.
10. OECD Guidelines for Testing of Chemicals (Organization of Economic Cooperation and Development, Paris, France), 1981.
11. Geating, J.; "Literature study of the biodegradability of chemicals in water"; USEPA Report **EPA-600/2-81-175**, 1981.
12. Paris, D. F.; Wolfe, N. L.; "Relationships between properties of a series of anilines and their transformation by bacteria"; Appl. Environ. Microbiol. **1987**, 53, 911-16.
13. Paris, D. F.; Wolfe, N. L.; Steen, W. C.; "Structure-activity relationships in microbial transformation of phenols"; Appl. Environ. Microbiol. **1982**, 44, 153-58.
14. Wolfe, N. L.; Paris, D. F.; Steen, W. C.; Baugham, G. L.; "Correlation of microbial degradation rates with chemical structure"; Environ. Sci. Tech. **1980**, 14, 1143-44.
15. Paris, D. F.; Wolfe, N. L.; Steen, W. C.; "Microbial transformation of ester of chlorinated carboxylic acids"; Appl. Environ. Microbiol. **1984**, 47, 7-11.

16. Banerjee, S.; Howard, P. H.; Rosenberg, A. M.; Dombrowski, A. E.; Sikka, H.; Tullis, D. L.; "Development of a general kinetic model for biodegradation and its application to chlorophenols"; Environ. Sci. Tech. **1984**, 18, 416-22.
17. Vaishnav, D. D.; Boethling, R. S.; Babeu, L.; "Quantitative structure-biodegradability relationships for alcohols, ketones and alicyclic compounds"; Chemosphere **1987**, 16, 695-703.
18. Pitter, P.; "Correlation between the structure of aromatic compounds and the rate of their biological degradation"; Collection Czechoslovak Chemical Comm. **1984**, 49, 2891-96.
19. Dearden, J. C.; Nicholson, R. M.; "The prediction of biodegradabilities by the use of quantitative structure-activity relationships : Correlation of biological oxygen demand with atomic charge difference"; Pestici. Sci. **1986**, 17, 305-10.
20. Govind, R.; "Treatability of toxics in wastewater systems"; Hazardous Substances **1987**, 2, 16-24.
21. Boethling, R. S.; "Application of molecular topology to quantitative structure-biodegradability relationships"; Environ. Tox. Chem. **1986**, 5, 797-806.
22. Sugatt, R. H.; O'Grady, D. P.; Banerjee, S.; Howard, P. H.; Gledhill, W. E.; "Shake flask biodegradation of 14 commercial phthalate esters"; Appl. Environ. Microbiol. **1984**, 47, 601-06.
23. Wolfe, N. L.; Burns, L. A.; Steen, W. E.; "Use of linear free energy relationships and an evaluative model to assess the fate and transport of phthalate esters in the aquatic environment"; Chemosphere **1980**, 9, 393-402.
24. Visser, S. A.; Lamontagne, G.; Zoulalian, V.; Tessier, A.; "Bacteria active in the degradation of phenols in polluted waters of St. Lawrence river"; Arch. Environ. Contam. Toxicol. **1977**, 6, 455-69.
25. Pandey, R. A.; Kaul, S. N.; Kumaran, P.; Badrinath, S. D.; "Determination of kinetic constants for bio-oxidation of some aromatic phenolic compounds"; Asian Environ. **1986**, 8, 4-8.
26. Urano, K.; Kato, Z.; "Evaluation of biodegradation ranks of priority organic compounds"; J. Hazardous Mtl. **1986**, 13, 147-59.

RECEIVED November 10, 1989

Chapter 10

Removal of Fluoride Ion from Wastewater by a Hydrous Cerium Oxide Adsorbent

J. Nomura, H. Imai, and T. Miyake

Shin-Nihon Chemical Industry Company, Ltd., 1-1-1, Uchisaiwaicho, Chiyodaku, Tokyo 100, Japan

A new adsorption process has been developed and shown practical for removal of fluoride ion from waste water to effect concentrations of 1 mg/L or less. The adsorbent developed for this process consists of polyolefinic resin beads carrying hydrous cerium oxide podwer with a large specific area. It is high in selectivity for fluoride ion and exhibits an adsorption capacity of 8 to 25 mg-F/mL-adsorbent at fluorine concentrations of 0.2 to 20 mg/L, with adsorption conforming to the Freundlich isotherm, and is readily regenerated by alkalis for repeated, long-term use.
This process consists essentially of adjustment to PH 2 to 5, fluorine adsorption, adsorbent regeneration by NaOH and precipitiation of CaF_2 from used regenerant by addition of $CaCl_2$.

Effluent fluorine has long been found in the waste water of industrial facilities which use or process mineral resources, such as coal-fired power plants, aluminum smelteries, and phosphorus and fluorine manufacturing plants. In recent years, the use of hydrogen fluoride and other inorganic fluorine compounds as fluorinating agents has increased rapidly in the manufacture of chlorofluorocarbon gases and fluoroplastics, and in the cleaning or surface treatment processes of semiconductor, metal, and glass production. This has led to increasing concern about the ecological effects of water-borne fluoride ions, in view of their extremely strong chemical activity. It has been reported that long-term intake of drinking water containing as little as 2 - 4 mg/L of fluoride ion tends to cause dental and skeletal fluorosis (1).

Waste water containing dissolved fluorine and fluorides is

0097–6156/90/0422–0157$06.00/0

commonly treated by coagulating precipitation, in which calcium hydroxide or calcium chloride is added to obtain calcium fluoride as a precipitate. However, a lower limit on the fluoride ion concentration that can be achieved in this way is imposed by the solubility of calcium fluoride, ≈ 10 - 15 mg/L. To obtain lower fluoride concentrations in industrial effluents, adsorption processes utilizing activated alumina have been conventionally employed, and processes have also been developed which employ chelating resins for formation of metal-fluoride complexes (2). In both cases, however, practical utilization has been adversely affected by performance degradation and other problems associated with coexistent ions and long-term use.

In extensive investigations on inorganic ion exchangers, we have found that hydrous oxides of rare earth elements exhibit a highly selective adsorptivity for fluoride ions (3-4), and developed a high-performance adsorbent (READ-F) incorporating hydrous cerium oxide as the ion exchanger. This paper describes the fluoride ion adsorption characteristics of various hydrous rare earth oxides, the basic properties of the READ-F adsorbent, and a process based on this adsorbent for treatment of fluorine-containing waste water.

Experimental

Preparation of Fine-Powder Hydrous Oxides. Ammonia water was added to an aqueous solution of the metal or rare earth salt to precipitate the hydrous metal or rare earth oxide, which was then subjected to filtration, a water wash, and drying. For example, hydrous cerium oxide was prepared by first adding hydrogen peroxide followed by ammonia water to aqueous cerium(III) chloride solution, to effect precipitation of hydrous cerium(IV) oxide. The precipitate was filtered, washed with water, and dried to obtain a fine light-yellow powder.

Determination of Adsorption Capacity. The hydrous metal or rare earth oxide, or else a commercially available activated alumina, was suspended at a concentration of 1 g/L in an aqueous solution containing fluoride (F^-) and other ions, and then stirred at 20°C and pH 5 (pH 6.5 for activated alumina) until the F^- concentration in the liquid phase reached equilibrium. The amount of ion adsorbed was determined from the initial and final liquid-phase ion concentrations as measured by ion chromatography. The crystallinity and the specific surface area of the hydrous cerium oxide were determined by X-ray diffraction and BET gas adsorption, respectively.

Preparation of Hydrous Cerium Oxide Adsorbent. Adsorbent in bead form (READ-F) consisting of polymer-borne hydrous cerium oxide was prepared by uniformly dispersing the fine-power hydrous cerium oxide in dimethyl sulfoxide solution of ethylene-vinyl alcohol copolymer, subjecting the dispersion to centrifugal atomization, coagulation of the resulting droplets in a water bath, and removal of the solvent. The beads thus obtained were 60% void, and had an apparent specific gravity of 1.55, an average particle diameter of 0.65 mm, and a hydrous cerium oxide content of 0.30 g/mL-adsorbent.

Adsorption and Desorption Characteristics. Aqueous solution of pH 3 (pH 6 for activated alumina) with a fluorine content of 100 mg/L was injected at a space velocity of 20 h^{-1} into a column of 450 mm inner diameter packed with 200 L of READ-F or activated alumina, and the breakthrough curves were plotted. READ-F column regeneration was performed by passage of 1N NaOH solution.

READ-F performance was also examined by similar treatment of waste water from a semiconductor cleaning process, containing F^- and various acids.

Results and Discussion

F^- Adsorptivity. Table I shows the F^- adsorption capacities as measured for the fine-powder hydrous metal and rare earth oxides. The water content of each was determined from its ignition loss, and is indicated in the table by the value of n in the formula $M_aO_b \cdot nH_2O$. As shown, the hydrous oxides of the rare earth elements Sm, Nd, Gd, and Ce adsorbed more than 1 mol of F^- per mol of metal -- far higher than the values obtained for the activated alumina and the hydrous oxides of Zr, Ti, and Fe, whose adsorptivities also have been reported by other investigators (5-8).

Figure 1 shows the relation between the amount of F^- adsorbed per mol of metal and the metal's ionic potential ($\phi = z/r$; where z = valency, r = ionic radius). The metals having low ionic potentials, and thus high basicity, tended to show large adsorption capacities, in agreement with the correlation between adsorptivity and ionic potential reported by Abe (9).

Hydrous cerium oxide, with its high adsorption capacity and a water solubility lower than any other hydrous rare earth oxide, was chosen for further investigation and ultimately for utilization as the ion exchanger in the READ-F adsorbent, as described below.

Properties and Ion Adsorptivity of Hydrous Cerium Oxide. The water content and the ion adsorption capacities of hydrous metal oxides are generally known to vary with crystallization and drying conditions. Drying temperature is particularly influential; lower drying temperatures generally result in higher water content.

Table II shows the water content, specific surface area, and F^- adsorptivity at pH 5, for hydrous cerium oxide samples prepared at various drying temperatures.

The X-ray diffraction patterns of these samples are shown in Figure 2. The samples dried at -80°C and 50°C both exhibit a fluorite-structure diffraction pattern, similar to that of cerium oxide, but with much broader peaks, indicating considerable strain in their microcrystalline structure. The samples prepared at higher drying temperatures showed increased crystallinity and were also characterized by reduced specific surface area which may be attributed to increased particle size.

As shown in Figure 3, the adsorption capacity of hydrous cerium oxide tends to increase with increasing specific surface area, perhaps reaching a plateau at 100 m^2/g or above. Water content, on the other hand, does not clearly correlate with

Table I. F^- Adsorption Capacities of Hydrous Metal Oxides

Hydrous Metal Oxide [a]	Adsorption Capacity [b]	
	mg-F/g	mol-F/mol-metal
Sm_2O_3 $4.1H_2O$	120	1.33
Nd_2O_3 $4.7H_2O$	119	1.32
Gd_2O_3 $5.0H_2O$	106	1.26
CeO_2 $1.6H_2O$	105	1.13
La_2O_3 $3.0H_2O$	93	0.93
Y_2O_3 $2.1H_2O$	97	0.77
ZrO_2 $3.3H_2O$	65	0.64
TiO_2 $6.0H_2O$	59	0.41
Fe_2O_3 $2.9H_2O$	49	0.27
Al_2O_3 [c]	69	0.20
Al_2O_3 [d]	65	0.19
MnO_2 $0.5H_2O$	7	0.04

a) Dried at 50°C
b) Initial F^- concentration 150 mg/L
c) Mizusawa Chemical "Neobead D3"
d) Wako Chemical "Woelm"

SOURCE: Reprinted with permission from ref. 4. Copyright 1987 Chemical Society of Japan.

adsorption capacity. The samples dried at -80°C, 50°C, and 120°C, which were found to have similar specific surface areas, also showed nearly equal adsorption capacities even though the latter two were found to have water contents 50% higher and 40% lower than the first, respectively. The higher water content of the sample dried at 50°C would appear to consist largely of free water (as opposed to water of hydration), which plays no role in adsorption. The smaller water content of the sample dried at 120°C was apparently compensated by rehydration of its adsorption sites upon its dispersion in aqueous solution. Drying at 500°C or above, in

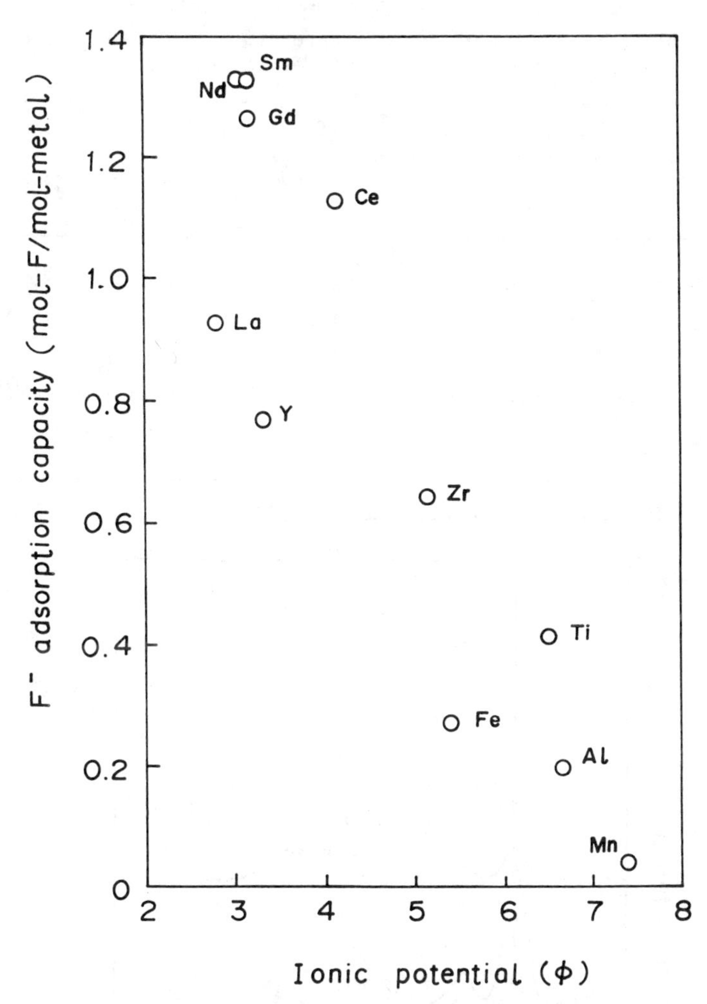

Figure 1. Dependence of F^- adsorption on ionic potential of central metal atom. (Reprinted with permission from ref. 4. Copyright 1987 Chemical Society of Japan.)

contrast, results in a far smaller adsorption capacity, which may be attributed to the substantially reduced specific surface area and also presumably to a greatly diminished capacity for rehydration.

Hydrous cerium oxide is insoluble in alkaline solutions. It is also only slightly soluble in acidic solutions, as indicated by a measured solubility of less than 0.1 mg/L in 0.1N aqueous hydrochloric acid (pH 1), for the sample prepared by drying at 120°C. It may therefore be expected to be highly durable in practical applications at pH 2 and above.

Table II. Effect of Drying Temperature on F^- Adsorption Capacity of Hydrous Cerium Oxide

Drying Temperature	Water Content after Drying (%)	Specific Surface Area (m^2/g)	F^- Adsorption Capacity a) (mol/g-hydrous cerium oxide)
-80°C	9.8	117	4.35×10^{-3}
50°C	14.3	116	4.40×10^{-3}
120°C	5.9	108	4.17×10^{-3}
500°C	0.9	70	1.80×10^{-3}
900°C	0	6	0.05×10^{-3}

a) Initial F^- concentration 5×10^{-3} mol/L

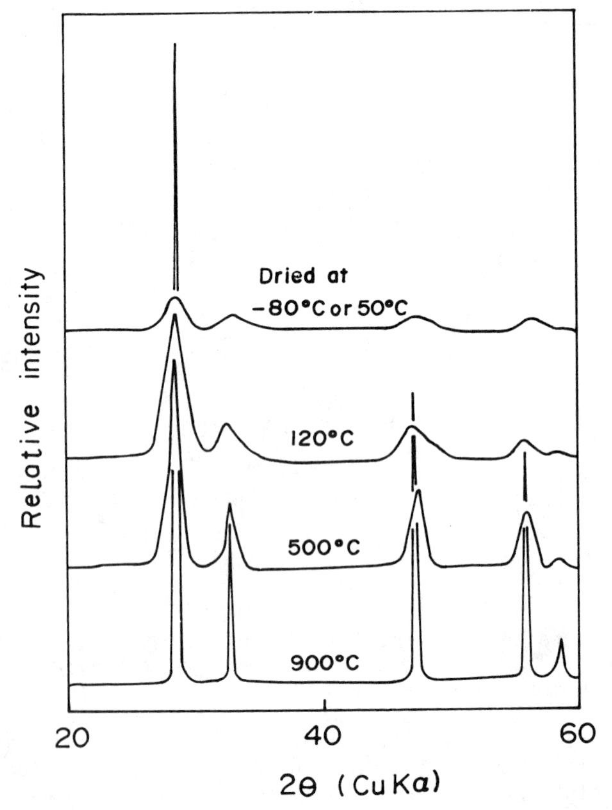

Figure 2. X-ray diffraction patterns of hydrous cerium oxide powders. (Reprinted with permission from ref. 4. Copyright 1987 Chemical Society of Japan.)

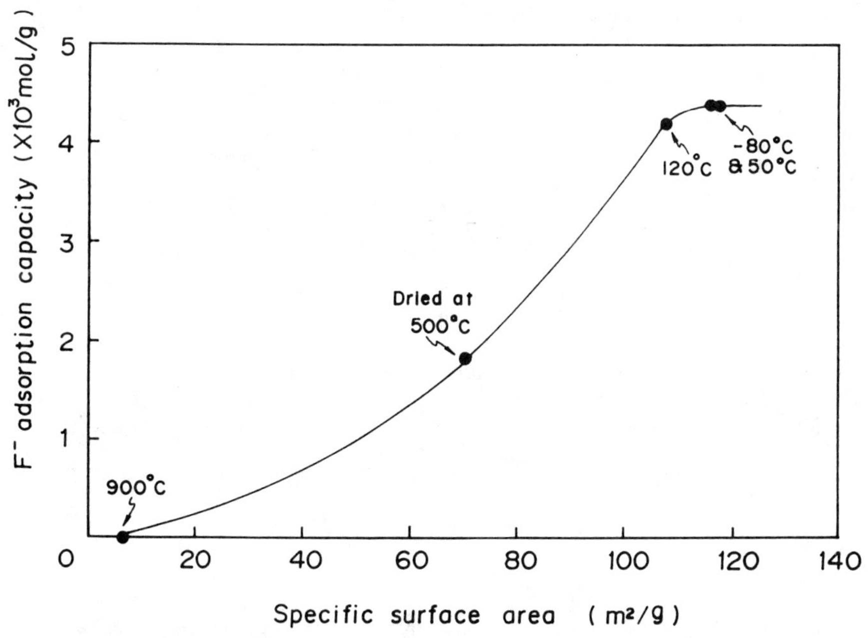

Figure 3. Dependence of F^- adsorption capacity on specific surface area.

Figure 4 shows the adsorptivity of hydrous cerium oxide for various anions as a function of the liquid-phase pH. The region of adsorption for F^- and SiF_6^{2-} is about pH 7 and lower, and that for SO_4^{2-}, Cl^-, and NO_3^- is about pH 5 and lower. The region of adsorption for $As(OH)_3$, and $B(OH)_4^-$ is neutral to moderately basic, and that for HPO_4^{2-} apparently extends throughout the entire pH range. The results also indicate that the adsorption capacity for F^- is at least twice as high as for any other ion.

Table III shows the adsorption selectivity of hydrous cerium oxide in a pH 5 solution containing F^- and other anions.

Table III. F^- Selectivity of Hydrous Cerium Oxide

Anion [a]	Adsorbed Anion (mol/g)	Residual Anion (mol/L)	F^- Selectivity (αx)
F^-	1.74	0.26	——
HPO_4^{2-}	0.98	1.02	7
SO_4^{2-}	0.26	1.74	45
Cl^-	ND	2.0	1000
Br^-	ND	2.0	1000
NO_3^-	ND	2.0	1000

a) Initial concentration of each anion 2×10^{-3} mol/L

SOURCE: Reprinted with permission from ref. 4. Copyright 1987 Chemical Society of Japan.

The F^- selectivity relative to each of the other X anions (αx) was calculated as:

$$\alpha x = \frac{\text{Adsorbed } F^-/\text{Residual } F^-}{\text{Adsorbed anion X/Residual anion X}}$$

It is evident that at this pH level hydrous cerium oxide adsorbs virtually no Cl^-, Br^-, or NO_3^- in the presence of F^-. In other studies, we have found that hydrous cerium oxide exhibits a similarly high level of F^- selectivity throughout the range pH 5 to 2.

The adsorption characteristics of hydrous cerium oxide are presumably attributable to the tendency for its OH^- ions, with an ionic radius of 1.45 Å and lying in a strained microcrystalline structure which may be classified crystallographically as a fluorite, to undergo replacement by ions having a smaller ionic radius, as is the case with F^- ions (1.33 Å).

Adsorption and Desorption Characteristics of READ-F. The F^- adsorption isotherm of the polymer-borne hydrous cerium oxide adsorbent, READ-F, is shown in Figure 5 in comparison with those of chelating resin and activated alumina. All three adsorbents exhibit Freundlich adsorption for F^-, with a linear relation

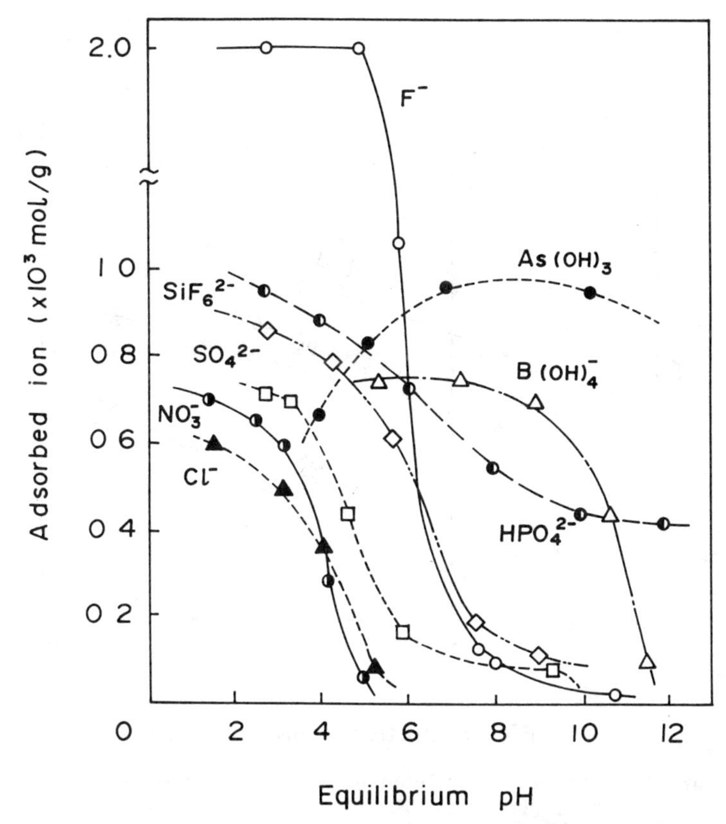

Figure 4. Effect of pH on anion adsorption of hydrous cerium oxide. (Reprinted with permission from ref. 4. Copyright 1987 Chemical Society of Japan.)

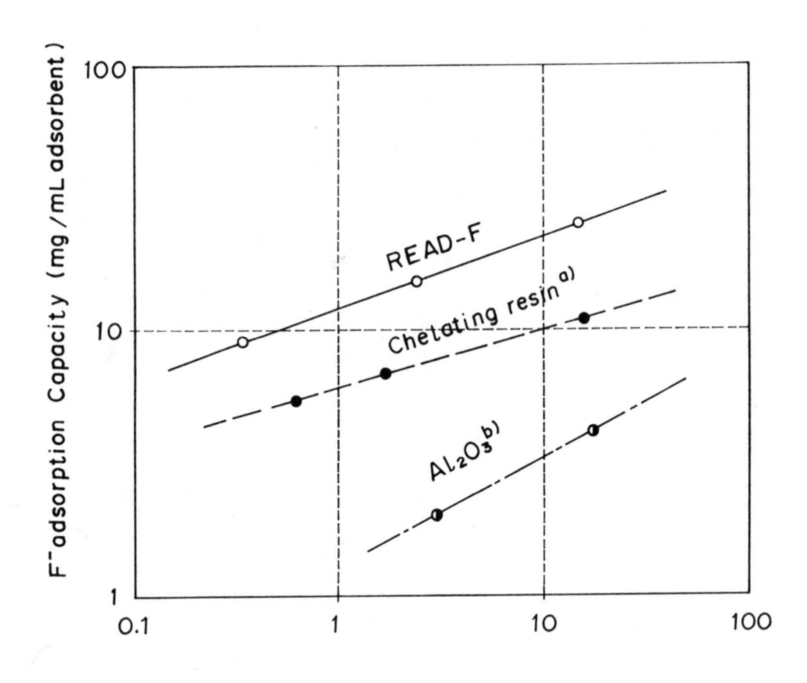

Figure 5. Comparison of adsorption equilibrium isotherms.
a) Unitika "Uniselec UR 3700"
b) Mizusawa Chemical "Neobead D3"

between adsorbed quantity and liquid-phase concentration in log-log plots. As indicated, the F^- adsorption capacity of READ-F is more than twice that of the chelating resin and about 6 times that of the activated alumina, and F^- adsorptivity is high even at F^- concentrations as low as several mg/L.

Figure 6 shows breakthrough curves for fixed-bed adsorption by READ-F and activated alumina, which indicate that 100 mg/L of F^- in aqueous solution can be effectively reduced to 1 mg/L or less by READ-F.

READ-F releases adsorbed ions upon contact with an alkali solution. A regeneration curve obtained by passage of aqueous 1N-NaOH as regenerant is shown in Figure 7.

Basic Process of Waste Water Treatment with READ-F. The essential components of the READ-F process for removal of fluorine from waste water are shown in Figure 8 and Figure 9. For waste water containing more than 100 mg/L of F^-, it is usually economically advantageous to employ calcium precipitation as a preliminary step, for reduction of the F^- concentration to 20 - 15 mg/L, before READ-F adsorption.

The waste water is passed through one READ-F adsorption column to obtain the end F^- concentration, while a second READ-F adsorption column undergoes regeneration by passage of NaOH solution. Used regenerant, containing NaF at high concentration, is treated with a calcium salt for precipitation and separation of CaF_2. The effective reaction in each step is essentially as follows.

Adsorption:

$$CeO_{2-x}(OH)_{2x} + yF^- \longrightarrow CeO_{2-x}F_y(OH)_{2x-y} + yOH^-$$

Regeneration:

$$CeO_{2-x}F_y(OH)_{2x-y} + yNaOH \longrightarrow CeO_{2-x}(OH)_{2x} + yNaF$$

Precipitation:

$$2NaF + CaCl_2 \longrightarrow 2NaCl + CaF_2$$

where $0 < x \leq 1.6$ and $0 < y \leq 1.1$, based on experimental data.

Treatment of Semiconductor Cleaning Effluent Containing Fluoric Acid. Semiconductor cleaning operations generally involve the use of nitric, sulfuric, and other acids, along with fluoric acid. The resulting waste water therefore contains various ions. Figure 10 shows the results of F^- adsorption by the READ-F process, conducted on such an effluent after preliminary treatment by calcium precipitation. As shown, the F^- content is reduced by READ-F adsorption from 23 mg/L to 0.3 mg/L, with virtually no change in the concentration of the other ions. Stable, high-level performance continues to the designated breakthrough eluate concentration of 1 mg/L after some 200 hours in operation and adsorption of about 25 mg-F/mL-adsorbent.

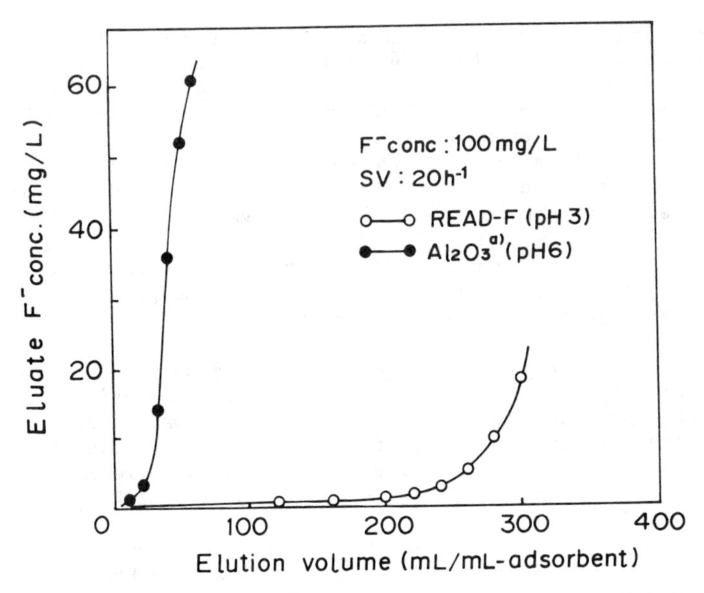

Figure 6. Breakthrough curves of F^- adsorption on READ-F and activated alumina.
a) Mizusawa Chemical "Neobead D3"

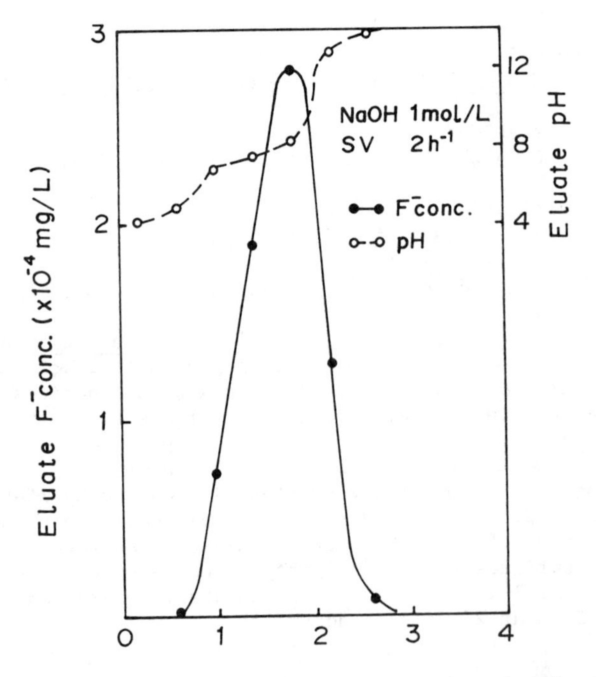

Figure 7. Regeneration of READ-F packed column.

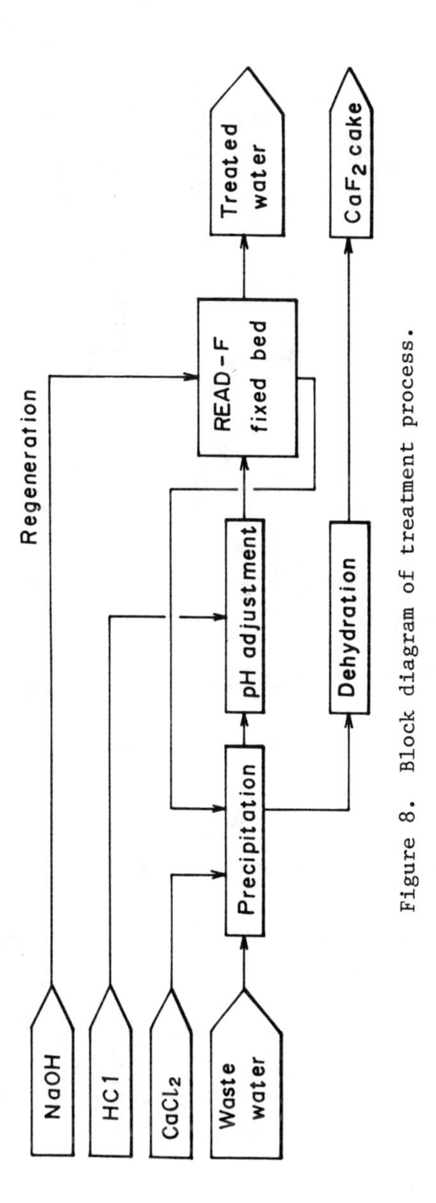

Figure 8. Block diagram of treatment process.

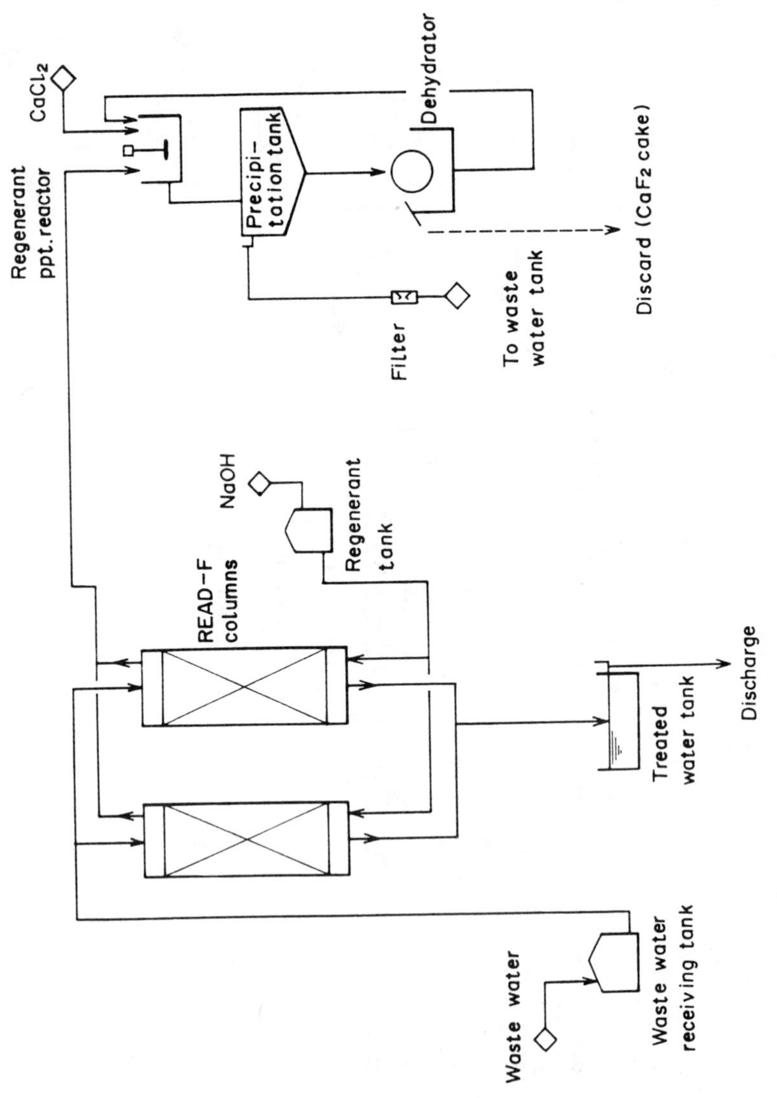

Figure 9. Flow diagram of a typical READ-F process.

Ion content before and after treatment (mg/L)

	F^-	Cl^-	NO_3^-	SO_4^{2-}	Ca^{2+}
Before	23	780	200	900	370
After	0.3	780	200	800	350

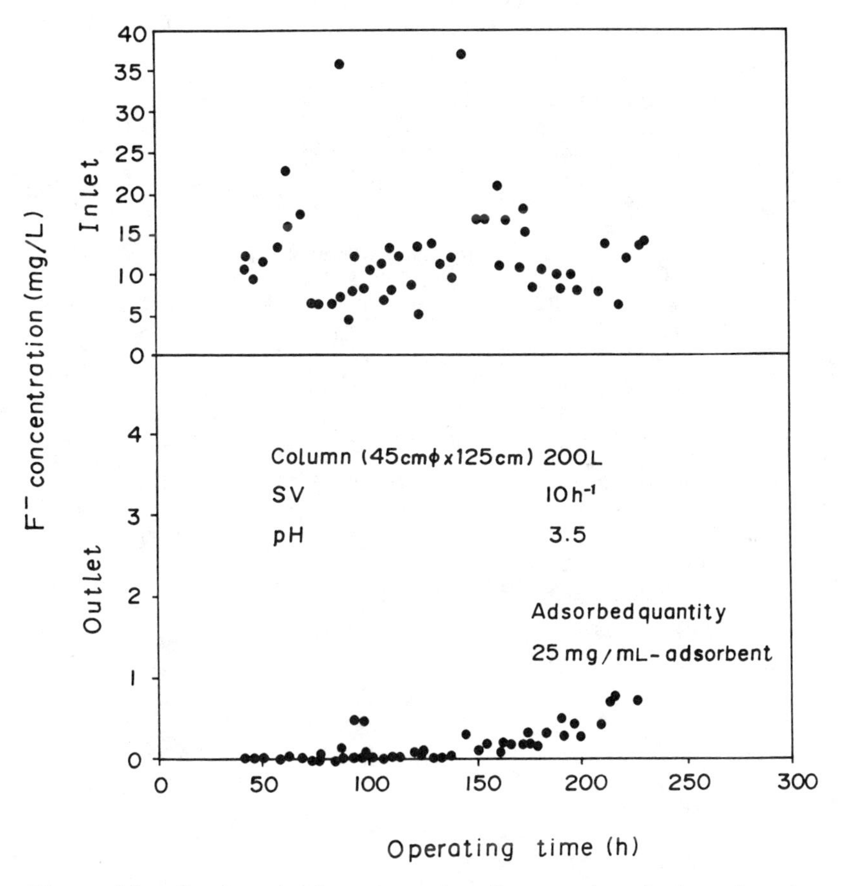

Figure 10. Treatment of waste water from semiconductor cleaning process.

Summary

(1) Hydrous rare earth oxides exhibit high adsorptivity for anions, corresponding to the low ionic potentials and thus the highly basic nature of their rare earth metals. Fluoride ions, in particular, readily undergo adsorption by ion exchange on these oxides, apparently because their ionic radius is slightly smaller than that of the hydroxide ions in the microcrystalline structure.
(2) READ-F adsorbent, consisting of polyolefinic resin beads carrying hydrous cerium oxide powder with a large specific surface area, exhibits highly selective adsorptivity for fluoride ions and an adsorption capacity about 6 times that of activated alumina. These characteristics, together with its capability for ready regeneration by alkalis, make possible a highly efficient, effective removal of dissolved fluorine from waste water.
(3) Effluents containing fluorine, such as those from semiconductor cleaning processes, can be effectively treated with an adsorption process employing such an adsorbent, to obtain fluorine concentrations of 1 mg/L or less.

References

1. Hileman, B. Chem. Eng. News 1988, August 1, 26.
2. Etigo, R.; Ishikura, M. Japan Kokai 36632, 1983.
3. Nomura, J.; Imai, H.; Ishibashi, Y.; Konishi, T. U.S. Patent 4 717 554, 1988.
4. Imai, H.; Nomura, J.; Ishibashi, Y.; Konishi, T. Chem. Soc. Japan 1987, 807
5. Kraus, K. A.; Philips, H. O. et al. Proc. 2nd Int. Conf. Peaceful Uses At. Energy, Geneva, 1958, 28, p3.
6. Hayano S. Special Research Project on Enviromental Science, Report by S-618 group (Japan) No. B135, 1982, 8-13.
7. Kulbi, H. Helv. Chim. Acta 1947, 30, 453.
8. Lewandowski, A.; Idzikowski, S. Chem Anal. (Warsaw), 1969, 14, 77.
9. Abe, M. Jpn. Analyst 1974, 23, 1254.

RECEIVED November 10, 1989

Chapter 11

Evaluation of a Cell–Biopolymer Sorbent for Uptake of Strontium from Dilute Solutions

J. S. Watson, C. D. Scott, and B. D. Faison

Chemical Technology Division, Oak Ridge National Laboratory, P.O. Box 2008, Oak Ridge, TN 37831

Immobilization of *Micrococcus luteus* within beads of bone gelatin results in a material which is able to adsorb significant quantities of strontium from dilute aqueous solutions analogous to some nuclear industry wastewaters. The mechanism appears to be principally an ion-exchange phenomenon. Both the bone gelatin and the microbial cells contribute to strontium removal; the principal contribution from the cells appears to be sorption onto cell wall material. This particular biosorbent may not be an immediate replacement for conventional ion-exchange materials currently used to remove strontium from wastewaters. However, the study does indicate that relatively inexpensive biological materials can be incorporated into particulate forms such as gel beads and used for the removal of dissolved metal ions from aqueous solution.

Metal ions are sorbed by a number of biological materials (1-5). In some cases, this phenomenon can be used to remove and concentrate ions from dilute waste solutions. Several forms of biological materials can be used, including microorganisms such as bacteria, algae, fungi, and portions of larger organisms. These biological materials may be effective sorbents as live, intact cells, although metal binding by killed cells has also been shown. In some cases, subcellular fractions, such as microbial cell walls, retain the cells' original metal-binding activity. Parts of larger organisms, such as plant roots or other tissue, can also be used to sorb metals.

Several of the important metals that have been shown to be concentrated by biological materials are listed in Table I. As many of these metals are environmental contaminants, biosorbents have potential application in pollution control (i.e., the treatment of industrial effluents prior to their release to the environment). Alternatively, biosorption may be a viable approach to the remediation of aquatic environments contaminated with such toxic metals. In some cases, the metals can be eluted (recovered) from the biosorbent. In other cases, the relatively low cost of the biosorbent could make combustion a practical way to concentrate sorbed metals prior to their reuse or ultimate disposal.

0097–6156/90/0422–0173$06.00/0

Table I. Metals Known to be Concentrated by Biological Materials[a]

Aluminum	Nickel
Cadmium	Plutonium
Cesium	Radium
Cobalt	Silver
Copper	Strontium
Gold	Thorium
Lead	Uranium
Manganese	Zinc

[a]Source: References 1, 2, and 5.

The most effective forms for any adsorbent are those which can be used in columns or other systems which give high adsorption rates and also permit the liquid to flow through the sorbent without significant backmixing (dispersion) of the fluid. These are usually various forms which can be packed into columns, generally immobilized within an inert support. The use of active sorbent (e.g., sorptive biopolymers) as support material has not been explored extensively.

This paper will consider the use of a microbial biosorbent that is incorporated within a support material which itself possesses sorptive capability and can be used within columnar bioreactors. The biosorbent consists of whole microorganisms that are immobilized within gelatin beads. The beads can be prepared with relatively uniform diameters so that the pressure losses through the column are moderate. The diffusion of metal ions through the gel particles is very high, resulting in correspondingly high adsorption rates. This combined biosorbent system is studied for the removal of strontium from a synthetic wastewater from the nuclear industry (6).

The Wastewater of Interest

The Oak Ridge National Laboratory (ORNL) processes considerable quantities of wastewaters which contain small but environmentally significant quantities of radioactive materials. A typical analysis of this wastewater is shown in Table II. The chemical composition of this wastewater is not significantly different from that of the process water found in many facilities where the water contains high calcium concentrations. The principal cation in the water is calcium, but there are significant concentrations of magnesium and sodium as well. The primary anion is bicarbonate, but sulfate, chloride, and nitrate anions are also present in significant concentrations.

The radioactive content of the water is summarized in Table II. Strontium-90 is by far the most important component in the water, making up approximately 2/3 of the total beta activity. The next most important component is ^{137}Cs, which constitutes approximately 1/15 of the beta activity. The compositions given in Tables I and II are only typical, as they can vary somewhat with operations at ORNL.

At the present time, this water is processed with zeolite ion exchangers to remove the radioactive components before it is discharged to the environment. The zeolite system is effective in protecting the environment and reducing the radioactivity

Table II. Composition of Reference ORNL Wastewater

Component	Concentration ppm (mg/L)
Ca^{2+}	40
Mg^{2+}	8
Na^{+}	5
K^{+}	2
Si^{3+}	2
Sr^{2+}	0.1
Al^{3+}	0.1
Fe^{2+}	0.1
Zn^{2+}	0.1
HCO_3^-	93
SO_4^{2-}	23
Cl^-	10
NO_3^-	11
CO_3^{2-}	7
F^-	1
Major Radiochemical Components	
^{90}Sr	4000 Bq/L
^{137}Cs	400 Bq/L
--------------	--------------
Gross beta	6000 Bq/L

(Reproduced with permission from Ref. 7. Copyright 1989 The Humana Press Inc.)

in the discharged water to levels well within accepted regulatory ranges, but regeneration of the zeolites produces significant volumes of solid wastes. The use of biosorbents is one approach which could reduce the volume of solid wastes either by permitting more efficient regeneration or by combustion (rather than regeneration) of the loaded sorbent. Ultimately the strontium (Sr) must be concentrated and fixed (immobilized) into a solid form, such as a cement grout-filled drum, for long-term storage/disposal. The more concentrated the Sr stream sent to immobilization, the lower the fixation and disposal costs. The studies described in this paper were aimed at establishing an improved understanding of the capabilities of biosorbent technology for treatment of this wastewater.

Selection of Biosorbents

A number of microorganisms have been surveyed to determine their ability to sorb Sr from dilute solutions (7). The results of this analysis are summarized in Table III. The initial Sr concentration in these equilibration experiments was 10 mg/L (ppm). This concentration, which is considerably higher than that in the wastewater of interest, permitted the use of atomic absorption spectrometry to measure dissolved Sr concentrations before and after equilibration. Based on the results of these screening experiments, the common bacterium *Micrococcus luteus* was chosen for further study. (The difficulties associated with immobilizing a filamentous fungus such as *Rhizopus* within gelatin beads discouraged further work with that organism.)

Strontium sorption by free cells of *M. luteus* was essentially instantaneous: an initial concentration of 10 mg/L Sr (as $SrCl_2$) was quantitatively removed from solution within 0.5 h (data not shown). However, a small percentage of bound Sr was released over the course of a 14-d incubation (7). This release paralleled a bleaching of the cells, which were originally yellow in color. This chromophoric material that binds Sr is apparently associated with the cell surface. In immobilized cell systems, no gradual Sr release occurred. Apparently, this material was retained within the gel beads after its release from the cells, probably because the material could not diffuse through the gel's mesh-like structure (7). This retention of all cell components/products is an important advantage of the immobilized adsorbent. An additional advantage of the present system is the use of a support which itself binds Sr.

Immobilization of Biosorbents

The *M. luteus* cells were immobilized within a bone gel to form a suitable material for use in a packed adsorption column. The gel particles were prepared using 15 wt % gelatin, 2 wt % propylene glycol alginate, and 20 to 40 wt % dry cells.

In many cases, it has been possible to form gel particles with exceptionally good uniformity by feeding the gel through a nozzle which vibrates at a fixed frequency (8). For the greatest uniformity, the vibrating frequency is adjusted to correspond to the frequency at which the jet becomes unstable to breakup into droplets. The bone gelatin, when loaded with the high cell concentrations desired for sorbent preparation, had a high viscosity and was thus difficult to pump through the small nozzle. Under these conditions, the gel particles formed also exhibited notable elasticity.

The most effective way to produce gel particles heavily loaded with cells involved rapid stirring of a suspension of gel in sodium hydroxide (for cross-linking). This approach produced a range of particle sizes. Particles of the desired size were obtained by size fractionation (i.e., screening). In larger scale operations, high yields of the desired gel particle sizes probably could be obtained by modified preparation methods. For instance, the very viscous gel suspension could be extruded from a relatively small nozzle, and the desired particle sizes could be obtained by cutting the extrusion rapidly with rotating knives. The further development of techniques for producing uniform gel particles from viscous suspensions was delayed pending the evaluation of the properties and performance of the gel particles which were prepared by the present method.

Behavior of Packed Columns with Cell-Filled Gel Particles

The cell-filled gel particles were packed into columns with 9- and 50-mL capacities. The packing procedure was relatively simple, since a slurry of the particles was simply poured into the column and the particles settled to the bottom of the column. Additional suspension was added until the settled volume of particles reached the desired height. There was no effort to compress the particles into the column or to reduce the void fraction otherwise. Care was taken to prevent the packed column from drying.

Some samples of gel particles were used over periods of up to a month and underwent several adsorption/desorption cycles (see below) without detectable changes. Microbial (hydrolytic) degradation of the gel particles was occasionally observed but was completely eliminated by boiling of both the sorbent and the wastewater feed solution prior to use. This treatment probably killed the organisms that were immobilized within the gel. However, no apparent effect on biosorbent activity was observed.

Strontium Uptake from Dilute Solutions. Experimental loading curves for Sr sorption by gelatin, with and without immobilized *Micrococcus* cells, are shown in Figure 1. The position of the breakthrough indicates the adsorption capacity of the biosorbent. The slope of the breakthrough gives an estimate of the rates of adsorption.

The midpoint of the breakthrough curve for the combined biosorbent occurred after approximately 70 column volumes of solution. Since the void fraction in the column is approximately 50%, this result indicates an effective distribution coefficient of approximately 150, based upon wet gel. That is a lower value than those reported for cells alone (Table III). Although there may be several reasons for this difference, the principal reason could be that concentrations in dry cells were used in the equilibrium experiments. The effective distribution coefficients obtained from column breakthrough measurements also represent the combined biosorbent system, which is composed of gelatin as well as cells. The swollen gel certainly contains more water than the packed, centrifuged cells used in the scouting batch equilibrations, resulting in less biosorbent per unit volume (or mass). Also, the column experiments correspond to adsorbent loadings in equilibrium with 10 mg/L Sr, while the batch experiments corresponded to loadings in equilibrium with much lower concentrations after equilibration.

A breakthrough curve for similar gel particles without cells (not shown) is essentially the same as that shown in Figure 1. At first glance, these results may suggest that the cells do not contribute significantly to Sr sorption. However, the cell-loaded gel particles used in the experiment shown in Figure 1 contain much less gelatin than the gel used in the experiment without cells because a significant portion of the volume (and mass) of gelatin is replaced with cells. The masses of dry gelatin

Table III. Rank of Organisms for Sr Adsorption from 100 mg/L Solution after 14 d

Organism	Distribution Coefficient[a]
Rhizopus	26,240
Micrococcus	9,980
Anabaena	8,200
Streptococcus	5,490
Bacillus	5,240
Chlamydomonas	3,750
Coelastrum	2,990
Penicillium	2,910
Scenedesmus	2,170
Citrobacter	1,800
Zooglea	1,540
Ashbya	1,380
Escherichia	1,230
Paecilomyces	1,080
Chlorella	1,070
Candida	980
Pseudomonas	870
Saccharomyces	760
Streptomyces	740
Caulobacter	450

[a] Distribution coefficient:

$$\frac{\mu g \text{ Sr bound to cells/g cells (dry wt basis)}}{\mu g \text{ Sr in solution/g solution}}$$

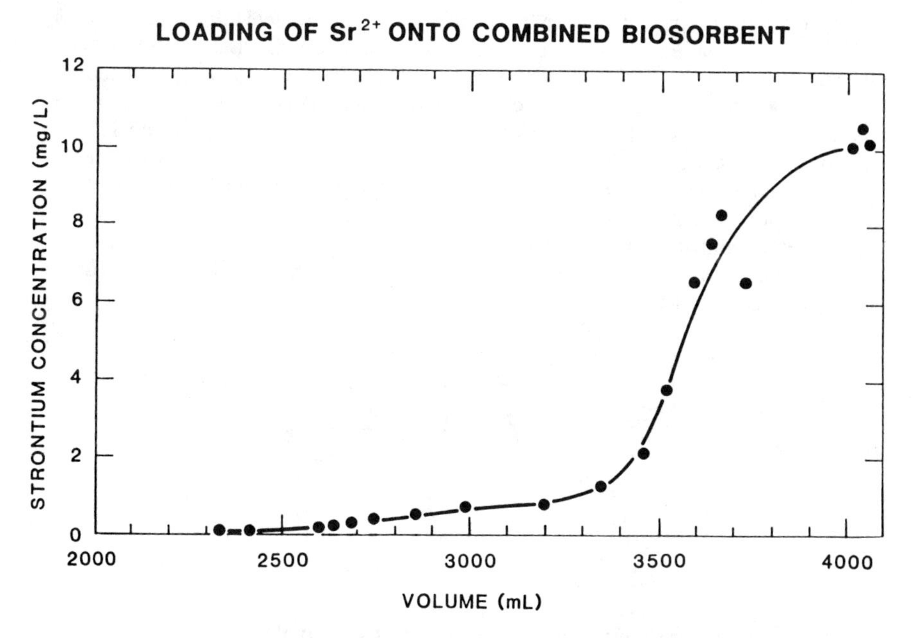

Figure 1. Loading of Sr^{2+} in a 50-mL column of bone gelatin with *Micrococcus luteus* cells. (Reproduced with permission from Ref. 7. Copyright 1989 The Humana Press Inc.)

and dry cells in the loaded gel are approximately equal; the mass of dry gelatin per unit volume of gel particles without cells is approximately twice that in the loaded particles. Thus, approximately half of the Sr adsorption shown in Figure 1 results from the cells and half results from the gelatin itself. Bone gelatin itself is relatively effective in removing Sr from dilute solutions, but inexpensive cells of *M. luteus* are also just as effective and can be used to displace significant fractions of the gelatin.

The Rate of Uptake. Estimates for the rate of strontium uptake can be obtained from breakthrough curves through examination of the portion of the curve near the midpoint of breakthrough. The shape of the curve in this region depends on the shape of the equilibrium isotherm, the kinetics (or mass transfer phenomena), and dispersion (and/or nonuniform flow distribution).

To estimate the rate of adsorption from the data in Figure 1, three assumptions were made. First, a linear equilibrium isotherm (constant distribution coefficient) was assumed for convenience. To be conservative, the effects of axial mixing (dispersion) were assumed to be zero. This latter assumption lumped both the dispersion and kinetics effects into the kinetics term. For this initial analysis, all of the kinetic resistance was assumed to be in the diffusion through the gel particles themselves. This last assumption gave a conservative estimate of the diffusion resistance since all other kinetic resistances added to the diffusion resistance and indicate a lower apparent diffusion coefficient within the gel particles.

The simplified (long bed) solution of Rosen (9) was then used to estimate the mass transfer resistance.

$$\frac{C_{out}}{C_{in}} = \frac{1}{2}\left[1 + \mathrm{erf}\left(\frac{\frac{3y}{2x} - 1}{2/\sqrt{5x}}\right)\right] \quad (1)$$

where

y/x	=	$(2\epsilon/3K)(vt/z - 1)$,
x	=	$3DKz/\epsilon vr^2$,
K	=	distribution coefficient (slope of the equilibrium curve),
v	=	flow velocity,
D	=	diffusion coefficient in the solid particle,
ϵ	=	void fraction in the bed,
z	=	bed length, and
t	=	time.

The diffusion coefficient can be evaluated from this equation by fitting the equation at two points. We chose to fit the equations at points where C_{out}/C_{in} was 0.2 and 0.8. The equation was evaluated for different values of x, and a plot was prepared of $(y_{.8} - y_{.2})/y_{.2}$ vs x. The y's represent dimensionless time (or solution volume) on the breakthrough curves when the effluent concentrations are 80% and 20% of the inlet concentration. Since a ratio of the y's is used, the units used in the breakthrough curve are not important. Once x is known, the diffusion coefficient can be easily calculated.

For the breakthrough curve shown in Figure 1, $y_{.8}$ is 3700 mL, and $y_{.2}$ is 3450 mL. Then x is found from the plot described previously to be approximately 4.5. This corresponds to a diffusion coefficient within the particle of approximately 0.3 x 10^{-6} cm^2/s, a value that is only approximately tenfold less than the diffusion coefficient in water.

The presence of gel polymer in the biosorbent could restrict the motion of ions and thus reduce the effective diffusion coefficients. The experimental findings, however, indicate that these gel particles have very open structures which are easily penetrated by the Sr ions. Qualitatively, it indicates that excellent adsorption kinetics are available for this system. Because essentially all of the kinetic resistances are lumped into this one parameter, the intraparticle diffusion coefficient, this result shows that none of the kinetic resistances are large.

One limitation to this treatment of breakthrough data needs to be mentioned. The Rosen equations apply to cases with linear isotherms, and any curvature in the isotherm will make the result less accurate. For adsorption processes one would expect many dilute systems to be approximated well by linear isotherms. However, the next section will show that this is an ion exchange system, and the linear isotherm will not be accurate. These results must be viewed as approximate, but reasonable approximations to use in the absence of full equilibrium information.

Mechanism of Strontium Adsorption

The mechanism for Sr adsorption by the combined biosorbent was not studied in detail. Routine experiments, however, suggest that the principal Sr uptake occurs via an ion-exchange mechanism. Evidence supporting the ion-exchange mechanism is presented in Figures 2 and 3. Strontium is shown to be replaced quantitatively by calcium (Ca) ions, and the molar quantities exchanged are equal. Gel beads without cells behaved in essentially the same manner. Batch tests with free cells of *M. luteus* yielded the same result.

Selectivity of the Biosorbent for Sr. If Sr adsorption is principally an ion-exchange phenomenon, the selectivity of the adsorbent for Sr over Ca is important, at least for applications such as the ORNL wastewater, which contains considerably more Ca than Sr.

There are indications of slight selectivity of the cell/gelatin adsorbent for Sr over Ca, as indicated in column loading experiments (Figures 4 and 5). Calcium ion breakthrough occurred first; however, the difference in the breakthrough was very small, suggesting that the selectivity for Sr was relatively small. A second qualitative indication of the selectivity for Sr is the overshoot of the Ca concentration during the later stages of the breakthrough (i.e., the increase in the Ca efflux concentration relative to the inlet concentration of Ca). This phenomenon arises when the least preferred ion (Ca) loads on the column in front of the moving band of the most preferred ion (Sr). The Sr band displaces this Ca, which adds to the Ca concentration in this portion of the breakthrough curve. However, the small size of this overshoot suggests that the selectivity for Sr is minimal. For most practical purposes, one would probably ignore the selectivity.

By using the mixture of gelatin and cells, it is not possible to determine if the slight selectivity for Sr comes from the gelatin or from the cells. Scouting experiments with cells alone, however, suggested that the cells actually prefer Ca; but, again, the difference in selectivity is believed to be small.

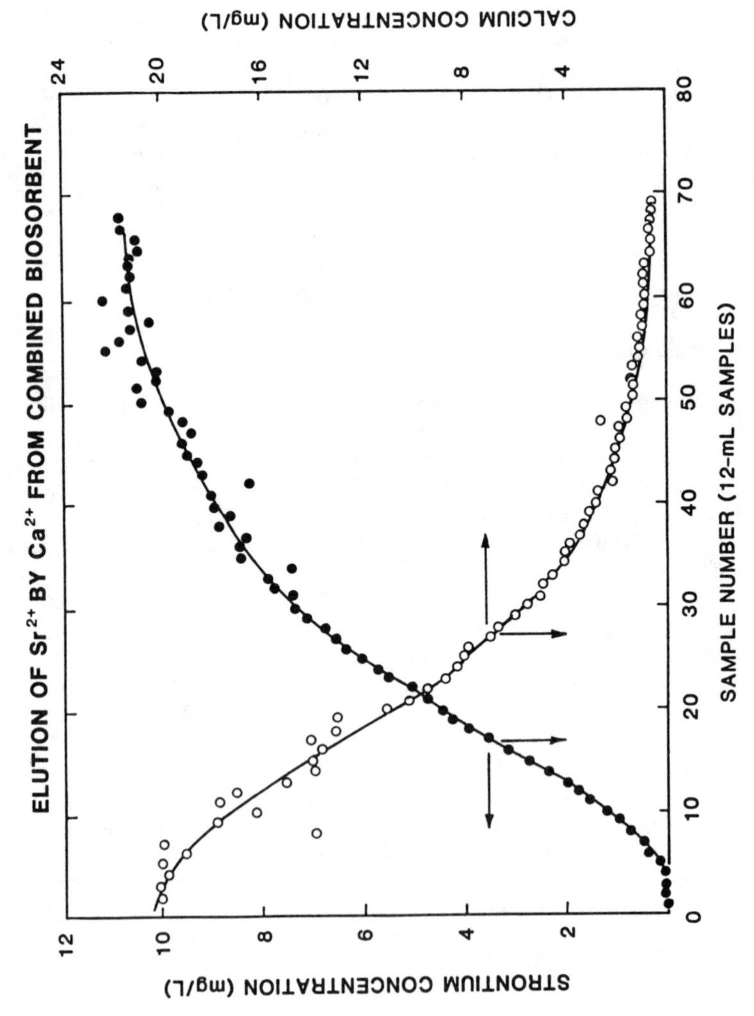

Figure 2. Elution of Sr^{2+} from a 9-mL column of bone gelatin with cells.

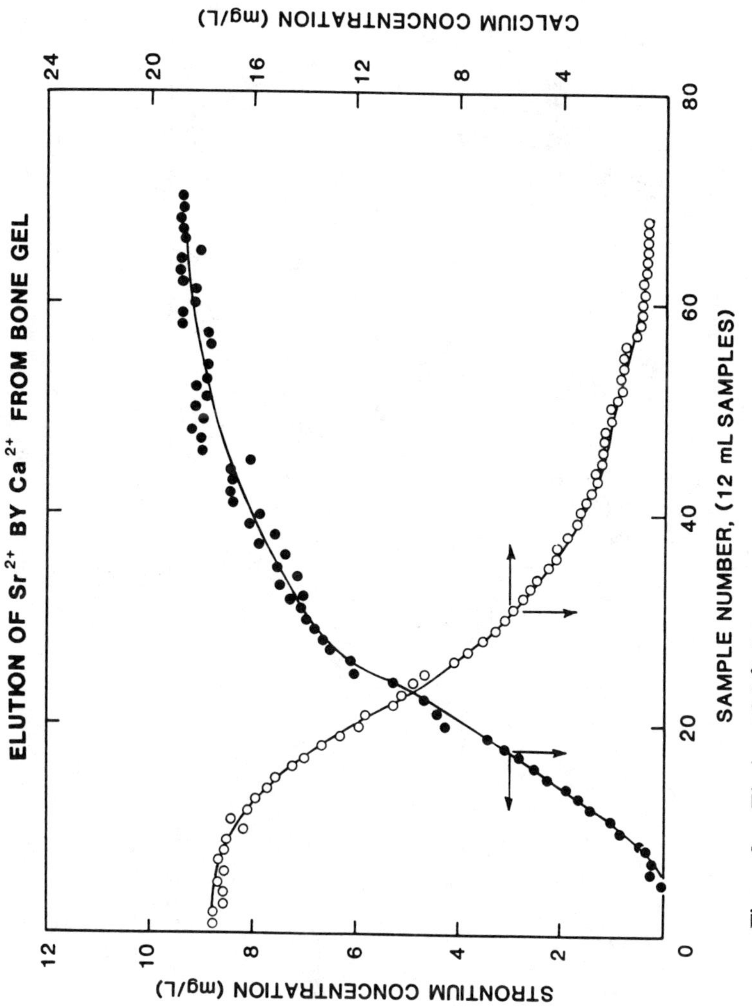

Figure 3. Elution of Sr^{2+} from a 9-mL column of bone gelatin without cells.

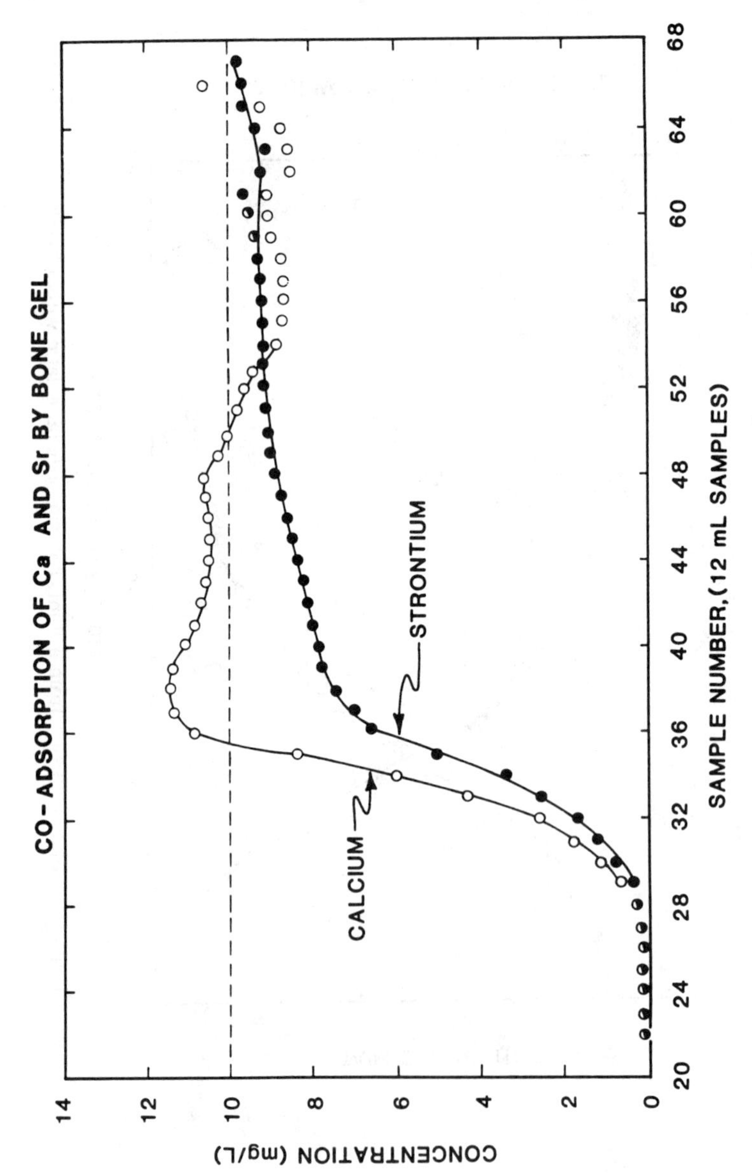

Figure 4. Coadsorption of Ca^{2+} ions (10 mg/L) and Sr^{2+} ions (10 mg/L) on a 9-mL column filled with bone gelatin and cells.

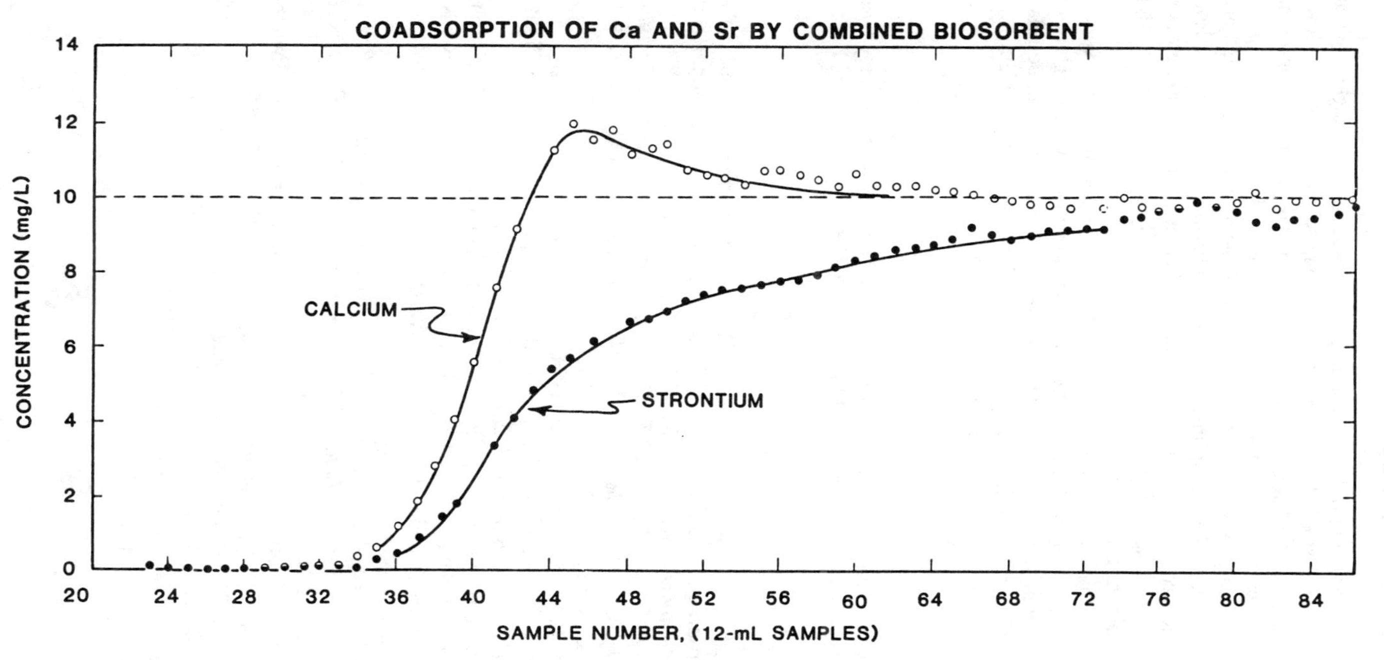

Figure 5. Coadsorption of Ca^{2+} ions (10 mg/L) and Sr^{2+} ions (10 mg/L) on a 9-mL column filled with bone gelatin without cells.

Conclusions and Recommendations

This study has focused upon only one cell-biopolymer system and a single application. The particular material studied does not show a strong selectivity for Sr. However, the study does suggest that the combined biosorbent has significant ion-exchange capacities. The capacity per unit volume is certainly smaller than the capacity of commercial ion-exchange resins, but the cost of the biological materials could also be much less. Whole microorganisms can be especially inexpensive adsorbent components. A less costly adsorbent/exchanger could be especially desirable when the material can only be used for a single cycle; the metal could be concentrated by burning the adsorbent/exchanger.

The gelatin that was used to immobilize the cells provided very high adsorption/exchange rates, and this can be important in cases in which high removals are required. The high rates indicate that the breakthrough curves can be relatively sharp, and good sorbent utilization can be obtained along with high removal efficiency.

Acknowledgments

The Oak Ridge National Laboratory is operated by Martin Marietta Energy Systems, Inc., under contract DE-AC05-84OR21400 with the U.S. Department of Energy.

Literature Cited

1. Shumate, S. E. et al.; "Separation of Heavy Metals from Aqueous Solutions Using Biosorbents - Development of Contacting Devices for Uranium Removal," *Biotechnol. and Bioeng. Symp. No. 10* **1980**, 27-34.
2. Strandberg, G. W.; Shumate, S. E.; *Accumulation of Uranium, Cesium, and Radium by Microbial Cells -- Bench-scale Studies*, ORNL/TM-7599, Oak Ridge National Laboratory, Oak Ridge, TN, 1982.
3. Macaskie, L. E.; *Biotechnol. Lett.* **1984**, *6*, 71-76.
4. Macaskie, L. E.; Dean, A. C. R.; "Strontium Accumulation by Immobilized Cells of *Citrobacter* sp.," *Biotechnol. Lett.* **1985**, *7*, 627-630.
5. Macaskie, L. E.; Dean, A. C. R.; "Microbial Metabolism, Desolubilization, and Deposition of Heavy Metals: Metal Uptake by Immobilized Cells and Applications to Detoxification of Liquid Wastes," *Biological Waste Treatment* (to be published by Alan R. Liss, Inc.), 1989.
6. U.S. Nuclear Regulatory Commission; *Calculation of Releases of Radioactive Materials in Gaseous and Liquid Effluents from Pressurized Water Reactors (PWR-GALE Code)*, NUREG-0017, 1976.
7. Watson, J. S.; Scott, C. D.; Faison, B. D.; "Adsorption of Sr by Immobilized Microorganisms," *Appl. Biochem. Biotechnol.* **1989**, *20/21*, 699-709.
8. Scott, C. D.; "Techniques for Producing Monodispersed Biocatalyst Beads for Use in Columnar Bioreactors," *Ann. NY Acad. Sci.* **1987**, *501*, 487-493.
9. Rosen, J. B.; *Ind. Eng. Chem.* **1954**, *46*, 1590.

RECEIVED November 10, 1989

Chapter 12

Selective Removal of Metals from Wastewater Using Affinity Dialysis

Shangxu Hu[1], Rakesh Govind[2,4], and James C. Davis[3]

[1]Department of Chemical Engineering, Zhejiang University, Hangzhou, People's Republic of China

[2]Department of Chemical Engineering, University of Cincinnati, Cincinnati, OH 45221

[3]British Petroleum Research, British Petroleum of America, Inc., 4440 Warrensville Center Road, Cleveland, OH 44128–2837

In this paper the application of an Affinity Dialysis Process has been studied for waste water treatment. The basic technique requires a solution of macromolecular agents in water which are capable of rapidly complexing metal ions. The macromolecular solution flows through the tube side of a hollow fiber membrane unit, with wastewater flowing through the shell side. The macromolecular solution can be regenerated by changing the pH of the solution. Experimental studies and a detailed mathematical model for the affinity dialysis system has been presented.

There is a growing need for an efficient and low-cost process for industrial wastewater treatment [1]. One of the problems associated with wastewater treatment is the recovery of metals from ore leachates and removal of toxic metals from wastewater. Various kinds of separation methods have been developed for this problem and are currently in industrial use [2]. The goal of our research is the application of membrane processes to this problem.

Membrane processes offer several benefits including lower energy utilization, easier to automate, lower maintenance cost due to less moving parts, and higher final product quality [3]. Membrane processes can be classified into four basic catagories:

1) presssure driven processes, such as: ultrafiltration, microfiltration, and reverse osmosis [4];
2) electromotive force driven process, such as electodialysis [5];
3) concentration driven process, such as dialysis [6];
4) chemical driven processes, such as: facilitated transport and coupled transport processes [7].

[4]Address correspondence to this author.

0097–6156/90/0422–0187$07.75/0

Pressure driven processes have the disadvantages of high pressure [8]. In industrial wastewater treatment, the usual case is to separate different species of metal ions with same polarity, and therefore electromotive force driven processes are not well suited. Chemical driven processes such as facilitated tranport use a microporous membrane impregnated with a liquid carrier agent. Problems associated with such processes include membrane instability due to loss of carrier, and difficulties in carrier regeneration[9].

Dialysis process has the advantage of physico-chemical stability, simplicity in equipment, and smoothness in operation; however, it generally exhibits inadequate productivity and selectivity[10].

Recently, new membrane processes have been developed, including the membrane-based solvent extraction[11] and the diafiltration process [12].

The membrane-based solvent extraction process utilizes microporous membranes as an interface between the aqueous solution and organic solvents containing liquid ion exchangers. Metal ions are transported from the aqueous solution to the organic phase at the interface created in the pores of the membrane. The organic solvent, which is loaded with metal ions in an extraction unit, is regenerated in contact with the stripping solution in a stripping unit.

This process features better membrane stability than facilitated transport process and higher selectivity than conventional dialysis process. However, in order to facilitate the metal ion transport and reduce the loss of organic solvent, larger pore sizes and thinner membranes must be used and the the organic solvent side must be kept at a lower pressure than the wastewater feed side. These conditions unavoidably cause a certain amount of water transport to the organic phase. The fine water droplets, which can not be completely separated from the organic phase, will carry stripping solution back to reduce the removal efficiency and adversely affect the selectivity of the process and its overall performance [11].

The difiltration process consists of a reactor unit, where macromolecular chelating agents or emulsified liquid ion exchange materials are used to selectively bind certain metal ions, a ultrafiltration unit to separate the complex from unreacted metal ions, and a stripping unit to regenerate the metal loaded macromolecular materials. Fouling, which results from concentration polarization and pore plugging, or a slow consolidation of the gel layer, can be a problem with the ultrafiltration unit[5].

Parallel to the development of the above mentioned processes, another new process named affinity dialysis was proposed [13], which is designed to keep the advantage of dialysis process while overcoming its shortcomings.

In affinity dialysis a chelating agent, which selectively reacts with certain metal ion species to form a complex in solution, is used in combination with a semi-permeable membrane designed to retain the polymer together with the complex but allow the unreacted metal ion species to freely permeate [14].

In a previous paper, the authors reported the successful application of affinity dialysis to the selective removal of metal ions from wastewater and presented the advantages of the process[15]. In this work, special emphasis has been given to the theoretical analysis of experimental data and a systematic numerical simulation of the process, such that its fundamental characteristics could be explored in depth.

Theoretical Modeling

The basic idea of affinity dialysis process used for separating metal ion species A and B in feed can be illustrated schematically in Figure 1, where x_A and x_B are the concentrations of metal ions A and B respectively in feed stream, y_A and y_B are those in polymer solution stream, y_P are that of polymer, and y_C and y_D are those of metal-polymer complexes in dialysate respectively.

Assuming that axial diffusion is negligible, the concentrations of components in the bulk stream along the radial direction are nearly

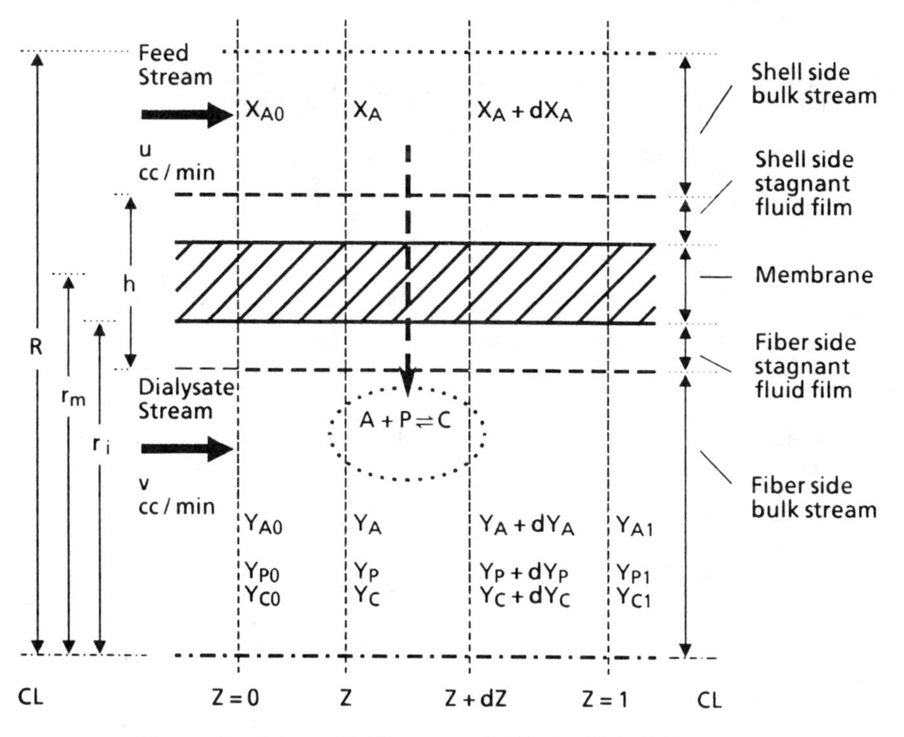

Figure 1 Schematic Diagram of Affinity Dialysis Process for Extracting Metal Species A

constant, and the chelating reaction is reversible with second order forward and first order backward kinetics [15], the reaction kinetic equations are:

$$\frac{dy_A}{d\theta} = -k_{AP} \cdot y_A \cdot y_P \tag{1}$$

$$\frac{dy_B}{d\theta} = -k_{BP} \cdot y_B \cdot y_P \tag{2}$$

$$\frac{dy_C}{d\theta} = -k_C \cdot y_C \tag{3}$$

$$\frac{dy_D}{d\theta} = -k_D \cdot y_D \tag{4}$$

The governing equations for affinity dialysis process as shown schematically in Figure 1 can then be derived as follows:

Defining dimensionless variables:

$$X_A = \frac{x_A}{x_{A0}},\ X_B = \frac{x_B}{x_{A0}},\ Y_A = \frac{y_A}{x_{A0}},\ Y_B = \frac{y_B}{x_{A0}}, \tag{5}$$

$$Y_P = \frac{y_P}{x_{A0}},\ Y_C = \frac{y_C}{x_{A0}},\ Y_D = \frac{y_D}{x_{A0}},\ and\ Z = \frac{z}{l}, \tag{6}$$

model equations in dimensionless form will be obtained, i.e.,

$$\frac{dX_A}{dZ} = P_A(Y_A - X_A) \tag{7}$$

$$\frac{dX_B}{dZ} = P_B(Y_B - X_B) \tag{8}$$

$$\frac{dY_A}{dZ} = P_C(X_A - Y_A) - K_{AP} \cdot Y_A \cdot Y_P + K_C \cdot Y_C$$

$$\frac{dY_B}{dZ} = P_D(X_B - Y_B) - K_{BP} \cdot Y_B \cdot Y_P + K_D \cdot Y_D$$

$$\frac{dY_P}{dZ} = K_C \cdot Y_C - K_{AP} \cdot Y_A \cdot Y_P + K_D \cdot Y_D - K_{BP} \cdot Y_B \cdot Y_P$$

$$\frac{dY_C}{dZ} = K_{AP} \cdot Y_A \cdot Y_P - K_C \cdot Y_C$$

$$\frac{dY_D}{dZ} = K_{BP} \cdot Y_B \cdot Y_P - K_D \cdot Y_D \tag{9}$$

with the conditions:

$$X_A = X_{A0} = 1.0\,,\ X_B = X_{B0},\ Y_A = Y_{A0},\ Y_B = Y_{B0},$$

$$Y_P = Y_{P0},\ Y_C = Y_{C0},\ and\ Y_D = Y_{D0},\ at\ Z = 0 \tag{10}$$

There are two sets of dimensionless groups in equation system (9). The first of them are:

$$P_A = \frac{p_A \cdot n \cdot 2\pi r_m \cdot l}{u \cdot h}$$

$$P_B = \frac{p_B \cdot n \cdot 2\pi r_m \cdot l}{u \cdot h}$$

$$P_C = \frac{p_A \cdot n \cdot 2\pi r_m \cdot l}{v \cdot h} \tag{11}$$

$$P_D = \frac{p_B \cdot n \cdot 2\pi r_m \cdot l}{v \cdot h}$$

In general, this set of dimensionless groups can be defined by:

$$P = \frac{p \cdot A}{w \cdot h} \tag{12}$$

where p is the permeability of membrane used, A the total effective area of the membrane device, h the effective thickness of the resistance to the permeation, which is approximately equal to the thickness of membrane, and w the flow rate of stream pasing through the particular side of the membrane device.

Physically P measures the specific amount of mass-permeation through the device in question under given conditions, so we name this dimensionless group "SPECIFIC PERMEATION" (SP).

The second set of dimensionless groups are:

$$K_{AP} = \frac{x_{A0} \cdot k_{AP} \cdot n \cdot \pi r_i^2 \cdot l}{v}$$

$$K_{BP} = \frac{x_{A0} \cdot k_{BP} \cdot n \cdot \pi r_i^2 \cdot l}{\upsilon}$$

$$K_C = \frac{k_C \cdot n \cdot \pi r_i^2 \cdot l}{\upsilon} \tag{13}$$

$$K_D = \frac{k_D \cdot n \cdot \pi r_i^2 \cdot l}{\upsilon}$$

In general, this set of dimensionless groups can be defined as:

$$K = \frac{k \cdot V \cdot x_{A0}^{(j-1)}}{w} \tag{14}$$

in which k is the reaction rate constant for corresponding reaction, V the effective volume where the reaction takes place, j the order of reaction, and w the flowrate of stream passing through.

Chemically K measures the specific amount of mass reacted within the device in question, so we name this dimensionless group "SPECIFIC CONVERSION" (SC).

Verification of Numerical Methods

General analytical solution to the mathematical model equations is difficult and hence numerical methods have been used. However, numerical solution should be verified with analytical solution of reduced model equations. These analytical solutions serve as a standard for evaluating the relative errors of numerical methods.

Assuming that there is only one species of metal ion A, which reacts with the polymer reversibly due to the excessive amount of polymer usually present in the solution stream, the concentration of polymer can be taken as constant. Therefore the reaction rate is affected only by the concentration of A, and a set of reduced equations was obtained, i.e.:

$$\frac{dX_A}{dZ} = P_A(Y_A - X_A) \tag{15}$$

$$\frac{dY_A}{dZ} = P_C(X_A - Y_A) - K_A \cdot Y_A + K_C \cdot Y_C \tag{16}$$

$$\frac{dY_P}{dZ} = K_C \cdot Y_C - K_A \cdot Y_A \tag{17}$$

$$\frac{dY_C}{dZ} = K_A \cdot Y_A - K_C \cdot Y_C \tag{18}$$

From equation (15) we have:

$$Y_A = \frac{1}{P_A} \cdot \frac{dX_A}{dZ} + X_A \tag{19}$$

By substituting (19) in (16), it follows:

$$\frac{d}{dZ}\left(\frac{1}{P_A} \cdot \frac{dX_A}{dZ} + X_A\right)$$

$$= P_C\left(X_A - \frac{1}{P_A} \cdot \frac{dX_A}{dZ} - X_A\right) - K_A \cdot \left(\frac{1}{P_A} \cdot \frac{dX_A}{dZ} + X_A\right) + K_C \cdot Y_C$$

or,

$$\frac{d^2 X_A}{dZ^2} + (P_A + P_C + K_A)\frac{dX_A}{dZ} + (K_A P_A X_A - K_C P_C Y_C) = 0 \tag{20}$$

Substitute (19) in (18) and obtain:

$$\frac{K_A}{P_A}\frac{dX_A}{dZ} + K_A X_A = K_C Y_C + \frac{dY_C}{dZ} \tag{21}$$

From (20) we have:

$$Y_C = \frac{1}{K_C P_A}\frac{d^2 X_A}{dZ^2} + \frac{P_A + P_C + K_A}{P_A K_C}\frac{dX_A}{dZ} + \frac{K_A}{K_C} X_A \tag{22}$$

By differentiating (22) with respect to Z, it gives:

$$\frac{dY_C}{dZ} = \frac{1}{K_C P_A}\frac{d^3 X_A}{dZ^3} + \frac{P_A + P_C + K_A}{P_A K_C}\frac{d^2 X_A}{dZ^2} + \frac{K_A}{K_C}\frac{dX_A}{dZ} \tag{23}$$

Substitute (22) and (23) in (21) to eliminate Y_C and obtain:

$$\frac{d^3 X_A}{dZ^3} + (P_A + P_C + K_A + K_C)\frac{d^2 X_A}{dZ^2} + (P_A K_A + P_A K_C + P_C K_C)\frac{dX_A}{dZ} = 0 \tag{24}$$

in which the only unknown dependent variable is X_A. Equation (24) is a third-order linear homogeneous differential equation, and can be written and solved in the following standard form:

$$\frac{d^3 X_A}{dZ^3} + a \frac{d^2 X_A}{dZ^2} + b \frac{dX_A}{dZ} = 0$$

where

$$a = P_A + P_C + K_A + K_C \tag{25}$$

and

$$b = P_A K_A + P_A K_C + P_C K_C \tag{26}$$

by making use of its auxiliary equation:

$$s^3 + a s^2 + b s = 0 \tag{27}$$

the three roots of which are:

$$s_1 = \frac{-1}{2} (a + \sqrt{(a^2 - 4b)})$$

$$s_2 = \frac{-1}{2} (a - \sqrt{(a^2 - 4b)})$$

$$s_3 = 0$$

Thus we have the solution of X_A:

$$X_A = C_1 \cdot \exp(s_1 Z) + C_2 \cdot \exp(s_2 Z) + C_3 \tag{28}$$

By differentiating (28) with respect to Z, it gives:

$$\frac{dX_A}{dZ} = C_1 s_1 \cdot \exp(s_1 Z) + C_2 s_2 \cdot \exp(s_2 Z) \tag{29}$$

Substitute (28) and (29) in (19) and get a solution of Y_A:

$$Y_A = C_1 (\frac{s_1}{P_A} + 1) \cdot \exp(s_1 Z) + C_2 (\frac{s_2}{P_A} + 1) \cdot \exp(s_2 Z) + C_3 \tag{30}$$

By substituting (30) in (18), it follows:

$$\frac{dY_C}{dZ} = K_A C_1 (\frac{s_1}{P_A} + 1) \cdot \exp(s_1 Z) + K_C C_2 (\frac{s_2}{P_A} + 1) \cdot \exp(s_2 Z) \tag{31}$$

$$+ K_A C_3 - K_C Y_C$$

Equation (31) is a typical first-order differential equation of the following form:

$$\frac{dY_C}{dZ} + K_C Y_C = \Phi(Z)$$

the analytical solution of which is:

$$Y_C = \frac{\int \Phi(Z) \exp(K_C Z)\, dZ}{\exp(K_C Z)}$$

By integrating $\Phi(Z) \exp(K_C Z)$, it gives:

$$Y_C = \frac{K_A C_1}{r_1 + K_C}\left(\frac{s_1}{P_A} + 1\right) \cdot \exp(s_1 Z) + \frac{K_A C_2}{s_2 + K_C}\left(\frac{s_2}{P_A} + 1\right) \cdot \exp(s_2 Z) + \frac{K_A C_3}{K_C} \quad (32)$$

Substitute (30) and (32) in (17), we obtain:

$$\frac{dY_P}{dZ} = \frac{K_A C_1 (-s_1)}{s_1 + K_C}\left(\frac{s_1}{P_A} + 1\right) \cdot \exp(s_1 Z) + \frac{K_A C_2 (-s_2)}{s_2 + K_C}\left(\frac{s_2}{P_A} + 1\right) \cdot \exp(s_2 Z) \quad (33)$$

which can be simply integrated and gives:

$$Y_P = C_4 - \frac{K_A C_1}{s_1 + K_C}\left(\frac{s_1}{P_A} + 1\right) \cdot \exp(s_1 Z) + \frac{K_A C_2}{s_2 + K_C}\left(\frac{s_2}{P_A} + 1\right) \cdot \exp(s_2 Z) \quad (34)$$

The integration constants C_1 through C_4 can be evaluated by applying the complementary conditions at $Z = 0$, i.e.:

$$X_{A0} = C_1 + C_2 + C_3 \quad (35)$$

$$Y_{A0} = C_1\left(\frac{s_1}{P_A} + 1\right) + C_2\left(\frac{s_2}{P_A} + 1\right) + C_3 \quad (36)$$

$$Y_{C0} = \frac{K_A C_1}{s_1 + K_C}\left(\frac{s_1}{P_A} + 1\right) + \frac{K_A C_2}{s_2 + K_C}\left(\frac{s_2}{P_A} + 1\right) + \frac{K_A C_3}{K_C} \quad (37)$$

$$Y_{P0} = C_4 - \frac{K_A C_1}{s_1 + K_C}\left(\frac{s_1}{P_A} + 1\right) - \frac{K_A C_2}{s_2 + K_C}\left(\frac{s_2}{P_A} + 1\right) \quad (38)$$

Equations (35) through (38) are a set of simultaneous linear equations, which can easily be solved and gives:

$$C_1 = \frac{f_2(Y_{A0} - X_{A0}) - (e_2 - 1)(\frac{K_C}{K_A} Y_{C0} - Y_{A0})}{(e_1 - 1) f_2 - (e_2 - 1) f_1} \tag{39}$$

$$C_2 = \frac{Y_{A0} - X_{A0} - (e_1 - 1) C_1}{(e_2 - 1)} \tag{40}$$

$$C_3 = \frac{K_C}{K_A}(Y_{C0} - d_1 e_1 C_1 - d_2 e_2 C_2) \tag{41}$$

$$C_4 = Y_{P0} + d_1 e_1 C_1 + d_2 e_2 C_2 \tag{42}$$

where

$$f_1 = \frac{d_1 e_1 K_C}{K_A} - e_1 \tag{43}$$

$$f_2 = \frac{d_2 e_2 K_C}{K_A} - e_2 \tag{44}$$

$$e_1 = \frac{s_1}{P_A} + 1 \tag{45}$$

$$e_2 = \frac{s_2}{P_A} + 1 \tag{46}$$

$$d_1 = \frac{K_A}{s_1 + K_C} \tag{47}$$

$$d_2 = \frac{K_A}{s_2 + K_C} \tag{48}$$

Equations (28),(30), (32), (34) are a complete set of analytical solutions to the reduced model equations (15), (16), (17), (18), and will be used as a standard to evaluate the errors of the numerical algorithm.

Since the mathematical model is essentially a set of ordinary differential equations (ODES), it can be solved in several ways. Practically, it is important to make a judicious selection of the many available numerical algorithms. Taking into account the precision, robustness, simplicity, and other criterion, after a series of systematic numerical experiments, we chose a category of single-step, automatically error controlled, variable step-size explicit integration algorithms, amongst which the 4th-order one is usually good enough for general usage.

Experimental Studies

As a typical case which may occur in many mineral processing plant, the removal of calcium ion from wastewater was selected for our study using affinity dialysis. After a selection amongst a variety of water soluble polymers, Poly-2-acrylamido-2-methyl-1-propanesulfonic acid (PAMP) was used as the chelating agent in this study.

The experimental system was designed to be operated in a continuous counter-current mode. As shown schematically in Figure 2, it mainly consists of an extraction unit, a strip unit, an acidificator and an alkalificator.

In the extraction unit, the calcium ions in the feed stream permeat through the membrane, react with PAMP, and stay in dialysate in the form of calcium-polymer complex. After pH adjustment through acidificator, the dialysate is pumped to strip unit, where the acidified PAMP is regenerated, and calcium ions permeate through the membrane. The regenerated PAMP is pumped through the alkalificator for a readjustment of pH and then flow back again to the extraction unit.

Hollow fiber artifical kidneys obtained from HOSPAL. 69-Meyzieu, France, designated DISSCAP 80 and DISSCAP 140 were used for the extraction unit. In the strip unit of the system, a product obtained from Amicon Corp., Danvers, Massachusetts, was used for the metal recovery due to the superior chemical stability of its hollow fiber.

All chemicals were reagent grade or better. Concentrated HCl and 50% NaOH were used for pH adjustment in the reactors to minimize the amount of water added to the system. PAMP was used as received from Aldrich.

Some of the original data, with relatively large experimental error, which is checked by making an overall material balance, was not used for further analysis in this study. The data used in this study is shown in Table 1, in which x_{A1} is the inlet concentration of calcium ion in feed waste, y_{P0} the inlet concentration of PAMP in dialysate, x_{A0} the outlet concentration of calcium ion in treated waste, and y_{C1} the outlet concentration of metal-polymer complex.

By using the proposed mathematical model and the numerical algorithm as verified in the previous section, a series of primary simulation runs were made, where the average physico-chemical parameter values were identified in our previous work[15]. A typical numerical solution is as shown in Figure 3. The experimental data points are also plotted on the same Figure and it shows that the simulation results agree fairly well with the experimental data. Other numerical solutions are listed in Table 2 in comparison with experimental data. In general, the relative errors are moderate and it can be concluded that the mathematical model is acceptable.

Simulation Studies

After both the validity of the proposed mathematical model and the presicion of numerical methods employed had been verified, extensive

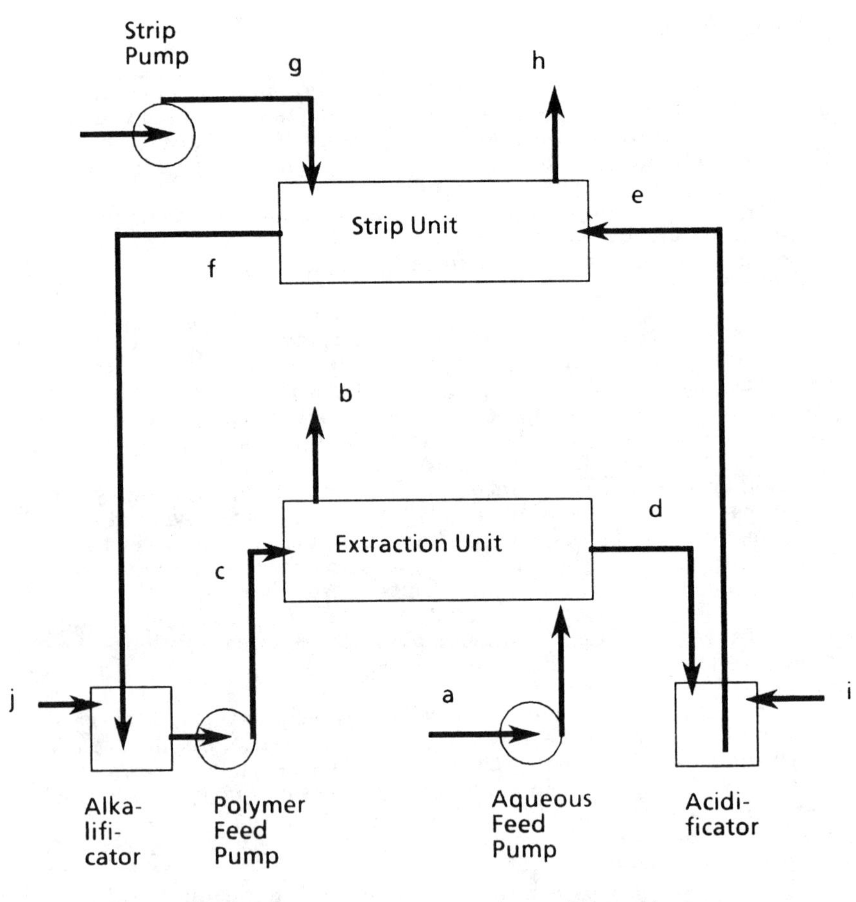

a : Feed stream
c : Lean polymer
e : Acidified polymer
g : Strip stream
i : Acid

b : Treated stream
d : Metal loaded polymer
f : Stripped polymer
h : Product recovery
j : Alkaline

ASDA
FIGURE 2 Experimental Affinity Dialysis System

Table 1 Experimental Data

Exp. No.	Feed Stream Flow Rate u cc / min	Feed Stream Ca^{++} Conc. x_{A1} ppm	Dialysate Flow Rate v cc /.min	Dialysate PAMP Conc. y_{P0} %	Outlet Conc. x_{A0} ppm	Outlet Conc. y_{C1} ppm
1	1.04	400	0.06	2.5	76	4600
2	1.00	360	0.12	2.5	25	2600
3	1.03	400	0.14	2.5	32	3750
4	1.07	400	0.21	2.5	12	2700
5	2.04	400	0.11	2.5	104	4910
6	2.10	360	0.23	2.5	35	3190
7	2.10	400	0.36	2.5	76	2700
8	2.04	400	5.20	0.0	104	------

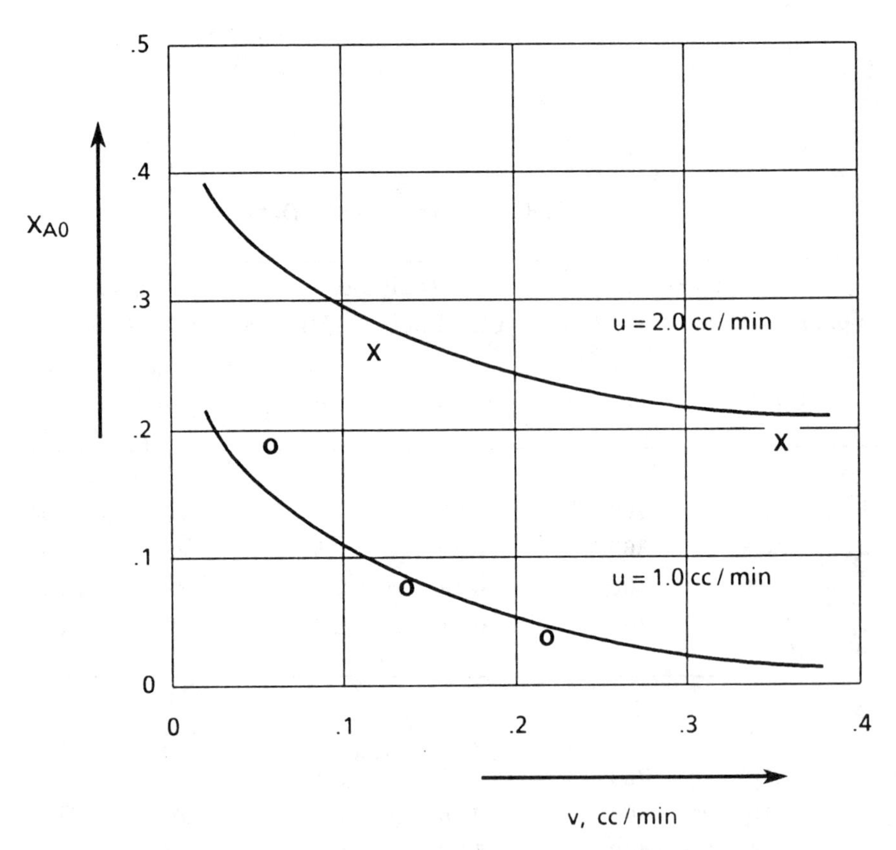

Figure 3: Experimental Data versus Simulation Results

Numerical Solution: -----------
Experimental Data: o o o u = 1.0 cc / min
x u = 2.0 cc / min

Table 2 Experimental Data versus Simulation Results

No.	Inlet Concentration x_{A1} ppm	Inlet Concentration PAMP %	Flow Rate Feed u cc / min	Flow Rate PAMP v cc / min	Dimensionless Conc. of X_A at Outlet Exper.	Dimensionless Conc. of X_A at Outlet Simul.	Error
1	400	2.5	1.04	0.06	0.19	0.16	0.03
2	360	2.5	1.00	0.12	0.07	0.09	0.02
3	400	2.5	1.03	0.14	0.08	0.08	0.00
4	400	2.5	1.07	0.21	0.03	0.04	0.01
5	400	2.5	2.04	0.11	0.26	0.29	0.03
6	360	2.5	2.10	0.23	0.10	0.08	0.02
7	400	2.5	2.10	0.36	0.19	0.22	0.03
8	400	0.0	2.04	5.20	0.26	0.28	0.02

simulation studies were conducted on the computer to explore the effect of system parameters.

Effect of Specific Permeation while no Chelating Reaction Occurs. The first series of simulation were run under the following conditions:

1. Counter-current flow,
2. Single ion species, $X_B = 0$,
3. No chelating reaction, $K_A = 0, K_C = 0$,
4. Fixed flow rate ratio at $R_p = (u/v) = -30.0$, and
5. The SP value varied from $P_A = -0.1$ to -12.5.

The negative values of R_P and P_A correspond to the Counter-current flow mode, i.e., v is taken as a positive value and u negative. The simulation results are shown in Figure 4, which shows that increasing the value of SP causes an increase of Y_A at outlet, while the X_A at outlet does not change much when there is no chelating reaction. Hence it is clear that the conventional dialysis process is not good for wastewater treatment if a relatively lower X_A at outlet is required.

Effects of Chelating Reaction Rate. A second series of simulation were run under the following conditions:

1. Counter-current flow,
2. Single ion species, $X_B = 0$,
3. $P_A = -2.5$, and $R_p = (u/v) = -30.0$,
4. Forward chelating reaction rate varied from $K_A = 0$ to 100,
5. Reverse reaction rate is negligible, $K_C = 0$.

The simulation results are shown in Figure 5, which illustrates that the chelating reaction rate plays an important role in decreasing the outlet concentration of X_A. Therefore it is shown that the affinity dialysis is much more effective than the conventional dialysis process. This has also been shown mathematically in the next section.

Effect of Specific Permeation while a Certain Degree of Chelating Reaction Exists. A third series of simulation were run under the following conditions:

1. Counter-current flow,
2. Single ion species, $X_B = 0$,
3. Fixed flow rate ratio at $R_p = (u/v) = -30.0$,
4. Fixed chelating reaction rates at $K_A = 100$ and $K_C = 0$,
5. The S P value varied from $P_A = -0.1$ to -12.5.

The simulation results are shown in Figure 6, which indicates that under the existence of a certain degree of chelating reaction, the SP value does play a role in decreasing the outlet concentration of X_A in the treated waste. It is evident that when SP value is relative lower, the ion species will quickly react with chelating agent as soon as it permeates through the membrane, thus the permeation rate is the controlling factor as shown in Figure 6(a); while in case of Figure 6(d), the chelating reaction rate is the controlling factor.

Competition of Two Metal Ion Species. The fourth series of simulation were run under the following conditions:

1. Counter-current flow,
2. Two ion species with same inlet concentration, $X_A = X_B = 1.0$,
3. Two ion species with same S P value, $P_A = P_B = -2.5$, and $P_C = P_D = 75$,

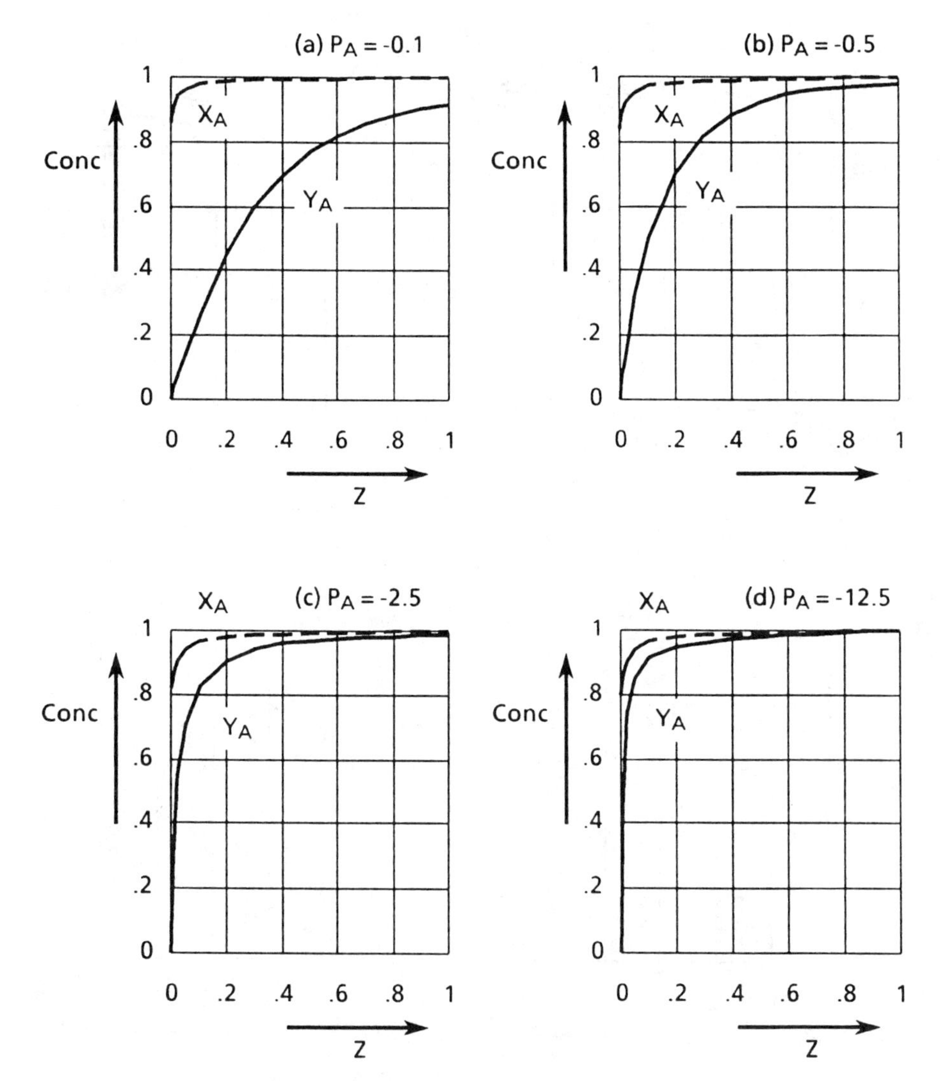

Figure 4: Effect of specific permeation while no chelating reaction exists.

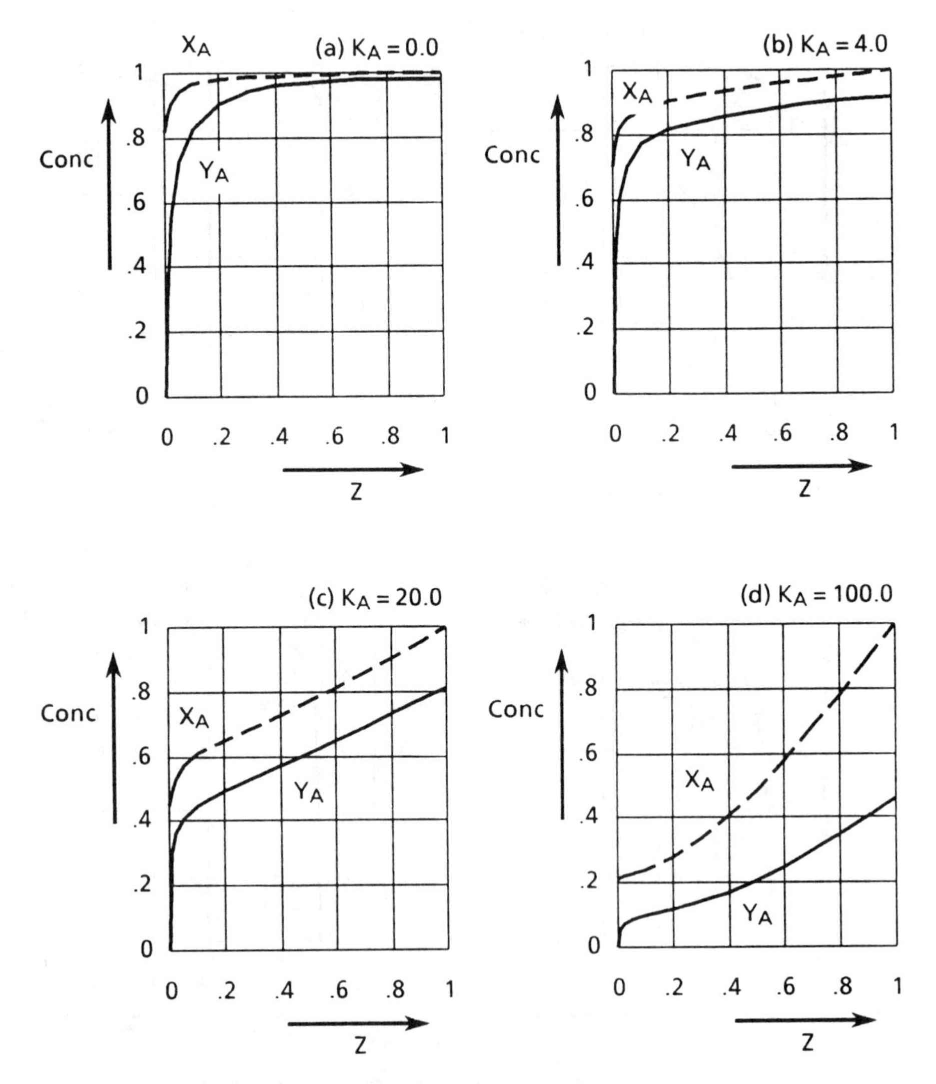

Figure 5: Effect of Chelating Reaction Rate

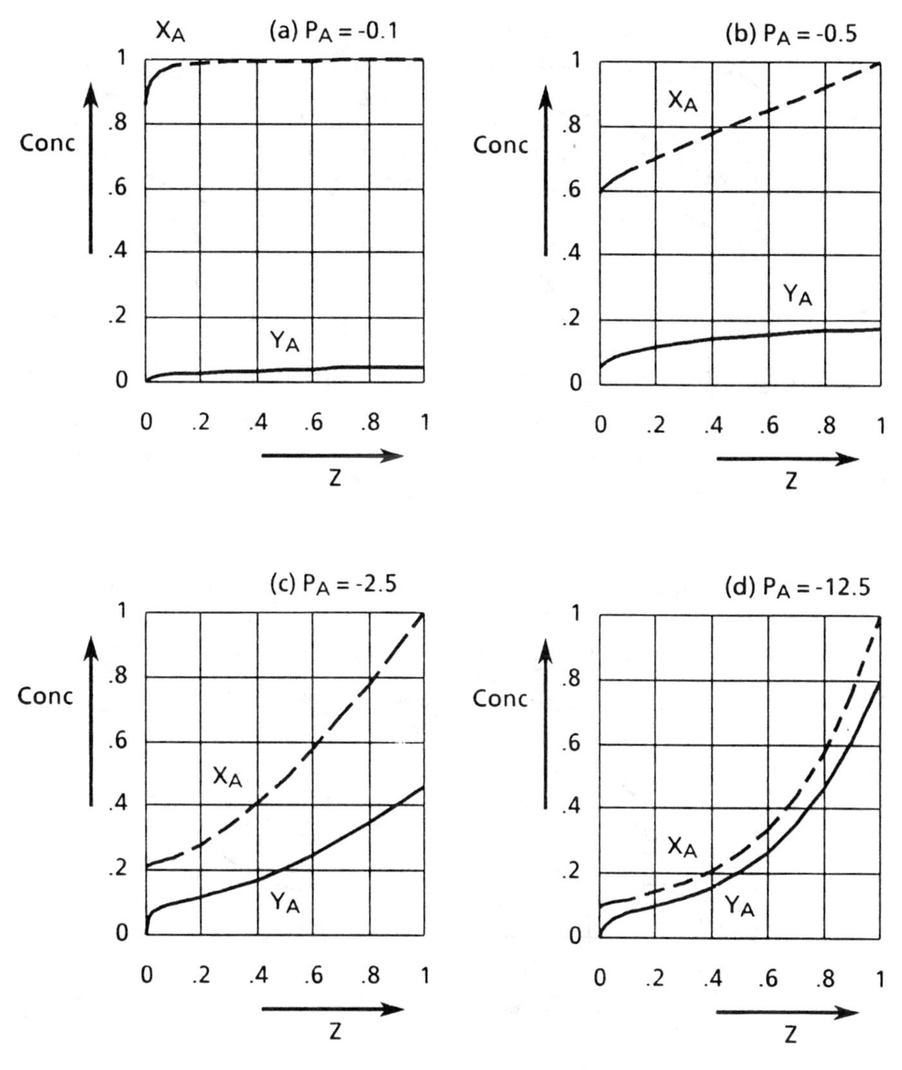

Figure 6: Effect of Specific Permeation While a Certain Chelating Reaction Exists

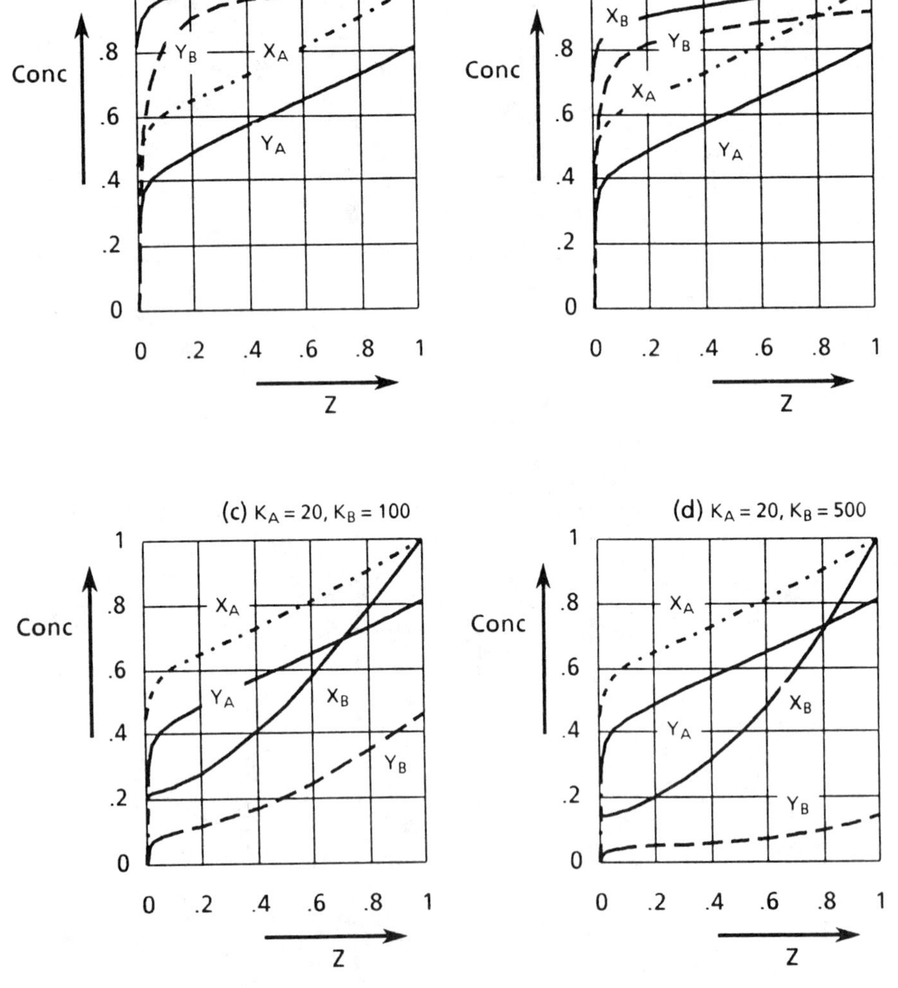

Figure 7: Competition of two ion species with different chelating reaction rates.

4. Chelating reaction of both ion species are irreversible $K_C = K_D = 0$,and

5. The SC value of first ion species has a fixed value of $K_A = 20$, while that of the second varied from $K_B = 0$ to 500.

The simulation results are shown in Figure 7, which shows that the chelating reaction rate affects the outlet concentrations of two species substantially even if they have the same permeation rate. This provides the basis for separating metal ion species through affinity dialysis.

Effects of Flow Pattern. A fifth series of simulation were run under the following conditions:

1. Single ion species, $X_B = 0$,
2. Irreversible chelating reaction ,$K_B = 0$,
3. The SP value is fixed at $P_A = -2.5$,
4. The flow rate ratio is fixed at $R_P = (u/v) = -30.0$, and
5. A comparison of co-current versus counter-current flow pattern at two chelating reaction rate, $K_A = 4$ and $K_C = 20$ respectively, were made.

The simulation results are shown in Figure 8, which illustrates that the flow pattern does affect both the outlet concentration of the feed stream, X_A, and that of polymer stream, Y_A, while the difference in the outlet concentration of polymer stream Y_A is more evident. In general, counter-current flow pattern is more efficient in decreasing the outlet concentration of X_A.

Discussion

By making overall material balance around the affinity dialysis device, it gives

$$M = u(X_{A1} - X_{A0}) = v(Y_{A1} - Y_{A0})$$

$$\frac{1}{u} = \frac{X_{A1} - X_{A0}}{M} = \frac{D_X}{M} \tag{49}$$

$$\frac{1}{v} = \frac{Y_{A1} - Y_{A0}}{M} = \frac{D_Y}{M} \tag{50}$$

From equations (28) and (30), we have at Z = 0:

$$X_{A0} = C_1 + C_2 + C_3$$

$$Y_{A0} = C_1\left(\frac{s_1}{P_A} + 1\right) + C_2\left(\frac{s_2}{P_A} + 1\right) + C_3$$

and at Z = 1:

$$X_{A1} = C_1\,exp(s_1) + C_2\,exp(s_2) + C_3$$

$$Y_{A0} = C_1\left(\frac{s_1}{P_A} + 1\right) exp(s_1) + C_2\left(\frac{s_2}{P_A} + 1\right) exp(s_2) + C_3$$

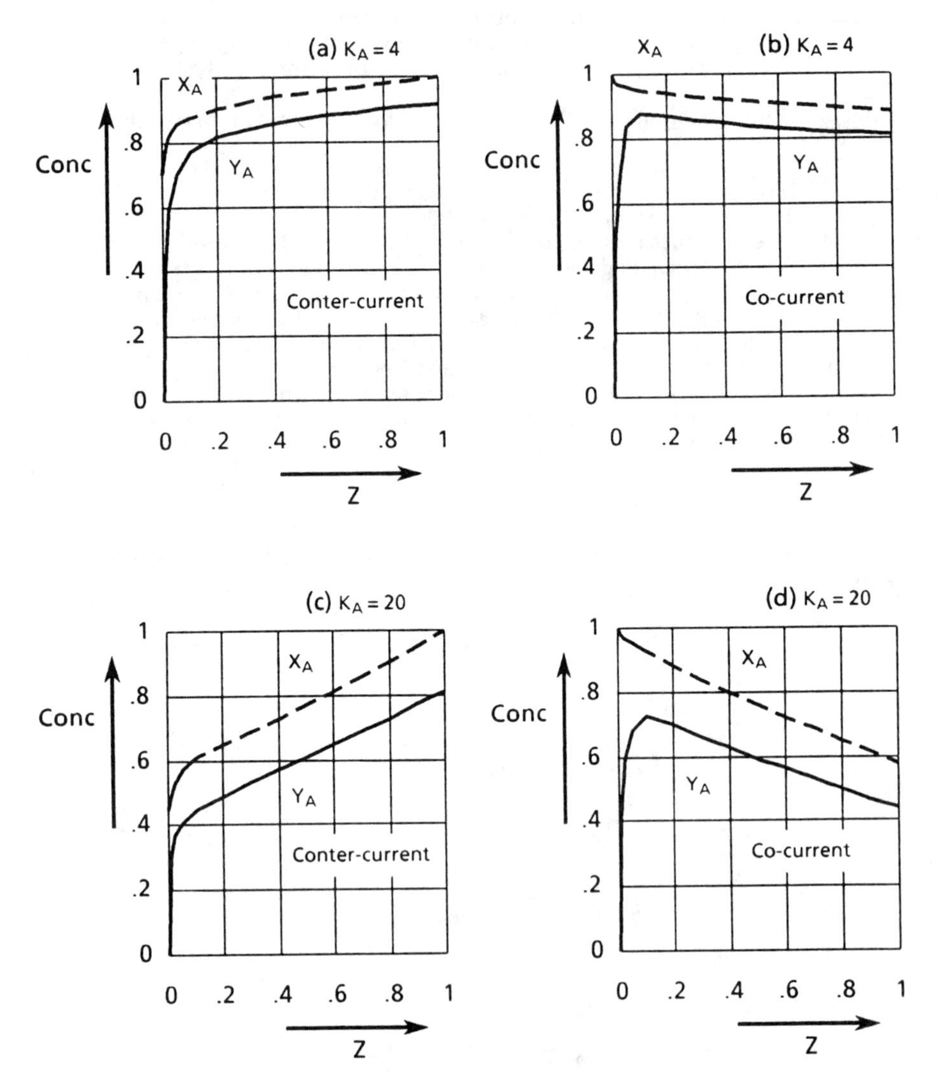

Figure 8: Effect of Flow Pattern on Concentration Profile

hence it follows that,

$$\frac{D_0}{D_1} = \frac{X_{A0} - Y_{A0}}{X_{A1} - Y_{A1}} = \frac{C_1 s_1 + C_2 s_2}{C_1 s_1 exp(s_1) + C_2 s_2 exp(s_2)} \quad (51)$$

Defining that

$$f = exp(a) / (\frac{C_1 s_1 + C_2 s_2}{C_1 s_1 exp(s_1) + C_2 s_2 exp(s_2)}) \quad (52)$$

$$K_S = K_A + K_C \quad (53)$$

$$J = K_S \cdot F \quad (54)$$

together with the definition of overall permeation rate

$$Q = q \cdot A = \frac{p}{h} \cdot A \quad (55)$$

and the dimensionless driving force

$$F = \frac{M}{Q} \quad (56)$$

then we have

$$a = ln(f \frac{D_0}{D_1}) \quad (57)$$

From definition it follows

$$a = P_A + P_C + K_A + K_C = \frac{Q}{M}(D_X + D_Y + J) \quad (58)$$

Combining equations (55), (56), (57), and (58), we obtain:

$$F = (\frac{D_X + D_Y}{ln(f \frac{D_0}{D_1}) - K_S}) \quad (59)$$

which is the driving force of affinity dialysis.

Evidently, for conventional dialysis process, $K_A = K_C = 0$, it leads to

$$a = P_A + P_C \,, \quad b = 0 \,, \quad s_1 = -a \,, \quad s_2 = 0 \,, \quad f = 0 \,, \quad K_S = 0$$

Under this condition, equation (59) is reduced to

$$F^o = \left(\frac{D_X + D_Y}{ln\left(\frac{D_0}{D_1}\right)} \right) = \frac{Lcg}{x_{A0}} \tag{60}$$

which is the dimensionless log-mean concentration gradient or the driving force of conventional dialysis process.

Define a new parameter

$$EF = \frac{F}{F^o} \tag{61}$$

named ENHANCEMENT FACTOR (EF), which quantitatively evaluates the efficiency of affinity dialysis in comparison with the conventional dialysis process. Figure 9 shows the EF increases with the chelating reaction rate K_A. It illustrates the fact that affinity dialysis exhibits much higher productivity than the conventional dialysis process.

CONCLUSIONS

In this paper the application of Affinity Dialysis Process to wastewater treatment has been studied. A mathematical model for the process was proposed. The numerical method employed for solving the mathematical model was checked using an analytical solution and showed good precision. The detailed model, solved numerically, has been compared with experimental data, and good agreement has been obtained. Extensive simulation studies showed the Affinity Dialysis Process exhibits much higher productivity and even better selectivity than the conventional dialysis process. Two new dimensionless groups proposed by the authors, the SPECIFIC PERMEATION and the SPECIFIC CONVERSION , were found very good in characterizing the nature of affinity dialysis process. A new criterion suggested by the authors, the ENHANCEMENT FACTOR , is good in quantifying the higher productivity of affinity dialysis process in comparison with the conventional dialysis process.

Legend of Symbols

a Coefficient in auxiliary equation (27) for linear homogeneous ordinary differential equation :
$a = P_A + P_C + K_A + K_C$

A Effective contact area of membrane fiber, [L^2] :
$A = 2n \cdot \pi r_m{}^2 \cdot l$

b Coefficient in auxiliary equqtion (27) :
$b = P_A K_A + P_A K_C + P_C K_C$

C Integration constants : C_1, C_2, C_3

D Dimensionless concentration difference :
$D_0 = X_{A0} - Y_{A0}$
$D_1 = X_{A1} - Y_{A1}$
$D_X = X_{A1} - X_{A0}$
$D_Y = Y_{A1} - Y_{A0}$

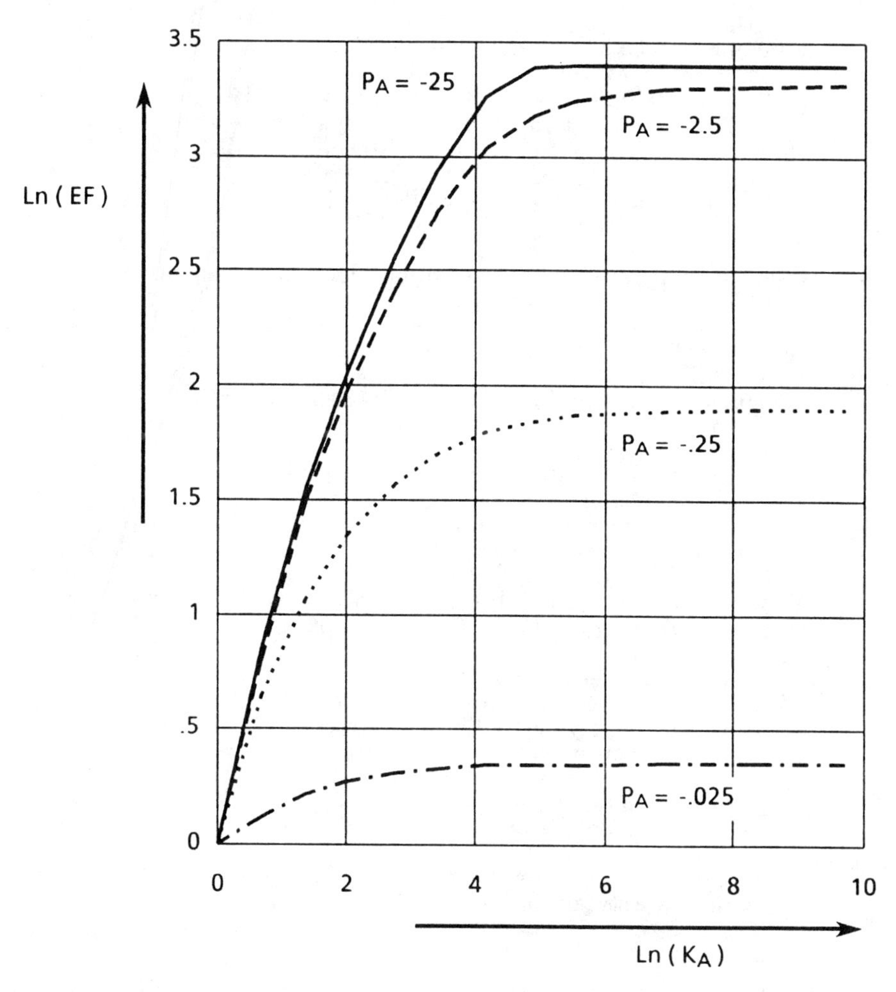

Figure 9 Enhancement Factor (EF) Versus Chelating Reaction Rate (K_A)
(Counter-current flow, single ion species, negligible reverse reaction)

EF Enhancement factor
f Function defined by equation (52)
F Dimensionless concentration gradient, or the Mass-transfer driving force:
$F = M / Q$
h Thickness of mass-transfer resistance, [L]
j Order of reaction
J Function defined by equation (56)
k Reaction rate constant
for 1st-order reaction, [1 / θ] : $dy_A / d\theta = - k_A \cdot y_A$
for 2nd-order reaction, [L3 / θ·W] : $dy_A / d\theta = - k_{AP} \cdot y_A \cdot y_P$
K Dimensionless group "SPECIFIC CONVERSION" (SC): $K = k V (x_{A0})^{j-1} / w$
l Effective length of membrane fiber, [L]
Lcg Log-mean concentration gradient, [W / L^3]
m Amount of overall mass-transport, [W / θ]
M Overall mass-transport based on dimensionless concentration, [L^3 / θ] :
$M = m / x_{A0}$
n Number of fiber used in the bundle
p Permeability, [L^2 / θ]
P Dimensionless group "SPECIFIC PERMEATION" (SP):
$P = p \cdot A / h \cdot w$
q Permeation rate, [L / θ] : $q = p / h$
Q Overall permeation rate, [L^3 / θ] : $Q = q \cdot A$
r Radius of membrane fiber, [L]
R Radius of shell side effective region, [L]
s Roots of auxiliary equation (27)
u Flow rate of feed stream, [L^3 / θ]
v Flow rate of polymer or dialysate stream, [L^3 / θ]
V Chelating reaction volume, [L^3] : $V = n \cdot \pi r_i^2 \cdot l$
w Flow rate in general, [L^3 / θ]
x Concentration in feed stream, [W / L^3]
X Dimensionless concentrationin feed stream: $X = x / x_A$
y Concentration inpolymer or dialysate stream, [W / L^3]
Y Dimensionless concentration in dialysate: $Y = y / x_{A1}$
z Linear distance along the length of membrane fiber from inlet of polymer stream, [L]
Z Dimensionless axial distance:
$Z = z / l$
at polymer stream inlet: $Z = 0$
at dialysate stream outlet: $Z = 1$

Subscripts

0 Polymer stream inlet: $Z_0 = 0$
$X_{A0}, Y_{A0}, Y_{C0}, Y_{P0}$
1 Dialysate stream outlet: $Z_1 = 1$
$X_{A1}, Y_{A1}, Y_{C1}, Y_{P1}$
A, B Metal ion:
$X_A, Y_A, p_A, P_A, k_A, K_A, q_A, Q_A; X_B, Y_B, p_B, P_B, k_B, K_B, q_B, Q_B$
C, D Metal-polymer complex: Y_C, Y_D

i Fluid film inside the membrane fiber: f_i, g_i, h_i, q_i
Inside radius of membrane fiber: r_i
Concentration at inlet of feed stream: X_{Ai}
for counter-current mode: $X_{Ai} = X_{A1}$
for co-current mode: $X_{Ai} = X_{A0}$
m Mean radius of membrane fiber: r_m
o Fluid film outside the membrane fiber: f_o, g_o, h_o, q_o
P Polymer: Y_P
S Summation
x With respect to concentration in feed stream: Dx
y With respect to concentration in dialysate stream: Dy

Literature Cited

[1] Cartwright, P. S. Desalination 1986 , 56, 17.
[2] Weber, W. F. ; Bowman, W. Chem. Eng. Prog. 1986 , 11 , 23.
[3] Baker, R. W. in Synthetic Membranes: Science, Engineering and Applications, Ed. by Bungay, P. M. et al., D. Reidel Publ. Co., Holland, 1986 , p437 .
[4] Bemberis, I. ; Neely, K. Chem. Eng. Prog. 1986 ,11 , 29.
[5] Applegate, L. E. Chem. Eng. 1984 , 11, 64.
[6] Jonsson, G. , in Synthetic Membrabes: Science, Engineering and Applications, Ed. by Bungay, P. M. et al., D. Reidal Publ. Co., Holland, 1986 , p625.
[7] Baker, R. W.; Tuttle, M. E.; Kelly, D. J.; Lonsdale, H. K. J. Membrane Sci. 1977 , 2, 213.
[8] Gooding, C. H. Chem. Eng. 1985 , 7, 56.
[9] Kim, B. M. J. Membrane Sci. 1984 , 21, 5.
[10] Prasad, R.; Sirkar, K. K. Separation Sci. and Techn. 1987 , 22 , 619.
[11] Lee, L. T. C.; Ho, W. S.; Liu, K. J. U. S. Patent 3,956,112.,1976
[12] Strathmann, H. Separation Sci. and Techn. 1980 , 15 , 1135.
[13] Ho, W. S.; Lee, L. T. C.; Liu, K. J.; U. S. Patent No. 3,957,504, 1976.
[14] Davis, J. C.; Valus, R. J.; Lawrence, E. G. Research Report to Corporate Research Laboratories, The standard Oil Co. 1987, No. 7107.
[15] Hu, S.; Govind, R.; Davis, J. C. ,Paper presented to the 18th Annual Meeting of the Fine Particle Society, 1987, Aug. 4/5, Boston.

RECEIVED January 12, 1990

Chapter 13

A New Approach for Treatment of Acid-Containing Waste Streams

A. M. Eyal[1], A. M. Baniel[1], and J. Mizrahi[2]

[1]Casali Institute of Applied Chemistry, School of Applied Science and Technology, Hebrew University of Jerusalem, 91904 Jerusalem, Israel
[2]Chiman, Development and Engineering, Ltd., P.O. Box 1494, 31012 Haifa, Israel

A new technology ("SEPROS") for treating acid-containing industrial waste streams is based on a novel family of extractants, composed of commercially available amines and organic acids. It provides for efficient, selective and reversible extraction of mineral acids and their salts. The extracted compounds are back-extracted by water, avoiding the consumption of auxiliary reagents such as neutralizing bases.

The new technology provides for recovery of acids and of salts in pure and concentrated solutions and greatly facilitates further treatments required to dispose of effluents and/or recover other valuable components. Of immediate industrial and ecological interest are the treatment of the waste streams from the titanium dioxide industry, the treatment of pickling liquors and the treatment of bleed streams from electrolytic zinc plants.

Mineral acids provide the H^+ activity required in processes such as leaching and pickling of metal surfaces. In most cases part of the acidity is not utilized due to strong dependence of reaction rate on H^+ concentration. Solutions containing part of the acid and its salts result. In electrowining of leach solutions acid is reformed, and a stream containing acid is bled to remove impurities. Acid-containing effluents are also formed in other processes such as those recovering acids from their salts by another acid.

Current practice of treating acid-containing effluents is neutralization by a low cost base thereby forming a sludge to be disposed of. Neutralization thus entails consumption of reagents, loss of acid values and yet does not avoid waste treatment. Enforcement of strict environmental protection laws poses a strong challenge and high economic motivation for development of improved treatments. When possible, it is preferred to separate acids and other valuable components from the effluent stream, thereby reducing the cost of waste management and recovering products, which can be recycled to the process or sold as such.

0097–6156/90/0422–0214$06.00/0

Volatile acids can be separated and recovered by evaporation. Processes were described, suggesting evaporating HNO_3 and HF from effluents of steel pickling (1), HCL from solutions containing also $ZnSO_4$ (2) and even H_2SO_4 from TiO_2 industries effluents (3). Corrosive solutions, high temperatures and possible precipitation of solids during evaporation impose high demands on materials of construction in addition to high energy requirements.

Methods such as membrane and chromatographic separations are used to separate small amounts of high value acids, but are not applicable for most industrial wastes.

For many separations, solvent extraction proved to be an attractive method, being selective, inexpensive, involving only the handling of liquids and operating at near-ambient temperatures. Solvent extraction of acids is being applied in many processes (4-6).

Acid Extraction and Couple Extractants

Extraction Mechanism and Current Extractants. Liquid-liquid extraction (LLX) processes consist of two basic operations: extraction and stripping (or back-extraction). In the former the extracted species is selectively transferred from a feed (usually aqueous) phase to an extractant (usually organic) phase. The two phases after extraction are denoted raffinate and extract respectively. In the stripping operation, the extracted species is transferred from the extract to another aqueous phase to form the product and to regenerate the extractant. Both operations depend on limited mutual miscibility of the two phases.

The simplest cases of acid separation from effluents, are those in which the acid-free solution can be discarded. The effluent is counter-currently extracted by an extractant, preferrably one that is selective and is efficient also at low acid concentrations, in order to achieve nearly complete acid removal. The raffinate is discarded and the extract is treated for extractant regeneration in a second battery.

Extractant regeneration can be simply achieved through treating the extract by a base (Fig. 1) to form a salt of the acid. This operation consumes a base and leads to loss of acidity. It is justified only when economic recovery of the acid is not possible or when the salt formed is valuable.

In many cases the mineral acid in the effluent is concentrated enough to be of value, if separated as such and recovered without considerable loss in concentration. In these cases, the preferred flowsheet is that of Fig. 2, where acid is recovered from extract by washing with water. Extractants suitable for such processes should provide reversibility in addition to selectivity and high efficiency at low acid concentrations.

For a better understanding of extractant properties and the possibilities of adjusting them to these requirements, driving forces and acid extraction mechanisms are briefly reviewed.

In most cases, the driving force for the total operation of transferring species from the feed to the product, is concentration difference. The extracted molecule is transferred into the extractant due to interaction with an active component in it. (In the following, extractant and extract will also be referred to as the organic phase.) In order to recover the extracted molecule, energy must be introduced

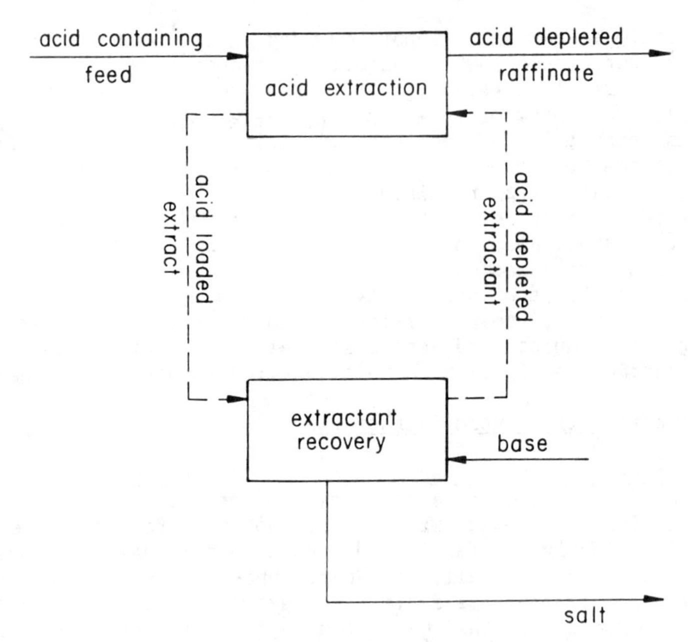

Fig. 1: Flowsheet of acid extraction followed by stripping through neutralization.

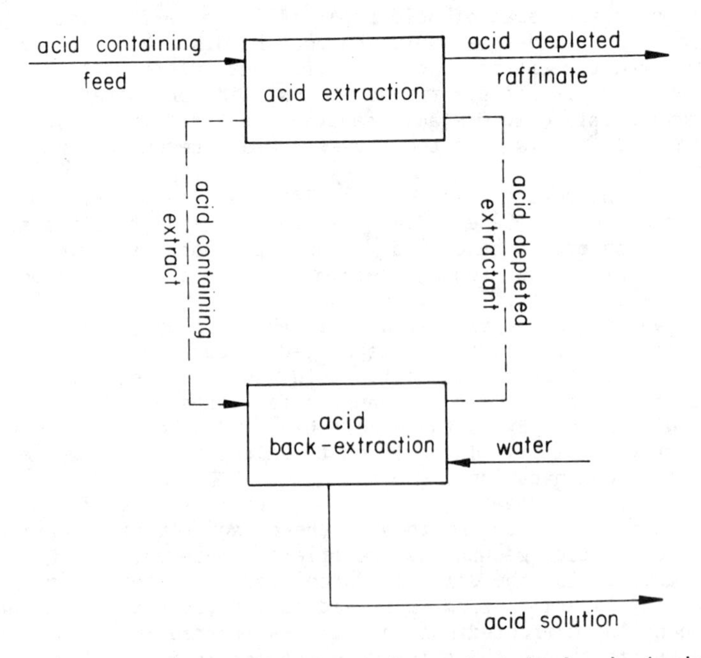

Fig. 2: Flowsheet of acid extraction and stripping by backwash with water.

(to overcome the interaction). When transferred to an aqueous stream (product), dilution energy might be applied. The degree of interaction with the active component, therefore, affects both the extraction and the stripping.

Extractants for recovery of acids contain an active component with unshared electron pairs, which can interact with the proton of the acid. The availability of these electron pairs determines the extraction mechanism. Weak bases containing oxygen atoms (alcohols, esters, ether, etc.) or equivalent sulfur containing compounds, interact with mineral acids to form hydrogen bonds. This mechanism is referred to as solvation mechanism, as the extractant replaces part (or most) of the water molecules solvating the acid molecule in the aqueous phase. In cases where the unshared electron pairs are more readily available, as with amines, the protons of the acid are co-ordinatively bonded to form cations such as R_3NH^+ (equivalent to ammonium ion), which form ion-pairs with the anions of the acid.

Amine-based extractants have several advantages such as low solubility in aqueous solutions and good chemical stability (both reducing extractant losses) and high selectivity. Amines have a strong extracting power enabling acid extraction from solutions of very low H^+ activity (see curve A in Fig. 3). This strong binding power, however, has an adverse effect on stripping - extraction is virtually not reversible. Amine-based extractants are thus considered suitable for processes according to Fig. 1, but not for those according to Fig. 2.

The binding strength of solvating extractants is relatively small, enabling stripping the extractant by washing with water. Their drawbacks are manifested in the extraction step. They are less selective and inefficient in extraction from solution of low H^+ activity (see curves E and F in Fig. 3). An example is provided by Buttinelli's study of H_2SO_4 recovery from bleeds of zinc electrowining plants (7). The extractants used (isobutyl alcohol and isoamyl alcohol) allow only partial recovery. About 25% of the acid has to be neutralized prior to further treatment. In addition, the solubility of the extractant in aqueous streams requires its recovery from raffinate and product. Phosphate esters, also extracting acids by the solvation mechanism, are less soluble, but hydrolyse at high acid concentration. Their use thus imposes a choice between extractant losses and working at low temperatures, both increasing the cost of recovery (8).

The degree of freedom in controlling the properties of commercially available solvating or amine extractants is limited (the main parameter being the relative contribution of the hydrophobic part). A more sophisticated adjustment would require special synthesis of reagents.

Couple Extractants. Grinstead and his co-workers described a family of extractants composed of an amine and of an organic acid (both water immiscible) in a diluent. They found them suitable for extracting salts (9,10).

Our group has thoroughly investigated acid extraction by these extractants (referred to in the following as couple extractants). We have found that they form a "new generation" of acid extractants providing some very important advantages (11,12).

Couple extractants are simply prepared by mixing the amine (b) and the organic acid (Ha). A salt forms as evidenced by heat evolution:

$$Ha + b = bHa$$

A mineral acid (HA) can be extracted from its aqueous solution according to:

$$bHa(org) + HA(aq) = bHA(org) + Ha(org)$$
(aq - aqueous, org - organic)

This reaction involves displacement of the organic acid by the mineral acid and not neutralization as in the case of acid extraction by amine. The energy of extraction is therefore considerably smaller. As the organic acid stays in the organic phase, a reverse displacement may take place in contacting the extract with water thereby recovering the mineral acid and the extractant.

The equilibrium distribution at each concentration of the aqueous phase, (expressed as a distribution curve) is determined by several parameters:

- amine basicity
- relative acidity of the two acids (organic and mineral)
- amine to organic acid molar ratio in the extractant
- steric factors
- diluent and modifier effects.

Effects of these parameters on distribution curves are shown in Figs. 3-5. They provide much freedom in controlling extractant properties. They allow designing extractants simply by mixing commercially available components. Extractants can be adjusted to combine reversibility and selectivity with nearly complete recovery of the acid.

These properties allowed the design of processes for wet process phosphoric acid upgrading (13) and for KNO_3 production (14). Couple extractants appear to be especially suitable for treatment of mineral acid-containing waste streams.

Effluent Treatments Utilizing Couple Extractants

Acid Recovery. Selectivity, reversibility, efficient extraction and low extractant losses (low solubility in aqueous solution and chemical stability) make couple extractants suitable for acid recovery from effluents according to the flowsheet in Fig. 2. Typical examples are HCl recovery from ion-exchangers regeneration liquors, H_2SO_4 recovery from effluents of TiO_2 industries, HF and HNO_3 recovery from pickling liquors and HCl and H_2SO_4 recovery from effluents of metal surface treatments.

Reversibility of couple extractants avoids considerable dilution on recovery. As the concentration of recovered acid reflects its activity in the feed (rather than its concentration there) common-ion and salting out effects can be utilized. In several cases, presence of salts in solution strongly affects extraction (see Fig. 6) allowing recovery at concentrations considerably higher than those in the feed.

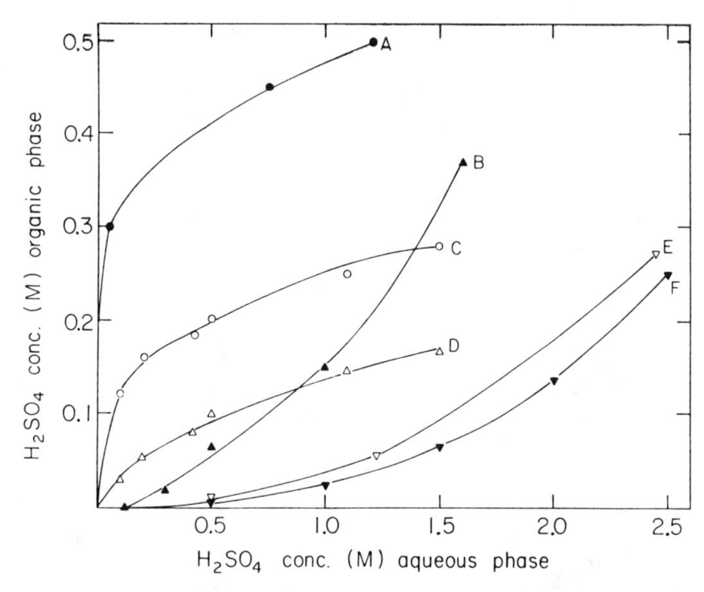

Fig. 3: H_2SO_4 distribution at ambient temperature between aqueous solutions and extractants composed of A. 0.5M TCA, B. 0.5 M TEHA + 0.5M DEHPA, C. 0.5M TCA + 0.5M DEHPA, D. 0.5M TCA + 1.0M DEHPA all of them in aromatic-free kerosene, E - TBP, F - isoamylalcohol.

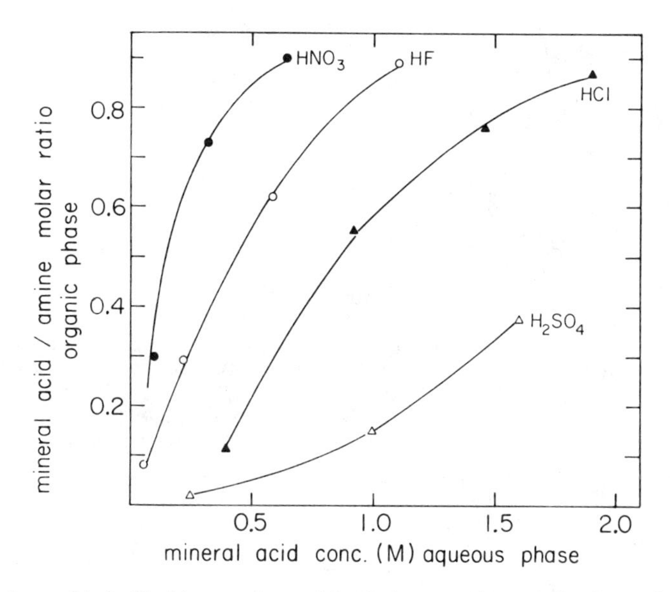

Fig. 4: Distribution at ambient temperature of mineral acids between aqueous solutions and an extractant composed of 0.5M TEHA + 0.5M DEHPA in aromatic-free kerosene.

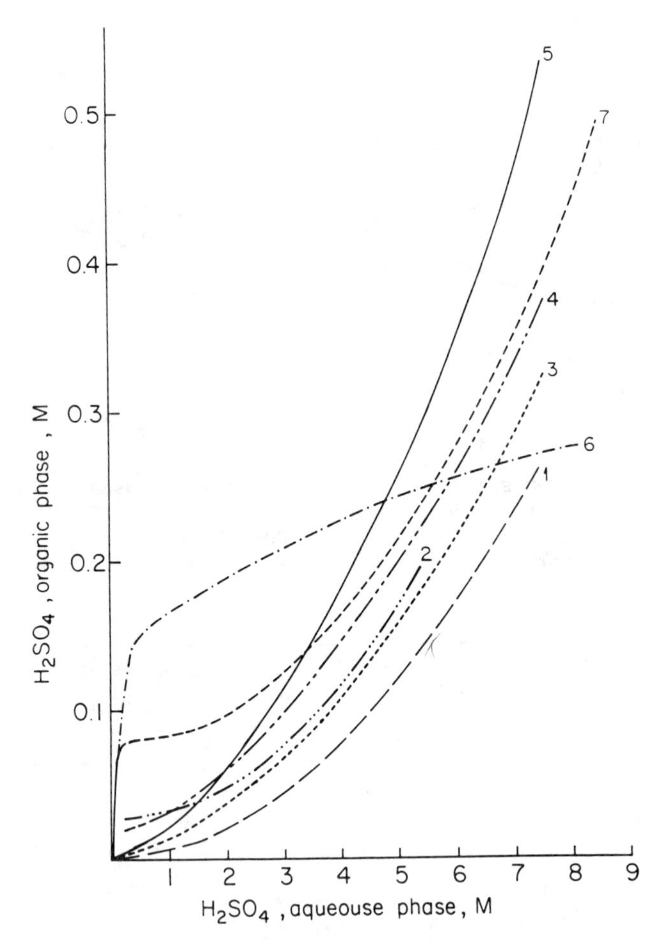

Fig. 5: H_2SO_4 distribution at ambient temperature between aqueous solutions and extractants composed of 1. 0.27M MTCA + 0.32M HDNNS, 2. 0.26M MTCA + 0.07M TLA + 0.30M HDNNS, 3. 0.31M MTCA + 0.31M HDNNS, 4. 0.34M MTCA + 0.30 HDNNS, 5. 0.5M MTCA + 0.5M HDNNS, 6. 0.25M TCA + 0.25M ABL, 7. 0.24M MTCA + 0.13M TLA + 0.29M HDNNS, all of them in aromatic-free kerosene.

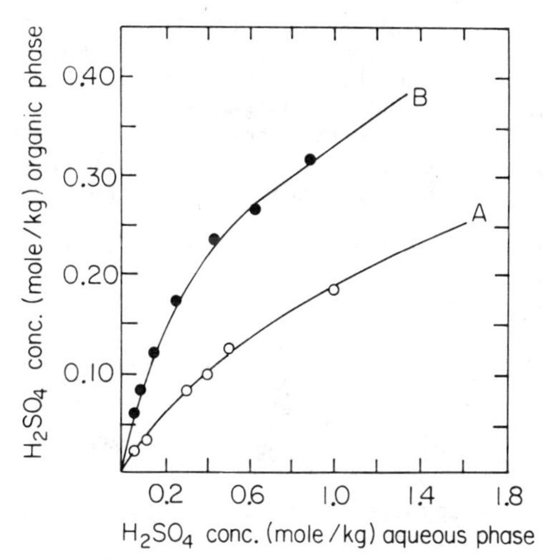

Fig. 6: H_2SO_4 distribution at ambient temperature between an extractant composed of 0.5M TCA + 1.0M DEHPA in aromatic-free kerosene and (A) salt-free aqueous solution or (B) aqueous solution containing $ZnSO_4$ + $MgSO_4$ + $MnSO_4$, total concentration of 2M.

Recovery of acids at still higher concentrations requires energy input. Some compositions of couple extractants show temperature effects (see Fig. 7). Performing of acid back-extraction at temperatures higher than those of extraction, allows concentration of the product without resorting to energy consuming water evaporation.

Valuable acids can be recovered from effluents containing their salts by addition of a cheaper acid. In cases where the salt of the added acid does not precipitate, a solution containing at least two anions, a cation and H^+ is formed. A method for selective separation of the more valuable acid is required.

A typical example is acid recovery from $HCl+FeCl_2$ containing effluents of surface treatments. Hydrolysis at high temperatures provides for separating the constituents of the solution into iron oxide and an azeotropic HCl solution (15). Alternatively, H_2SO_4 can be added and HCl evaporated, similarly to the process of $ZnCl_2$ conversion to $ZnSO_4$ (2). Both methods suffer from the disadvantages associated with acid evaporation and are not suitable for small operations. A much simpler process can be designed utilizing the selectivity of couple extractants to HCl over H_2SO_4 and over the corresponding salts. Concentration of the recovered hydrochloric acid will be determined by the concentration of the feed solution (adjustable by $FeSO_4 \cdot 7H_2O$ precipitation and by evaporation) and by the concentration of the sulforic acid added.

Similarly, the high selectivity of couple extractants to HCl in the system $H-NH_4-Cl-SO_4$ (Table I) allows conversion of NH_4Cl in solution to $(NH_4)_2SO_4$ and HCl according to:

$$2NH_4Cl(aq) + H_2SO_4(aq) + 2E(org) = (NH_4)_2SO_4(aq) + 2E\cdot HCl(org)$$

$$2E\cdot HCl(org) \overset{H_2O}{=} 2E(org) + 2HCl(aq)$$

(E - extractant, aq - aqueous phase, org - organic phase).

A challenge that often attracted process developers is the use of CO_2 for recovery of acids from their salts. We have developed a process (flow-sheet in Fig. 8) in which CO_2 is used to drive H_2SO_4 out of gypsum according to:

$$CaSO_4 + CO_2 + H_2O = CaCO_3 + H_2SO_4$$

As CO_2 is a weak acid, energy introduction is required. This process, described in a previous paper (16) recovers H_2SO_4 at a concentration of about 35%.

Recovery of an Acid and of a Salt. An additional operation is required when a salt should also be separated from the effluent prior to disposal. The discussion in the following will focus on precipitation and on solvent extraction as separation methods.

Precipitation. A serious environmental problem is posed by the effluent from TiO_2 pigment production via the sulfate-process. This solution, composed of about 25% H_2SO_4 and 12% sulfates, is generated in large volumes. The cost of effluent treatment caused eight older plants to shut down with the loss of 336,000 tons of capacity in U.S. and Europe. New plants operate in the chlorine route, which is more

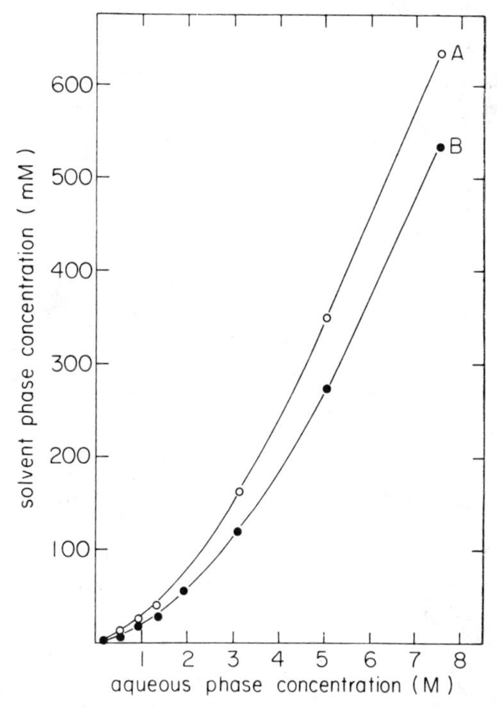

Fig. 7: H_2SO_4 distribution between aqueous solutions and an extractant composed of 0.5M MTCA + 0.5M HDNNS in aromatic-free kerosene at: A. 22°C and B. 80°C. (Reproduced with permission from Ref. 16. Copyright 1986 Marcel Dekker, Inc.).

Table I: Couple extractant selectivity in the H-NH_4-Cl-SO_4 system at ambient temperature. The couple is composed of 0.5M tricaprylylamine + 0.5M di-2ethylhexyl phosphoric acid in aromatic free kerosene. The extractant loaded with HCl (in a previous operation) was equilibrated with solutions of $(NH_4)_2SO_4$ and of $(NH_4)_2SO_4$ + NH_4Cl

Exp.	Aqueous phase initial comp. (mole/kg)		Aqueous/organic weight ratio	Organic phase mineral acid content (mole/kg)				Selectivity
				initial		final		
	$(NH_4)_2SO_4$	NH_4Cl		HCl	H_2SO_4	HCl	H_2SO_4	
A	1.52	1.71	1.4:1	0.70	-	0.30	0.040	5.2
B	2.27	0.85	1.4:1	0.70	-	0.28	0.045	11.3
C	3.03	-	1.4:1	0.70	-	0.21	0.065	24.7

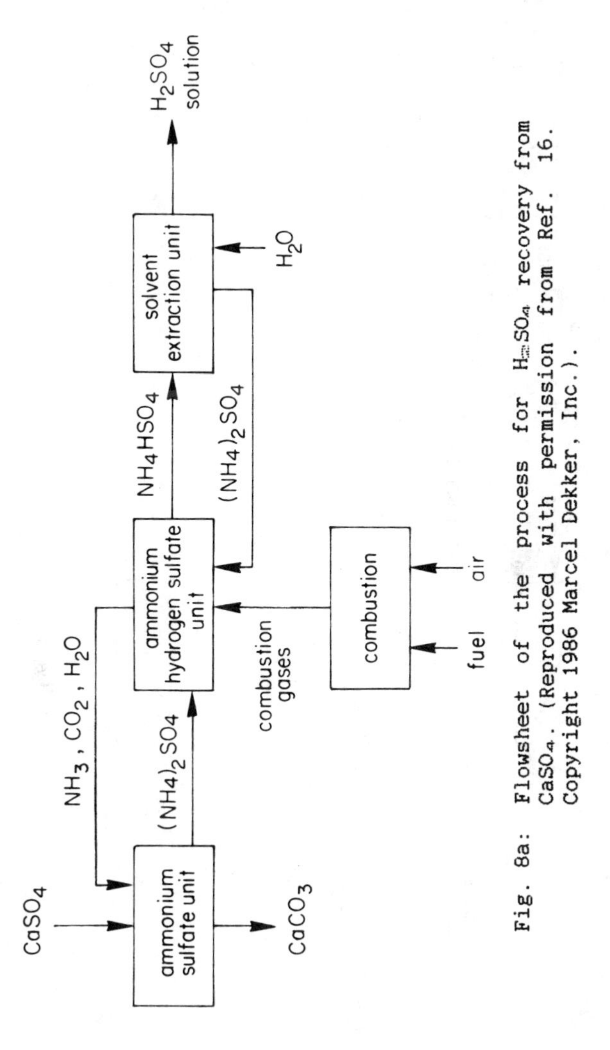

Fig. 8a: Flowsheet of the process for H_2SO_4 recovery from $CaSO_4$. (Reproduced with permission from Ref. 16. Copyright 1986 Marcel Dekker, Inc.).

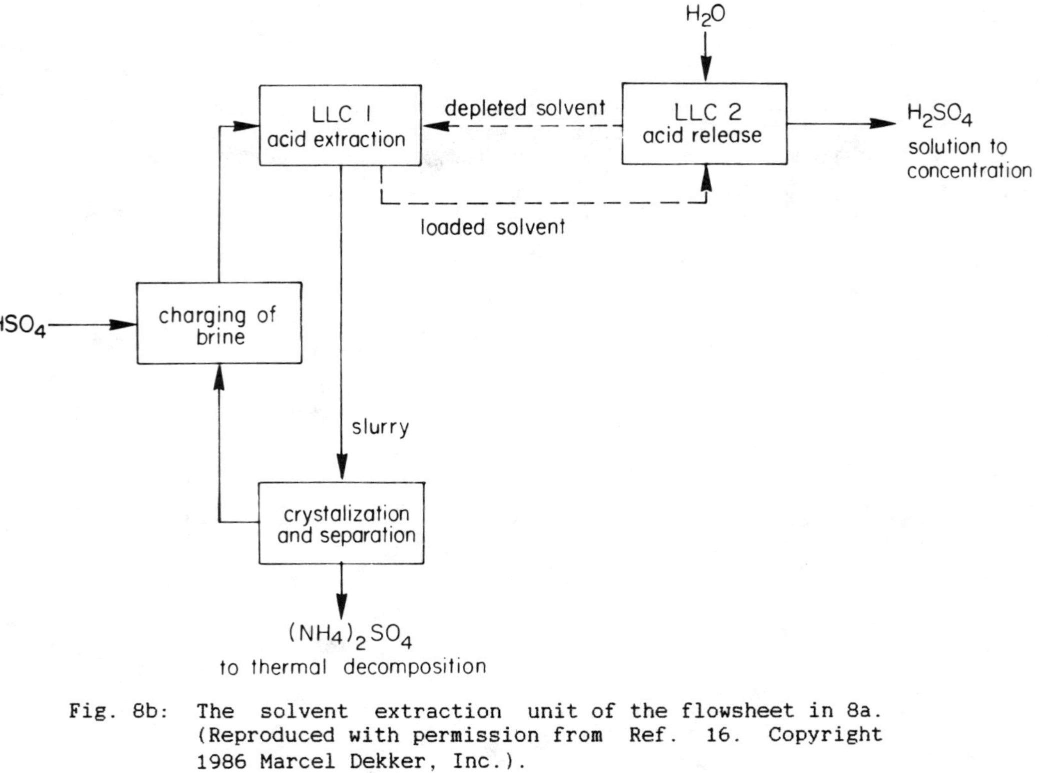

Fig. 8b: The solvent extraction unit of the flowsheet in 8a. (Reproduced with permission from Ref. 16. Copyright 1986 Marcel Dekker, Inc.).

expensive and still produces effluents to be treated (17). An economic technology for treatment of H_2SO_4 + $FeSO_4$ containing effluent is therefore of high importance.

If there is no sink for $FeSO_4$ solution, $FeSO_4$ has to be removed from the effluent in addition to H_2SO_4 recovery, as disposal in sea is limited (18). Processes based on evaporating both water and H_2SO_4 at high temperatures were suggested. They separate the solution into its consistituents - water, 98% H_2SO_4 and solid salts, which may be further treated (3). Alternatively, the solution is concentrated to 80-85% acid in order to precipitate the salts (17). These processes operate at very high temperatures at high acidity. Equipment cost and energy consumption are expected to be very high.

A much simpler and less demanding solution, based on couple extractants, can be applied where there is an outlet for H_2SO_4 solutions of intermediate concentration (40-60%). In this prospective process (Fig. 9) water is evaporated, thereby precipitating part of the salt. The salt-depleted solution is extracted to remove part of the acid. The raffinate is recycled and the acid is recovered from the extract. The concentration of the recovered acid is determined by the evaporation operation and can be adjusted to various needs. Recovered acid can be further concentrated by conventional methods since it is non scaling.

In this process variant, all the water in the effluent has to be evaporated similarly to the process discribed in (3) and in (17). Yet, the combination of the two operations - concentration and selective acid extraction - provides advantages for each operation:

- Acid/salt separation takes place in two operations. Complete separation in the concentration stage is therefore not required. High temperatures and high acidity can thus be avoided.
- As extraction raffinate is recycled through the concentration stage, complete extraction is not required. More freedom is provided in selection of extractant, resulting in higher concentration of the product acid.

Similar processes can also be applied for treatment of other effluents from which both acid and salt have to be separated. A typical example is effluents from industries utilizing MnO_2 and H_2SO_4 for oxidation. Thus, for example, oxidation of p-methoxy toluene to anisaldehyde or thiols to disulfides produce an aqueous effluent containing $MnSO_4$, H_2SO_4 and soluble organics. Current procedure - neutralizing by lime and disposal - entails the disadvantages of neutralization. A process similar to that in Fig. 9 can be applied if the organics are evaporated with the water. Recovery of solid $MnSO_4$, H_2SO_4 solution and possibly the organics and saving on waste treatment provide a strong economic motivation for this process.

Recoveries by Extraction of Both Acid and Metal Salt. Extraction of both acids and salts may, in some cases, be more practical and more economic than extraction of the acids and precipitation of the salts. Couple extractants provide several advantages in such operations as will be illustrated by a process we developed for treating the bleed of zinc electrowining.

Typical bleed composition is: 15-16% H_2SO_4, 8-10% $ZnSO_4$, 4-5% $MgSO_4$, 1-1.5% $MnSO_4$, 200 ppm chlorides, 300-400 ppm calcium and small amounts of Fe, Ni, Co, Cu and Cd. The main constituents - $ZnSO_4$ and

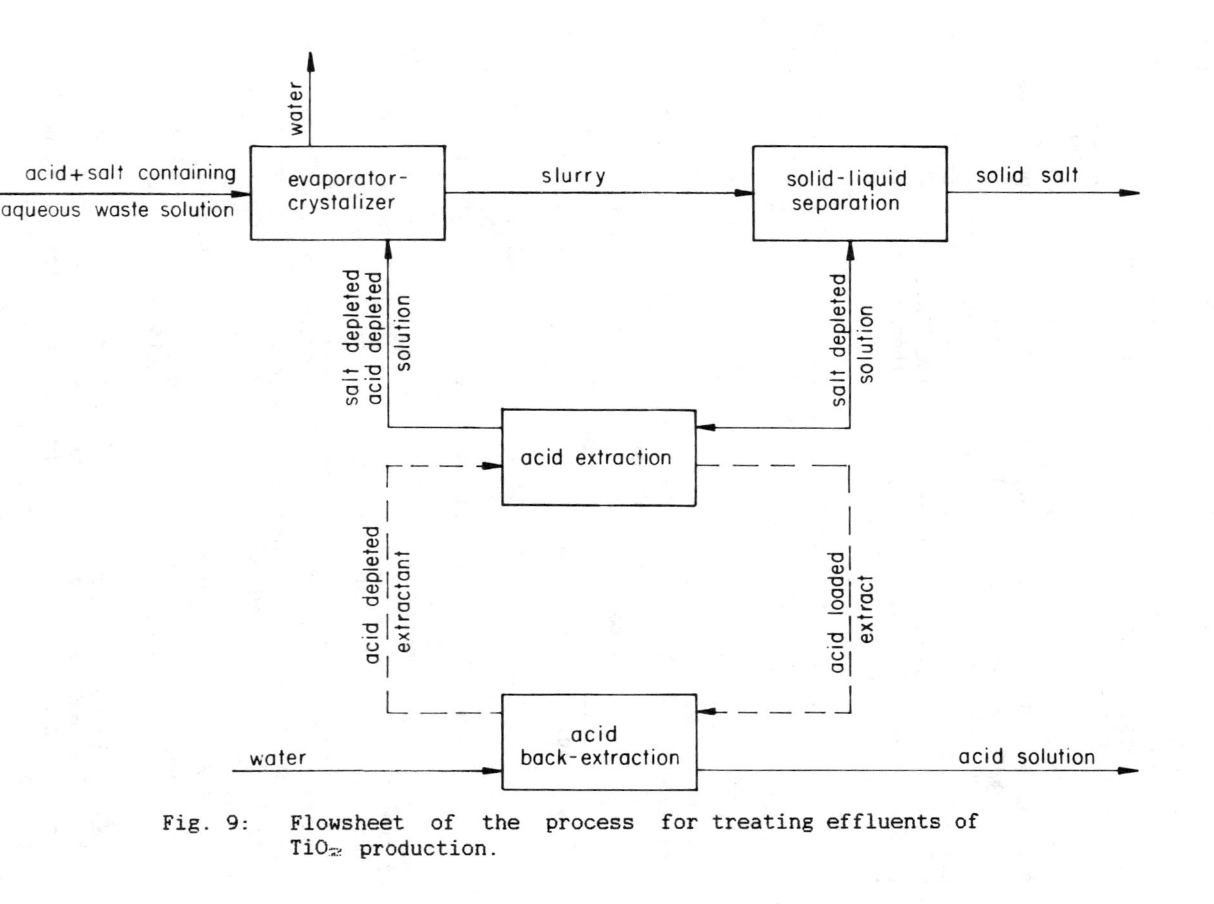

Fig. 9: Flowsheet of the process for treating effluents of TiO_2 production.

H_2SO_4 - are valuable and can be recycled into the process without even resorting to separation between them. An attractive approach would be to remove the impurities and recycle the rest. Yet the different characteristics of the various impurities would impose distinct separations resulting in an expensive complex treatment sequence.

Current practice consists of controlled addition of a base such as lime. After neutralization of the acid, $Zn(OH)_2$ is precipitated, separated and recycled to production. This treatment suffers from many drawbacks:

- acidity loss amounts to the sum of free H_2SO_4 content of the effluent and the acid required to convert the recycled $Zn(OH)_2$ back to $ZnSO_4$.
- reagent consumption - lime is required to neutralize the lost acid.
- formation of large volumes of gypsum that should be disposed of.

Reducing these burdens on zinc production poses an important challenge to process development. Several suggestions were made based on zinc extraction.

Zinc is not extractable by cation exchangers from solutions of high acidity. pH50 of 1.2 and 5.4 were found for extraction by 30% di-2 ethylhexyl phosphoric acid (DEHPA) in Shellsol and by 30% Versatic 911 in Shellsol respectively (19).

Several publications suggest extraction of zinc as an anionic complex. NaCl is added to the effluent and the chloride complex is extracted by tributyphosphate (TBP) (20) or by an amine (2). A chloride excess is required (19). Addition of chloride anion disturbs further treatment for H_2SO_4 recovery. Yet the main drawback is that zinc is recovered as a chloride that cannot be directly recycled to electrowining. Application of another electrowining process that can utilize $ZnCl_2$ was suggested (21), but recovery of $ZnSO_4$ is still more attractive.

$ZnCl_2$ can be converted to $ZnSO_4$ by addition of H_2SO_4 and evaporating HCl (2). Alternatively this conversion may use a liquid cation exchanger (2,15,19), such as di-2 ethylhexyl phosphoric acid, denoted here as HDEHP:

$$ZnCl_2(aq)+2HDEHP(org) = Zn(DEHP)_2(org)+2HCl(aq)$$

$$Zn(DEHP)_2(org) + H_2SO_4(aq) = ZnSO_4(aq) + 2HDEHP(org)$$

These two methods for converting $ZnCl_2$ to $ZnSO_4$ entail the addition of an operation. HCl has to be evaporated or recovered as a dilute solution through application of a second extractant.

An alternative approach consists of starting with H_2SO_4 removal from the effluent. The acid depleted solution can thus be extracted by a liquid cation exchanger for Zn recovery. Buttineli (7) studied H_2SO_4 pre-extraction from the effluent using alcohols. The low efficiency of these extractants in extraction of H_2SO_4 at the low concentration range, imposes the choice between losing an important part of the acid and recovering it as dilute solution. In addition, the extractants studied were quite soluble in aqueous solutions and should be recovered from them by evaporation. Another drawback is that zinc recovery through extraction by a cation exchanger does not avoid loss

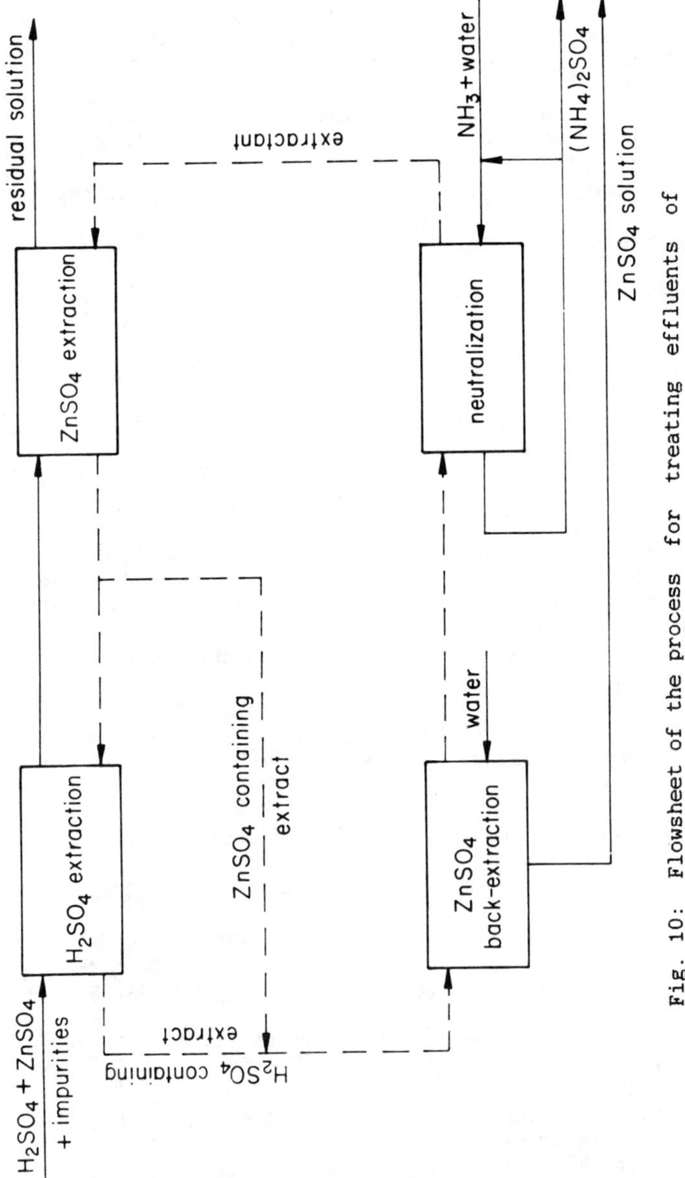

Fig. 10: Flowsheet of the process for treating effluents of zinc electrowining.

of acid in an amount equivalent to the recovered zinc (acid formed on extraction must be neutralized).

We have developed a process based on consecutive reversible extraction of both H_2SO_4 and $ZnSO_4$ by the same couple extractant (Eyal, A.M.; Baniel, A.M.; Hajdu, K.; Mizrahi J., submitted for publication in Solv. Ext. Ion. Exch.). One varient of process flow-sheet is shown in Fig. 10. The extractant is a kerosene solution of a tertiary amine (tricaprylyl amine - Alamine 336 - was found suitable) and an organic acid such as di-2-ethylhexyl phosphoric acid.

This process was designed for an effluent from a zinc production utilizing ammonia for iron removal as jarosite (a sodium base can also be used). This base in our case is partially introduced through the effluent treatment process and is in fact utilized twice; first for neutralizing the acid in the extractant, and secondly for the jarosite precipitation. (Use of reagents from the main process in effluent treatment and recycling with the recovered compound may open new possibilities in treatment of many other effluents.)

This process, with only small adjustments, can also operate without the use of ammonia. The acid will then be recovered from the extract by water or by $ZnSO_4$ solution. Somewhat more process water will be required and the recycle stream will be more dilute.

The advantages of the new process are clear:

- Zn is recovered as sulfate which can be recycled to the process.
- no concentration or acid evaporation is required.
- use of more than one extractant is avoided (no cross-contamination).
- lime consumption and gypsum formation are considerably reduced.
- about 95% of the acidity and 95% of the Zn are recovered in concentrations which allow direct recycle to the process.

Summary and Conclusions

Couple extractants extracting acids and salts efficiently, selectively and reversibly provide advantages in solvent extraction processes for treating mineral acid-containing waste streams. There is flexibility in adjusting their properties to various effluents. Recovery of acids and salts reduce treatment cost and decrease reagent losses. Of immediate industrial and ecological interest are the treatment of waste streams from titanium dioxide industry, the treatment of pickling liquors and the treatment of bleed streams from zinc electrowining plants.

Acknowledgments

We gratefully acknowledge the assistance of Mrs. B. Hazan and Mr. R. Bloch in the experimental work.

The proprietary know-how and patents relevant to the implementation of the new technology have been assigned to SEPROS Separation Processes Ltd., Kibbutz Ramat Rachel, 90900 Israel.

Legends of Symbols

ABL - alpha bromolauric acid (prepared in our laboratory by bromination of lauric acid).
aq - aqueous phase.
DEHPA - di-2 ethylhexyl phosphoric acid (by Pfaltz and Bauer).
HDEHP - di-2 ethylhexyl phosphoric acid (also denoted as DEHPA).
HDNNS - dinonyl naphthalene sulfonic acid (Synex DN-052 by King Industries).
LLX - liquid-liquid extraction.
MTCA - methyl tricaprylyl amine (Aliquat 336 by henkel).
org - organic phase.
TBP - tributylphosphate (by Riedel-de Haen).
TCA - Tricaprylyl amine (Alamine 336 by Henkel).
TEHA - tris-2 ethylhexyl amine (by Fluka).
TLA - trilaurylamine (alamine 304 by Henkel).

Literature Cited

1. Nyman, B.; Koivnen, T; "The Outokumpu Process for Pickling Acid Recovery"; In Proceedings of Int. Symp. on Iron Control in Hydrometallurgy, Dutrizac, J.E.; Monhemius, A.J., Eds.; Horwood: Chichester, UK, 1986, p. 520.
2. Anderson, S.O.S.; Reinhardt, H.; "Recovery of of Metals from Liquid Effluents"; In Handbook of Solvent Extraction; Lo, T.C.; Baird, M.H.I.; Hanson, C. Eds.; J. Wiley & Sons: New York, NY, 1983, p. 751.
3. Smith, I.; Cameron, G.M.; Peterson, H.C.; "Acid Recovery Cuts Waste Output"; Chemical Engineering Feb. 3, 1986, p. 44.
4. Blumberg, R.; "Miscellaneous Inorganic Processes"; In Handbook of Solvent Extraction, Lo, T.C.; Baird, M.H.I.; Hanson, C. Eds., J. Wiley & Sons: New York, NY, 1983, p. 826.
5. IMI Staff; "Development and Implementation of Solvent Extraction Processes in the Chemical Process Industries"; In Proc. Int. Solv. Ext. Conf., The Hague, 1971, 2, 1386.
6. Araten, Y.; Baniel, A.; Blumberg, R.; The Fertilizer Society of London, Proceedings No. 99, 1967.
7. Buttinelli, D.; Giavarini, C.; Mercanti, A.; "Pilot-Plant Investigation on H_2SO_4 Extraction by Alcohols from Spent Electrolytes" In Proc. Int. Solv. Extr. Conf., Denver Co., 1983, p. 422.
8. Fujiwaki, Y.; Doi, E.; Morimoto, Y.; "Recovery of Inorganic Acids by Solvent Extraction at Low Temperatures", Jpn. 55/18512, 8 Feb., 1980.
9. Grinstead, R.R.; Davis, J.C.; Lynn, S.; Charlesworth, R.K.; "Extraction by Phase Separation with Mixed Ionic Solvents"; Ind. Eng. chem. Prod. Res. Develop. 1969, 8, 218.
10. Grinstead, R.R.; Davis, J.C.; "Extraction by Phase Separation With Mixed Ionic Solvents, Recovery of Magnesium Chloride from Sea Water Concentrates"; Ind. Eng. Chem. Prod. Res. Develop. 1970, 9, 66.
11. Eyal, A.; Baniel, A.; "Extraction of Strong Mineral Acids by Organic Acid-Base Couples"; Ind. Eng. Chem. Proc. Des. Develop. 1982, 21, 334.

12. Eyal, A.; Hajdu, K.; Applebaum, C.; Baniel, A.: "Recovery of Acids and Reactions in Concentrated Aqueous Solutions Mediated by Acid-Base Solvents"; Proc. Int. Solv. Ext. Conf., Denver Co., 1983, p. 411.
13. Eyal, A.; Baniel, A.; "A Process for Defluorination and Purification of Wet process Phosphoric Acids Containing High Al Concentrations"; Solv. Extr. Ion Exch. 1984, 2, 677.
14. Eyal, A.; Mizrahi, J.; Baniel, A.; "Potassium Nitrate Through Solvent Separation of Strong Acids"; Ind. Eng. Chem. Proc. Des. Dev., 1985, 24, 387.
15. Tunley, T.H.; Kohler, P.; Sampson, T.D.; Econ, B.; "The Metsep Process for the Separation of Strong Acids"; J. of The South African Inst. of Mining and Metallurgy, May 1976, p. 423.
16. Eyal, A.; Applebaum, C.; Baniel, A.; "Sulfuric Acid Recovery Through Solvent Aided Decomposition of Ammonium Sulfate"; Solv. Extr. Ion. Exch., 1986, 4, 803.
17. Lazarko, L.; Short, H.; Johanson, E.; "TiO_2's Future is Keyed to New Technologies"; Chemical Engineering, Jan. 1989, p. 39.
18. Short, H.; "At-Sea Disposal: Many Questions Remain Unanaswered"; Chemical Engineering, Sept. 5, 1983, p. 30.
19. Thorsen, G.; "Commercial Processes for Cadmium and Zinc"; In Handbook of Solvent Extraction, Lo, T.C.; Baird, M.H.I.; Hanson, C., Eds.; J. Wiley & Sons: New York, NY, 1983, p. 709.
20. Buttinelli, D.; Giavarini, C.; "Development of a Zinc-Extraction Process by TBP from Sulphate Waste Liquors"; Chemistry and Industry, 1981, 6, 395.
21. Buttinelli, D.; Giavarini, C.; Lupi, C.; "Zinc Recovery from Bleed Off Streams by Solvent Extraction and Electrowining"; Proc. Int. Solv. Ext. Conf., Moscow, USSR, 1988, IV, p. 295.

RECEIVED November 10, 1989

SOILS, RESIDUES, AND RECYCLE TECHNIQUES

Chapter 14

Dechlorination of Organic Compounds Contained in Hazardous Wastes

Potassium Hydroxide with Polyethylene Glycol Reagent

T. O. Tiernan[1], D. J. Wagel[1], G. F. VanNess[1], J. H. Garrett[1], J. G. Solch[1], and C. Rogers[2]

[1]Department of Chemistry, Wright State University, Dayton, OH 45435
[2]Risk Reduction Engineering Laboratory, U.S. Environmental Protection Agency, Cincinnati, OH 45268

Laboratory and field tests have demonstrated that a chemical reagent prepared by reacting potassium hydroxide with polyethylene glycol (KPEG) effectively dechlorinates polychlorinated dibenzo-p-dioxins (PCDD) and polychlorinated dibenzofurans (PCDF) which are contained in liquid hazardous wastes or which are sorbed on spent activated carbon. The present paper is concerned primarily with the effects of treatment duration, reaction temperature, the presence of water and volatile materials, and the quantity of the KPEG reagent on the efficiency of the KPEG dechlorination reaction with PCDD/PCDF in such waste materials. The results of these studies indicate that the concentrations of PCDD/PCDF in waste liquids can be reduced to levels of less than one part-per-billion by direct treatment with KPEG. In the case of the contaminated carbon matrix, however, direct KPEG treatment is somewhat less effective, and reduction of PCDD/PCDF to such concentrations appears to be best achieved by first extracting the carbon matrix with an appropriate solvent, and then treating the extract with KPEG.

Chlorinated organic chemicals constitute a significant portion of the hazardous materials which are accumulating in the environment as the result of manufacturing processes and improper waste disposal practices. Among such chlorinated compounds, the polychlorinated dibenzo-p-dioxins (PCDD), polychlorinated dibenzofurans (PCDF), and polychlorinated biphenyls (PCB) are important because of their toxicity, and the degree of public concern over their presence and transport in water, air and soils. Since the toxicity of PCDD, PCDF, PCB and other chlorocarbon congeners can generally be significantly reduced by removing chlorine atoms from these

0097-6156/90/0422-0236$06.00/0

molecules, waste treatment processes which result in dechlorination of these compounds are of great interest. One such process entails treatment of wastes with appropriate chemical reagents. A chemical dechlorination process has several advantages. In principle, simple and relatively inexpensive reagents can be used, and the waste and the reagent can be combined at the treatment site. Consequently, the process can be scaled to a size which is appropriate to either small or large quantities of waste. Using recently developed reagents, dehalogenation reactions can be accomplished at moderate temperatures. This simplifies the design of reactors required for such treatment and reduces the need for controlling volatile emissions. Also, such processes are obviously economical in terms of the amount of energy required for the process. All of these factors suggest lower costs for chemical dechlorination processes than are required for incineration or other processes which necessitate large fixed plant investments, and are confined to a given site.

Dehalogenation reactions are well known in organic chemistry, but dechlorination is seldom used for preparative purposes. Traditionally, dechlorination reactions require strong bases or reducing agents, often require high temperatures, and are sometimes violent and difficult to control. Moreover, such processes frequently result in rearrangement of the chlorine substituents on the chlorinated species being treated, rather than complete dechlorination. However, a simple, low temperature dehalogenation process has recently been described (1,2). The dechlorinating reagent in this case is prepared by combining a polyethylene glycol (PEG) with solid potassium hydroxide (approximately two moles of KOH for each mole of PEG) and heating the mixture at 90-100°C for approximately one hour. The resulting active species is apparently a potassium alkoxide (KPEG), which may form an ether following displacement of chloride from the chlorinated molecule.

The effectiveness of the KPEG reagent has already been demonstrated in several studies. The initial investigations established that this reagent can dechlorinate some aliphatic chlorinated compounds, as well as certain aromatic chlorinated compounds, such as PCB (3). Further studies conducted in our laboratory demonstrated that KPEG can effectively dechlorinate 2,3,7,8-tetrachlorodibenzo-p-dioxin (TCDD) and other PCDD and PCDF at moderate temperatures, when these compounds are contained either in pure solutions or in complex chemical waste mixtures. These laboratory scale experiments were subsequently scaled up and utilized by the U.S. Environmental Protection Agency (EPA) in field tests to treat several thousand gallons of liquid wastes (as well as several tons of soils containing these wastes) which were contaminated with relatively high concentrations of various chlorocarbons, including PCDD, PCDF, PCB other chlorinated organics (4).

As testing of the KPEG reagent proceeds, several different applications are being explored. One problem of increasing concern is the generation of hazardous waste by small industries, research facilities, and laboratories accomplishing analyses of large numbers of samples containing hazardous compounds. The costs for disposal by these facilities of wastes contaminated with chlorinated organics is quite high, and disposal of wastes contaminated with PCDD/PCDF is always difficult, and at times impossible. Because the scale of the KPEG reaction can easily be adjusted to accommodate the volume of contaminated waste, on-site dechlorination of laboratory wastes could prove to be an important application of this KPEG treatment process.

The studies described herein were designed to yield a better understanding of the operating conditions which are most effective for dechlorination of PCDD/PCDF using the KPEG reagent. As noted above, effective procedures have previously been established for the treatment of several complex chemical wastes and soils contaminated with these wastes (4). The present paper is concerned mainly with experiments intended to define optimum reaction conditions for destruction of PCDD/PCDF and other chlorocarbons which are sorbed on spent activated carbon. While the latter is a difficult task, a viable chemical process for regeneration of activated carbon which has been used to remove toxic chlorocarbons from wastewater should markedly reduce the cost of wastewater treatment.

Experimental

Laboratory Reaction Vessel. The laboratory experiments described herein were conducted in controlled-access laboratories by personnel wearing safety garments and organic vapor removing breathing apparatus. The reactions were accomplished either in a glovebox with carbon and particulate filters on the outlet, or inside a high flow rate chemical hood. The all-glass reaction vessel used in these studies was a 500 mL, thick-wall reaction kettle which fits into a heating mantle. The glass cover of the kettle had four ground glass openings which provide access for a variable rate stirrer, a thermometer, a water-cooled condenser and a sample introduction port or a purge gas inlet. During the laboratory tests, the stirrer was operated at approximately 100 rpm, the mantle temperature was adjusted to meet the experimental requirements, and the condenser was operated either in a vertical position so as to continuously return materials volatilized from the reaction mixture to the reaction flask, or the condenser was positioned at a 30 degree angle downward with respect to the flask, in order to allow volatile materials to condense and be collected in a separate 250 mL collection flask. The temperature of the reaction mixture was monitored with a thermometer, the tip of which was immersed in the reaction mixture, and the heating mantle temperature was measured with a thermocouple placed between the glass kettle and the heating mantle. During the treatment process, the reaction vessel was sealed and an activated charcoal trap was attached to the condenser outlet.

Procedures for Analysis of Reaction Products. In some experiments, time-resolved studies were conducted. At selected intervals during these tests, and at the conclusion of each experiment, aliquots of the reaction mixture were removed and transferred to separate bottles. The aliquots were immediately weighed and a sufficient quantity of 50% sulfuric acid was added to produce an acidic solution, thus quenching the dechlorination reaction. These aliquots of the reaction mixture were spiked with ^{13}C-labelled PCDD/PCDF isomer standards representative of the tetra- through octachlorinated congener groups of these compounds. Depending on the analyses to be performed, other standards such as isotopically labelled PCB and chlorobenzenes were also added to the aliquots of the reaction mixture. The quenched reaction products, which included any solid material, were mixed with sufficient sodium sulfate to produce a freely flowing powder (typically, 8 g of sodium sulfate was mixed with each gram of sample) and the sample/sodium sulfate mixture was then subjected to Soxhlet extraction for a period of 16 hours using benzene/acetone (1/1) as the extraction solvent. It should be noted that Soxhlet extraction, rather than less rigorous extraction procedures, is required to effectively remove PCDD/PCDF from the carbon matrix. Liquid portions of the quenched reaction mixture aliquots (containing no solid material) were extracted repetitively with several portions of hexane. The Soxhlet extracts were then concentrated and rediluted with hexane, and these extracts, as well as the extracts of the liquid reaction products, were prepared using procedures which have been described in detail elsewhere (5,6). Briefly, these preparation procedures initially involved sequential extraction of the sample with 30 mL quantities of 20% potassium hydroxide, water, concentrated sulfuric acid, and water. The reaction mixture extracts were then chromatographed on a multi-layer silica column using hexane as the eluting solvent. Following concentration, the extracts were chromatographed on a three-gram activated basic alumina column, from which a fraction containing PCB was isolated by eluting with 8% methylene chloride in hexane, and a fraction containing PCDD/PCDF was then collected by eluting the column with 50% methylene chloride in hexane. The PCDD/PCDF fraction was introduced onto a PX-21 charcoal/celite 545 column, from which the PCDD/PCDF fraction was ultimately recovered by back elution with toluene. Another activated basic alumina column was then used to prepare the final PCDD/PCDF fraction which was concentrated to near dryness. Prior to analysis the extract was rediluted with a tridecane solution containing isotopically labeled 1,2,3,4-TCDD and 1,2,7,8-tetrachlorodibenzofuran (TCDF) (^{13}C- or ^{37}Cl-labelled compounds).

PCDD/PCDF in the samples were quantified using a high resolution gas chromatograph coupled to a mass spectrometer (GC-MS). The computer controlled mass spectrometer was operated in the selected ion mode, and two or more masses were monitored for each ^{13}C-labelled PCDD/PCDF internal standard and for each native PCDD/PCDF compound to be quantified. The PCDD/PCDF isomers were separated on a 60 m DB-5 fused silica capillary GC column which yielded data on the 2,3,7,8-TCDD isomer as well as on the total tetra- through octachlorinated congener groups of these compounds. Additional capillary GC columns were used to obtain isomer specific data on the 2,3,7,8-substituted PCDD/PCDF isomers, in cases where this was desired. These methods yielded low part-per-trillion detection limits for PCDD and PCDF.

The samples derived from the laboratory and field treatment tests were generally analyzed for tetra-, penta-, hexa-, hepta- and octa-CDD/CDF, and the success of the dechlorination process was gauged by the extent to which the concentrations of these compounds were reduced in the treated samples as compared to the untreated samples. Further studies are in progress to determine the ultimate destruction products of the KPEG reaction.

It should be mentioned that while the ultimate reaction products (other than residual PCDD/PCDF) from the studies described herein are still not known, the gross product mixtures resulting from KPEG treatment of pure 2,3,7,8-TCDD, as well as treatment of a hazardous waste containing 120 parts-per-billion (ppb) of 2,3,7,8-TCDD have been evaluated for toxicological effects. The conclusions from these tests indicated that the products of the reaction were not mutagenic and were not toxic to aquatic organisms or mammals (4). Studies involving the KPEG-treated 2,3,7,8-TCDD showed that the reaction products were less toxic to the guinea pig by a factor of 350 than the untreated 2,3,7,8-TCDD.

Waste Samples Used in Treatment Tests. The chemical waste samples subjected to KPEG treatment in the course of the studies described herein were obtained from actual hazardous waste sites. The pentachlorophenol (PCP)-petroleum oil waste sample was collected from a 9000 gallon tank at a former wood-preserving plant site located in Montana. This waste contained about 3% PCP, as well as high concentrations of the entire chlorinated spectrum of PCDD/PCDF (up to about 80 ppm), which made on-site treatment the only practical remedial procedure. The contaminated activated carbon tested here was obtained from a facility which treats aqueous leachate from a hazardous waste landfill. This carbon contained fine particulate material and was saturated with water (approximately 35% by weight).

Results and Discussion

KPEG Treatment of PCP Wood-Preserving Wastes in Laboratory and Field Tests. As noted above, a PCP-petroleum oil wood-preserving waste was one of the hazardous waste materials tested in the KPEG studies. The results of the analysis of a sample of the untreated waste from this site are presented in Table I. Laboratory KPEG-treatment tests on this waste were conducted at temperatures of 70°C and 100°C, with aliquots of the reaction mixture being withdrawn for analyses at 15 minute intervals. In these laboratory experiments, 20 g of waste were combined with 1 g of potassium hydroxide and 21 g of KPEG reagent. The KPEG reagent used here was prepared by mixing two moles of solid potassium hydroxide with each mole of PEG-400 and heating the mixture at 90-100°C for one hour. Portions of the data from these tests are shown in Table I and indicate that, at both of the reaction temperatures noted, nearly complete destruction of the PCDD/PCDF isomers was accomplished within 30 minutes. Obviously, there is some trade-off here between reaction time and temperature.

Treatment of 9000 gallons of this same waste was subsequently accomplished by U.S. EPA at the field site, using a truck-mounted stainless steel batch reactor. This 2700 gallon reactor was heated by a boiler which circulated steam through a heat exchanger. The field treatment entailed mixing 600 gallons of KPEG with 1400-2000 gallons of waste in the truck-mounted reactor. The temperature of

Table I. Concentrations of PCDD/PCDF in PCP-Petroleum Oil Wood Preserving Waste Before Treatment and Following Laboratory and Field KPEG Treatment Tests

Total Congeners	Untreated Waste	Laboratory Treatment 15 min, 70°C	Laboratory Treatment 30 min. 70°C	Laboratory Treatment 15 min. 100°C	Laboratory Treatment 30 min. 100°C	Composite Results from Five Batch Treatments at the Field Site
	Concentration (ppb)					
TCDD	420	ND (0.46)	ND (0.37)	ND (1.3)	ND (0.28)	ND (0.64)[a]
PeCDD	820	ND (1.0)	ND (0.92)	ND (2.1)	ND (0.41)	ND (0.64)
HxCDD	3000	ND (1.8)	ND (1.8)	ND (3.4)	ND (2.4)	ND (0.57)
HpCDD	21000	11	2.1	2.3	1.4	ND (0.33)
OCDD	84000	6.5	4.0	4.1	2.6	ND (0.84)
TCDF	150	330	16	110	ND (0.35)	ND (0.64)
PeCDF	500	ND (0.43)	ND (0.47)	ND (0.60)	ND (0.29)	ND (0.69)
HxCDF	3900	4.9	3.0	2.7	ND (0.76)	ND (0.98)
HpCDF	5400	5.8	2.6	ND (0.91)	ND (1.1)	ND (0.81)
OCDF	6200	ND (3.6)	ND (3.6)	ND (3.2)	ND (1.6)	ND (0.66)

[a] Average detection limit for 5 batches

this mixture was maintained at 150°C for a period of 1.5 hours. Following this treatment, portions of the treated waste were analyzed and no detectable levels of PCDD/PCDF were found in the treated waste. The results of these field test treatments are also summarized in Table I. The data presented for the latter are based on the average results of five separate treated batches of waste. The field treatment conditions used here were somewhat more rigorous than the laboratory conditions in terms of the temperatures used, and this resulted in complete dechlorination of the tetra- through octachlorinated PCDD/PCDF in the waste.

Laboratory Study of the Effects of Reaction Time on KPEG Treatment of Contaminated Carbon. The effects of reaction time on the KPEG dechlorination of PCDD/PCDF which had been trapped on activated carbon in the course of wastewater cleanup were also investigated. A 1:1 mixture by weight of carbon and KPEG was used in these tests, and the temperature of the reaction mixture was increased until water and other volatiles present in the contaminated carbon samples refluxed. The condenser attached to the reaction vessel was positioned in the vertical position in this case, so that the volatilized materials were condensed and returned to the reaction kettle continuously. The temperature for each test in this series was maintained at 100°C. Aliquots of the reaction mixture were removed from the reaction vessel at various time intervals and analyzed for PCDD/PCDF. Table II presents a selected portion of the data from these tests, and shows the concentrations of PCDD/PCDF present in the contaminated carbon before KPEG treatment and following treatment for periods of 4 hours and 72 hours, respectively. The total PCDD/PCDF concentrations (the sum of all congener groups) decreased by 14% during the four hour treatment, while treatment for 72 hours resulted in a 60% decrease. As can be seen from Table II, the decrease in total PCDD/PCDF concentrations during the 72-hour treatment was accompanied by an increase in the concentrations of the lower chlorinated (tetra- and penta-) congeners, as compared to the levels of these in the untreated sample. Since the overall replicability of these experiments was found to be on the order of ±30%, it can be seen that these increases are generally much larger than those which would result from the combined variability in the sample, the treatment procedures and the analyses. Apparently the higher chlorinated PCDD/PCDF (hexa-, hepta,- octa-) have been stripped of chlorine atoms by the KPEG reagent, yielding elevated concentrations of the lower chlorinated congeners, which have not yet been further dechlorinated, under these conditions, when the reaction is stopped at 72 hours. Similar results were observed in our earlier studies of KPEG treatment of PCP wastes containing PCDD/PCDF. Continuing the reaction for a longer period of time or using a higher reaction temperature should drive the dechlorination reaction toward completion. The presence of water in the carbon-KPEG mixture may be a factor which impedes these reactions. Clearly, the activated carbon is a very different matrix from the waste oils and contaminated soil samples containing PCDD/PCDF and PCB, for which earlier studies have already demonstrated that KPEG treatment can fully dechlorinate the tetra- through octachlorinated congeners.

Table II. Concentration and Percent Destruction of PCDD/PCDF in Contaminated Carbon Before and After KPEG Treatment Demonstrating Effect of Reaction Time

	Concentration (ppb)		
Total Congeners	Untreated Contaminated Carbon	Following Treatment at 100°C for 4 hours	Following Treatment at 100°C for 72 hours
TCDD	1500	770	1600
PeCDD	100	24	ND (60)
HxCDD	920	640	ND (39)
HpCDD	540	560	ND (13)
OCDD	ND (430)	13	ND (12)
TCDF	470	170	280
PeCDF	450	16	62
HxCDF	190	7.9	ND (5.0)
HpCDF	350	1600	ND (100)
OCDF	280	360	ND (18)
Total PCDD/PCDF	4800	4200	1900
% of Total PCDD/PCDF Dechlorinated	--	13%	60%

Further experiments with the contaminated carbon are in progress to investigate the effects of water and reaction temperature on the KPEG dechlorination reaction. The results described above demonstrate the need for careful laboratory tests to evaluate waste matrices prior to field treatment in order to define operating parameters which will result in full dechlorination of the toxic contaminants.

Laboratory Study of the Effects of Water and Other Volatiles on the KPEG Treatment of Contaminated Carbon. The contaminated carbon which had been treated with KPEG for a period of four hours in the experiment discussed above was treated a second time under different conditions. During this second test, the condenser attached to the reaction vessel was positioned so that the components which volatilized as the reaction mixture was heated were condensed into a separate collection vessel. After 1.5 hours, these volatiles were completely removed from the mixture and the temperature of the mixture rose to 146°C, with no increase in the temperature of the heating mantle. The mixture was maintained at 146°C for an additional 2 hours.

In a separate experiment, the volatile components present in another aliquot of the original contaminated carbon were removed by distillation prior to the addition of the KPEG reagent. Following removal of the volatiles, the KPEG was added and the reaction mixture was heated for four hours at a temperature of 135°C.

Table III compares the observed concentrations of PCDD/PCDF in the products from the two experiments just described to the initial concentrations of these compounds in the contaminated carbon and to the corresponding results obtained from the four hour, 100°C direct-treatment experiment which was described in the previous section. The results summarized in Table III indicate that the removal of the water and other volatiles from the sample which had been previously treated with KPEG for four hours, with a concurrent increase in the temperature of the reaction mixture from 100°C to 146°C, resulted in dechlorination of 81% of the total PCDD/PCDF congeners present in the carbon. Just as was observed for the 72-hour KPEG treatment experiment (with no removal of volatiles) which was discussed in the previous section, however, most of the residual PCDD/PCDF contamination consists of the lower chlorinated congeners, indicating that the reaction has not reached completion. The experiment in which the water was removed prior to addition of the KPEG resulted in the dechlorination of 60% of the total PCDD/PCDF present on the carbon. While the residual concentrations of lower chlorinated congener groups were not as high as in the experiment just discussed, the concentrations of higher chlorinated congeners was greater. The results of the latter test would also have been affected by the somewhat lower reaction temperature (135°C) which was achieved in the last volatile removal experiment described here. Analyses of the distillate from these two experiments indicated that most of the PCDD/PCDF congeners were non-detectable, ensuring that the PCDD/PCDF had actually been dechlorinated and not simply removed from the reaction mixture along with the volatiles. While these experiments clearly demonstrate that the removal of the volatiles from the reaction mixture results in more complete destruction in a

Table III. Concentration and Percent Destruction of PCDD/PCDF in Contaminated Carbon Before and After Treatment of KPEG Demonstrating Effect of Removing Water from Reaction Mixture

	Concentration (ppb)			
Total Congener	Untreated Contaminated Carbon	Water Not Removed, Treated at 100°C for 4 hours	Water Removed Following Addition of KPEG, Treated at 146°C for 3.5 hours	Water Removed Before Addition of KPEG, Treated at 135°C for 4 hours
TCDD	1500	770	740	75
PeCDD	100	24	ND (1.2)	43
HxCDD	920	640	ND (1.2)	950
HpCDD	540	560	ND (1.8)	21
OCDD	ND (430)	13	1.9	ND (13)
TCDF	470	170	120	42
PeCDF	450	16	15	29
HxCDF	190	7.9	15	75
HpCDF	350	1600	13	630
OCDF	280	360	12	ND (74)
Total PCDD/PCDF	4800	4200	920	1900
% destruction of Total PCDD/PCDF 115	--	13%	81%	60%

significantly shorter period of time, this is probably due to the concurrent increase in the temperature of the reaction mixture which occurs. These results do not indicate a clear choice between removal of the volatiles before or after the introduction of the KPEG in terms of efficiency of the dechlorination. The presence of water in the treatment mixture has been suggested as a possible explanation for the incomplete dechlorination of 2,3,7,8-TCDD in a contaminated (120 ppb) waste sample consisting of solvents, oil and water during larger scale KPEG field tests (4). In these field tests, retreatment of the initially treated waste with additional KPEG reduced the concentration of 2,3,7,8-TCDD to a non-detectable level.

Laboratory Study of the Effects of the Quantity of KPEG Reagent on the Dechlorination Reaction. Reducing the quantity of KPEG which is used to treat a waste sample also reduces the final volume of treated material and simplifies the neutralization of the reaction mixture which would ultimately be required to render the highly alkaline treatment products safe for conventional disposal. Using the contaminated carbon matrix, an experiment was conducted with a carbon to KPEG ratio of 20:1, much higher than that used in the experiments already described. An excess of potassium hydroxide was added to the carbon in this case, and the sample was heated at 120°C for 72 hours. The results of this test are presented in Table IV. These data indicate that, as with other treatments discussed earlier, dechlorination of the higher chlorinated PCDD/PCDF is essentially complete, but some of the tetra- and pentachlorinated congeners remain after the 72-hour treatment. These results demonstrate, however, that the reaction is not significantly impeded by using a relatively low proportion of KPEG as compared to carbon. Recent field treatment tests on PCB-contaminated soils containing up to 2200 ppb of PCB which were conducted by U.S. EPA have shown that effective dechlorination of PCB can be achieved even with soil-to-KPEG ratios as high as 100:1.

Laboratory Study of KPEG Treatment of the Extract of a Contaminated Carbon Sample. An alternative approach for dechlorinating PCDD/PCDF sorbed on activated carbon entails extraction of the PCDD/PCDF and other extractable materials present in the carbon using an appropriate solvent, followed by KPEG treatment of the extract. Such a test was conducted previously, using methylene chloride/acetone (1/1) to extract the PCDD/PCDF from the carbon. The resultant extract was then treated with KPEG at 145°C for four hours. The result of this experiment was a reduction of all PCDD/PCDF congeners to non-detectable concentrations (Tiernan, T. O.; Wagel, D. J.; VanNess, G. F.; Garrett, J. H.; Solch, J. G.; Rogers, C. J.; *Chemosphere*, in press). These results suggest that extractable organic material present on the contaminated carbon is not responsible for impeding dechlorination of the activated-carbon-bound PCDD/PCDF isomers. Possibly, inhomogeneity of the carbon matrix and/or incomplete mixing and contact of the KPEG reagent with the carbon may account for the incomplete destruction of the PCDD/PCDF which was observed in the

Table IV. Concentration and Destruction Efficiency for the PCDD/PCDF Congeners Before and After Treatment Using a 20:1 Ratio of Contaminated Carbon:KPEG

Total Congener	Concentration (ppb) Untreated Contaminated Carbon	Following Treatment with KPEG at 120°C for 72 hours
TCDD	1500	980
PeCDD	100	ND (60)
HxCDD	920	ND (6.0)
HpCDD	540	ND (4.1)
OCDD	ND (430)	ND (2.2)
TCDF	470	270
PeCDF	450	11.4
HxCDF	190	ND (7)
HpCDF	350	ND (100)
OCDF	280	ND (13)
Total PCDD/PCDF	4800	1260
% Destruction of Total PCDD/PCDF	--	74%

tests described earlier. Additional experiments were accomplished to explore this possibility.

Laboratory Study of the Effects of Improving the Mixing of the KPEG Reagent and the Contaminated Carbon During Treatment. The somewhat limited dechlorination of PCDD/PCDF congeners which was observed when contaminated carbon containing these compounds was directly treated with KPEG may result from the inability of the KPEG to penetrate the matrix and effectively contact the carbon-bound PCDD/PCDF. In an effort to more completely distribute the dechlorinating reagent through the carbon matrix, an experiment was conducted in which 100 g of the contaminated carbon was thoroughly mixed with 90 mL of water, 10 mL of PEG-400 and 12.5 g of sodium hydroxide. The water and other volatiles were then removed by distillation of the mixture and the residual carbon and KPEG were heated to a temperature of 170°C for a period of six hours. This procedure effectively incorporates both the removal of the volatile components and the higher reaction temperatures which were shown to enhance the dechlorination process in the previous experiments. The measured concentrations of the PCDD/PCDF in the reaction mixture, as determined by analyses following this treatment, are summarized in Table V. The extent of dechlorination of the total PCDD/PCDF congeners achieved in this test was 82%, which is quite similar to the best overall results obtained in the previous treatment experiments. As in these other earlier experiments, the TCDD and TCDF congeners sorbed on the activated carbon appear to resist the dechlorination process, the present results showing 49% destruction of the TCDD congeners and 77% destruction of the TCDF congeners.

It is well known that TCDD and TCDF bind quite strongly to an activated carbon matrix. In fact, in the course of analyses of samples to determine the content of these compounds, these isomers are separated from other interferences by sorption on a PX-21 charcoal chromatography column, as briefly noted in the experimental section. Once sorbed on the carbon, these isomers resist elution, even by relatively strong solvents, and are removed efficiently only when eluted with toluene. This suggests that in future experiments with the contaminated carbon, it may be useful to employ an unchlorinated co-solvent in the KPEG process, which might displace the PCDD/PCDF from the carbon sites and permit subsequent dechlorination of the TCDD and TCDF isomers in the KPEG solution. Experiments to evaluate this will be initiated shortly.

Effectiveness of KPEG Reagent on 2,3,7,8-Substituted PCDD/PCDF Isomers in Contaminated Carbon. It is also of interest to determine the effectiveness of the KPEG process for dechlorination of specific PCDD/PCDF isomers, in particular the 2,3,7,8-substituted isomers which are the more toxic compounds. Data which demonstrate the results of KPEG treatment in dechlorinating specific 2,3,7,8 isomers are presented in Table VI. These results were obtained following the laboratory experiment in which the contaminated carbon was treated directly with the KPEG reagent at 100°C for four hours. Dechlorination of the 2,3,7,8-substituted isomers of the penta-, hexa-, and hepta-chlorinated congener groups is seen to be essentially complete, while the destruction efficiency for the

Table V. Concentrations of PCDD/PCDF in Untreated and KPEG-Treated Contaminated Carbon--Carbon Was Treated by Dispersing KPEG in a Water Slurry and Then Removing Water and Volatiles by Distillation Prior to Elevating Mixture Temperature to Final Reaction Temperature

	Concentration (ppb)	
Total Congener	Untreated Contaminated Carbon	Treated Carbon 6 hours at 170°C
TCDD	1500	750
PeCDD	100	ND (2.2)
HxCDD	920	ND (2.8)
HpCDD	540	ND (2.5)
OCDD	ND (430)	ND (4.3)
TCDF	470	109
PeCDF	450	6.0
HxCDF	190	ND (1.3)
HpCDF	350	ND (22)
OCDF	280	ND (66)
Total PCDD/PCDF Congeners	4800	870
% Destruction	--	82%

Table VI. Comparison of Residuals of Total PCDD/PCDF Congener Groups to Residuals of 2,3,7,8-Substituted Isomers of Those Groups Following Direct KPEG Treatment of Spent Carbon Containing These Compounds

PCDD/PCDF Congener Groups and Specific PCDD/PCDF Isomers	Residual %	
Total TCDD	52	
2,3,7,8-TCDD		58
Total HxCDD	69	
1,2,3,4,7,8-HxCDD		≤0.79
1,2,3,6,7,8-HxCDD		≤0.16
Total HpCDD	--[a]	
1,2,3,4,6,7,8-HpCDD		≤0.02
Total TCDF	35	
2,3,7,8-TCDF		31
Total PeCDF	3.5	
2,3,4,7,8-PeCDF		10
Total HxCDF	4.2	
1,2,3,4,7,8-HxCDF		≤0.39
Total HpCDF	--[a]	
1,2,3,4,7,8,9-HpCDF		≤0.11

[a] Treatment resulted in an increase in the concentrations of the total congener group

2,3,7,8-tetrachlorinated isomers is approximately the same as that achieved for the corresponding congener group as a whole. These results are significant because most of the toxicity which results from PCDD/PCDF contamination is the result of the presence of the 2,3,7,8-substituted isomers. By removing these isomers, the toxicity of the treated material can be significantly reduced even if the concentrations of other PCDD/PCDF isomers in the various congener groups are not reduced to non-detectable levels.

Literature Cited

1. Kimura, Y.; Regen, S. L.; "Polyethylene Glycols Are Extraordinary Catalysts in Liquid-Liquid Two-Phase Dehydrohalogenation"; *J. Org. Chem.* **1982**, *47*, 2493-4.
2. Kimura, Y.; Regen, S. L.; "Poly(ethylene glycols) and Poly (ethylene glycol)-Grafted Copolymers Are Extraordinary Catalysts for Dehydrohalogenation Under Two-Phase and Three-Phase Conditions"; *J. Org. Chem.* **1983** *48*, 195-8.
3. Rogers, C. J.; Kornel, A.; U.S. Patent No. 4 675 464, 1987.
4. Tiernan, T. O.; Wagel, D. J.; Garrett, J. H.; VanNess, G. F.; Solch, J. G.; Harden, L. A.; Rogers, C. J.; "Laboratory and Field Tests to Demonstrate the Efficacy of KPEG Reagent for Detoxification of Hazardous Wastes Containing Polychlorinated Dibenzo-p-dioxins (PCDD) and Dibenzofurans (PCDF) and Soils Contaminated with Such Chemical Wastes"; *Chemosphere*, **1989**, *18*, 835-41.
5. Solch, J. G.; Ferguson, G. L.; Tiernan, T. O.; VanNess, G. F.; Garrett, J. H.; Wagel, D. J.; Taylor, M. L.; "Analytical Methodology for Determination of 2,3,7,8-tetrachlorodibenzo-p-dioxin in Soils"; In *Chlorinated Dioxins and Dibenzofurans in the Total Environment II*; Keith, L. H.; Rappe, C.; Choudhary, G.; Eds.; Butterworth Publishers: Boston, Massachusetts, 1985, pp. 377-97.
6. Tiernan, T. O.; "Analytical Chemistry of Polychlorinated Dibenzo-p-dioxins and Dibenzofurans: A Review of the Current Status"; In *Chlorinated Dioxins and Dibenzofurans in the Total Environment*; Choudhary, G.; Keith, L. H.; Rappe, C.; Eds.; Butterworth Publishers: Boston, Massachusetts, 1983, pp. 211-37.

RECEIVED November 10, 1989

Chapter 15

Evaluation of Wood-Treating Plant Sites for Land Treatment of Creosote- and Pentachlorophenol-Contaminated Soils

Hamid Borazjani, Brenda J. Ferguson, Linda K. McFarland, Gary D. McGinnis, Daniel F. Pope, David A. Strobel, and Jennifer L. Wagner

Mississippi Forest Products Laboratory, Mississippi State University, P.O. Drawer FP, Mississippi State, MS 39762

Creosote and pentachlorophenol contaminated soils at eight wood-treating plant sites in the southeastern United States were evaluated and found suitable for remediation by land treatment. The first phase of the study included characterization of the chemical, physical, and morphological properties of the soil, levels of pentachlorophenol and polynuclear aromatic hydrocarbons, and an evaluation of the history of treating operations at the site. The second phase included studies of the rates of degradation and soil transport of pentachlorophenol and polynuclear aromatic hydrocarbons.

Creosote and pentachlorophenol have been widely used to treat wood products to increase their resistance to decay. Before strict environmental regulations were instituted, common practices at wood treating plants resulted in the release of substantial amounts of creosote and pentachlorophenol wastes into the environment. Exudation of treating solutions from freshly treated wood, accidental spills, and disposal of waste treating solution sludges in earthen pits were the major contributors to pollution. There are thousands of wood-treating plant sites in the United States, and most of them have contaminated soil and waste sludges to be cleaned up. One method that could be used for remediation of these sludges and contaminated soil is land treatment. Land treatment uses the microbiological, chemical and physical processes occurring in soil to immobilize and transform organic wastes to simpler, less toxic forms. Although waste sludges could be applied to clean soil to be transformed, land treatment techniques are usually used to remediate previously contaminated soil.

In order to use land treatment for a particular waste, two questions must be answered: 1) Are the waste compounds transformed in the soil? and 2) Are the waste compounds retained in the active

0097–6156/90/0422–0252$06.00/0

treatment zone long enough for complete transformation? The second question concerns the tendency of the waste compounds to move down through the soil (migration) along with the flow of soil water. Most transformation takes place in the top 50 to 100 centimeters of soil, probably because of the greater supply of oxygen there. Transformation below this zone is often very slow, and may not be fast enough to prevent contamination of the deep subsoil. If transformation does occur and the waste compounds are retained in the active treatment zone long enough for acceptable treatment to take place, then land treatment can be considered a viable alternative for disposal of the waste in question.

This paper reports the results of the first and second phase of a three-phase study to evaluate the efficacy of land treatment as an on-site management alternative for waste sludges and contaminated soil containing pentachlorophenol and creosote from wood-treating facilities. During the first phase, the chemical and physical characteristics of the soil and waste sludges at eight wood-treating plant sites were evaluated. During the second phase, laboratory experiments were conducted to determine the potential for transformation of creosote compounds and pentachlorophenol in the site soils and to determine the potential for migration of these compounds in the site soils. Work on four of the sites is now complete and a summary of the results is reported here. The third phase (currently in progress) involves a field evaluation study at one of the wood-treating plant sites.

Methods

Site Studies. Eight wood-treating plant sites in the southeastern United States were selected for study (Table I). The sites were chosen so that they could be used for a field evaluation study, which required that: 1) An area of 1/2 to 1 acre be available for use in the field evaluation, 2) the site must have a source of waste sludges, 3) the site should have had a low level exposure to creosote or pentachlorophenol so that an acclimated microflora is available, but the site should be without high levels of contamination in the treatment zone, and 4) a means of disposal of runoff water be available. A general history of treating operations at each site was taken. A study of the chemical, physical, and morphological characteristics of each site soil was done using standard methods (1). Part of the results are shown in Table I. The soil and sludges were analyzed for seventeen creosote components (polynuclear aromatic hydrocarbons, PAHs), pentachlorophenol, and other chemical characteristics (Tables II-V) using standard EPA recommended techniques for extraction, cleanup and analysis - Method Nos. 3540, 3520, 3630, 8100, 8040 (2).

Transformation Studies. Fifty kilogram samples of soil from each site were transported in insulated (but open to air) containers to the laboratory where they were kept cool (20°C) and moist until used. The soils were sieved to remove rocks and coarse plant materials. Dried, ground chicken manure was added to all soils at 4% by weight. This addition accomplished several objectives. The manure furnished: 1) a carbon source for potential co-metabolism,

Table I. Characteristics of the Eight Sites Used in This Study

Site location	Size & age of plant	Preservative used	Major land resource areas	Soil	Weight %		
					Sand[a]	Silt[a]	Clay[a]
Grenada, MS	100 acres 78 years	Pentachlorophenol Creosote	Southern MS Valley silty uplands	Grenada silt loam	16.06	70.17	13.77
Gulfport, MS	100 acres 80 years	Pentachlorophenol Creosote	Eastern Gulf Coast flatwoods	Smithton sandy loam	57.04	28.88	14.08
Wiggins, MS	100 acres 15 years	Pentachlorophenol Creosote	Southern Coastal Plain	McLaurin sandy loam	72.55	24.16	3.29
Columbus, MS	--	Creosote	Southern Coastal Plain	Latonia loamy sand	80.03	16.42	3.55
Atlanta, GA	15 acres 63 years	Pentachlorophenol Creosote	Southern Piedmont	Urban land	--	--	--
Wilmington, NC	--	Pentachlorophenol Creosote	Atlantic Coast flatwoods	Urban land	91.5	6.0	2.5
Meridian, MS	125 acres 61 years	Pentachlorophenol Creosote	Southern Coastal Plain	Stough sandy loam	60.2	31.4	8.4
Chattanooga, TN	76 acres 62 years	Creosote	Southern Appalachian Ridges and Valleys	Urban land complex	13.01	46.77	40.22

[a]These samples were taken from the surface to a depth of 5 inches.

Table II. Soil Concentration of PCP and PAHs at the Proposed Field Evaluation Sites

Depth (inches)	Grenada	Gulfport	Wiggins	Columbus	Atlanta	Wilmington	Meridian	Chattanooga
	Pentachlorophenol (ppm in soil)							
0-10	ND[a]	0.112	0.389	ND	20.64	1.418	0.129	0.288
10-20	ND	ND	0.017	ND	0.088	0.218	0.090	0.099
20-30	ND	ND	ND	ND	0.130	0.209	0.096	0.090
30-40	ND	ND	ND	ND	0.147	--	0.104	0.074
40-50	ND	ND	ND	ND	0.319	--	0.053	0.057
50-60	ND	ND	ND	ND	--	--	--	--
	Total polycyclic aromatics in soil[b] (ppm)							
0-10	ND	1.78	0.33	195.9	110.81	193.3	ND	121.76
10-20	ND	ND	ND	27.45	ND	40.55	ND	ND
20-30	ND	ND	ND	ND	ND	43.94	ND	ND
30-40	ND	ND	ND	ND	ND	--	ND	ND
40-50	ND	ND	ND	ND	ND	--	ND	ND
50-60	ND	ND	ND	ND	ND	--	ND	--

[a]ND = not detected.
[b]The total concentration of 16 polycyclic aromatic hydrocarbons (naphthalene, 2-methylnaphthalene, 1-methylnaphthalene, biphenyl, acenaphthylene, acenaphthene, dibenzofuran, fluorene, phenanthrene, anthracene, carbazole, fluoranthene, pyrene, 1,2-benzanthracene, chrysene, benzo(a)pyrene, benzo(ghi)perylene.

Table III. Composition of the Sludges[a]

	Water content (%)	Inorganic solids (%)	pH	Total organic carbon (%)	Total phenolics (%)	Oil and grease (%)	Pentachlorophenol[e] (ppm)	Polycyclic aromatic hydrocarbons[e,f] (ppm)
Grenada	74.58	1.11	6.30	7.37	.0041	9.74	6,699	96,078
Gulfport	30.62	1.38	4.80	22.50	.0097	44.03	5,656	101,023
Wiggins #1[b]	36.07	23.35	3.00	37.85	.0045	15.86	29,022	20,463
Wiggins #2[c]	31.56	42.42	3.50	49.45	.0130	22.57	30,060	47,075
Wiggins #3[d]	36.52	35.68	5.70	36.03	.0171	17.90	1,893	114,127
Columbus	34.44	4.45	5.90	49.79	.0224	44.60	ND[g]	475,372
Atlanta	69.10	7.14	5.00	25.33	.0120	14.17	51,974	119,546
Wilmington	26.60	64.44	7.20	4.02	.0007	.044	ND	10,007
Meridian	48.27	1.73	4.00	31.96	.0114	35.34	13,891	119,124
Chattanooga	67.35	16.91	7.10	14.61	.0003	3.68	ND	72,346

[a]Values based on the starting weight of sludge.
[b]Lagoon contains mainly pentachlorophenol.
[c]Lagoon contains mainly pentachlorophenol in a heavy oil.
[d]Lagoon contains mainly creosote. This sludge was used in the transformation and migration studies.
[e]These values are the means of two replicates and are determined on a dry basis. All were determined by capillary column gas chromatography.
[f]Total of the 17 major polycyclic aromatic hydrocarbons found in creosote.
[g]ND = not detected.

Table IV. PAH Composition of Site Sludges

	N	2Mn	1Mn	Bi	Ac	Ace	Di	Fl	Ph	An	Ca	Flu	Py	1,2B	Ch	Bz	Bzg
	(ppb)																
Grenada	67000	24150	13250	5850	5250	21500	17000	18000	43000	15000	3450	27000	19500	3250	5850	3600	5050
Gulfport	13500	14000	7450	3000	2635	10150	9600	10250	30000	7200	2100	17000	12500	2050	3650	1050	ND[b]
Wiggins #1	3400	2450	1400	535	215	1550	1300	1750	5000	2550	570	2150	1500	185	495	75	ND
Wiggins #2	10200	7450	4000	1900	1050	5550	6050	7450	21000	8150	2650	11500	7500	1300	2400	355	ND
Wiggins #3	17500	12000	6350	3500	2000	13000	11500	14000	34000	14500	4250	22500	19000	3850	6000	580	ND
Columbus	70500	29500	16500	10500	7650	31000	32500	34000	53000	23000	12500	49500	38000	12500	17000	3500	6850
Atlanta	39400	23000	11500	6600	2800	16500	16000	18000	45000	24500	9550	23000	15500	3400	5800	1100	8050
Wilmington	350	330	185	ND	ND	400	425	585	1550	1525	190	840	430	150	150	ND	ND
Meridian	16500	5350	2700	1650	1800	5150	6850	7350	29500	6550	2050	20000	11700	2200	4800	1350	550
Chattanooga	1200	815	585	445	ND	1230	1150	1415	5400	2200	870	3550	2100	200	200	ND	ND

N = Naphthalene
2Mn = 2-Methylnaphthalene
1Mn = 1-Methylnaphthalene
Bi = Biphenyl
Ac = Acenaphthylene
Ace = Acenaphthene
Di = Dibenzofuran
Fl = Fluorene
Ph = Phenanthrene
An = Anthracene
Ca = Carbazole
Flu = Fluoranthene
Py = Pyrene
1,2B = 1,2-Benzanthracene
Ch = Chrysene
Bz = Benzo(a)pyrene
Bzg = Benzo(ghi)perylene
Pentachlorophenol

[a]These values were obtained by GC/MS.
[b]ND = Not detected.

Table V. Half Lives and 95% Confidence Limits of PAHs and PCP in Columbus Soil Loaded with Site Sludges at 0.33, 1.0, and 3.0% by Soil Dry Weight

Compounds	0.33% Loading			1.0% Loading			3.0% Loading		
	t 1/2 (days)	Lower limit	Upper limit	t 1/2 (days)	Lower limit	Upper limit	t 1/2 (days)	Lower limit	Upper limit
Naphthalene	1	1	1	14	10	28	14	11	21
2-Methylnaphthalene	1	1	1	7	4	29	24	19	33
1-Methylnaphthalene	1	1	1	3	3	4	39	28	61
Biphenyl	1	1	1	5	3	10	57	30	578
Acenaphthylene	1	1	1	9	5	NT[a]	112	89	147
Acenaphthene	4	2	8	25	17	50	124	96	169
Dibenzofuran	3	3	4	2	17	28	147	99	301
Fluorene	3	3	4	31	22	52	224	169	347
Phenanthrene	18	11	50	25	17	50	578	173	NT
Anthracene	46	35	68	NT	NT	NT	173	96	866
Carbazole	35	30	43	75	48	169	90	62	169
Fluoranthene	53	29	248	289	165	1155	107	67	267
Pyrene	231	100	NT	433	187	NT	99	62	248
1,2 Benzanthracene	347	122	NT	578	173	NT	315	82	NT
Chrysene	102	61	301	365	173	NT	98	47	NT
Benzo-a-pyrene	NT	NT	NT	NT	NT	NT	NT	NT	NT
Benzo-ghi-perylene	ND[b]	ND	ND	NT	NT	NT	158	61	NT
Pentachlorophenol	1087	334	NT	NT	NT	NT	385	248	758

[a]NT = no transformation observed.
[b]ND = not detected.

which has been found in at least some instances to be an important component of the transformation process (3); 2) both major and minor nutrients; and 3) a wide variety of microbes that were potentially important biodegraders. Also, added organic matter should markedly decrease mobility of hazardous constituents in applied organic wastes, which is highly desirable in a land treatment operation (4,5). Previous field experience in landfarming operations (unpublished data) indicated that addition of manures was beneficial, so the effect of adding manures was not specifically studied (i.e., with no-manure controls) in the research reported here. Although other animal manures might serve as well, chicken manure was chosen for study because it is readily available in many parts of the United States. A typical analysis of the chicken manure used in this study is given below:

Total organic carbon = 8.97%
Total nitrogen = 1.35%
Total phosphorous = 0.12%

Samples of the sludges from each site were taken from several locations in the sludge pits and mixed together to give a representative sludge from each site. Sludge was added to the soil at loading rates of 0.3%, 1.0%, and 3.0% (sludge wet weight/soil dry weight) with three replications of each site soil/loading rate combination. Refer to Table III for sludge composition. The experimental unit was a brown glass container with a lid, containing 500 g (dry weight) of soil mixed with sludge. The sludge from each site was tested in the soil from that site. For sludges with no pentachlorophenol, reagent grade pentachlorophenol in methanol was mixed with the soil at 200 ppm pentachlorophenol by weight. The soil moisture content was adjusted to 70% of water-holding capacity. The water holding capacity was determined by saturating the soil with water, allowing the soil to drain for 24 hours, and then drying at 105°C. Water was added weekly to each experimental unit as needed to maintain the 70% moisture content. All the units were kept in a constant temperature room (22° $\pm$ 2°C) for the duration of the study. The soil in each unit was mixed each week by stirring with a spatula, to simulate tillage operations commonly used in landfarming. At 0, 30, 60, and 90 days 40 gram samples of the soil in each dish were taken for chemical analysis. The analyses were conducted according to the standard EPA methods referred to earlier (2). Transformation rates and upper and lower confidence limits for the transformation rates were calculated with a linear regression using first order kinetics. Half lives were calculated from the transformation rates.

Migration Studies. Eighteen undisturbed soil cores were taken at each site for the migration studies. A stainless steel cylinder (22 cm diameter x 76 cm long) lined with a section (18 cm x 60 cm) of high density polyethylene pipe (Phillips Driscopipe) was driven into the soil. This produced an undisturbed soil core (18 cm x 51 cm) enclosed in the section of pipe.

In the laboratory the soil cores were placed on a rack with each core resting on a fiberglass mat supported by a stainless steel screen set in a large glass funnel. Each soil core was tested as follows: 1) A solution of 500 ml of 500 ppm NaCl in water was poured on the top of each core; 2) after this solution entered the soil, water was put on top of each core periodically; and 3) the water draining from the bottom of each core was collected in 500 ml increments and tested for Cl^- concentration. The twelve cores from each site showing the sharpest, most uniform Cl^- 'peak' were chosen for the migration studies.

Half the soil cores from each site were randomly chosen as controls to measure background levels of creosote compounds and PCP. All the cores had the top 15 cm of soil removed, pulverized and mixed with chicken manure (4% by weight). The soil was replaced in the control cores without further modification. In the remaining cores from each site the site sludge was mixed with the removed soil at 3% by weight, and the removed soil was replaced in the soil cores ('loaded cores'). For site sludges with no pentachlorophenol, pentachlorophenol was added to the removed soil at 200 ppm by weight. Water was added to each core at a rate equivalent to 5 cm rainfall each week (1300 ml/core/week). The water draining from each core was sampled and analyzed at monthly intervals. After three months, each soil core was sectioned into 6 equal portions and each portion was analyzed to determine how far the sludge components had moved down through the core.

Results and Discussion

Site Studies. A summary of the general site characteristics is given in Table I. Most of the sites are old sites and had extensive areas of contamination. All but two sites recorded use of both pentachlorophenol and creosote in their wood treating operations. Records at most plant sites were very sketchy. The soil types varied from a heavy clay at Chattanooga to an almost pure sand at Wilmington. The soils at the proposed field evaluation sites had low to moderate levels of contaminants except the Grenada site, where no contamination was detected (Table II). The more mobile pentachlorophenol was detected as deep as 50 inches in three soils, but creosote components (PAHs) were found only in shallow layers. Note that these results apply only to the area at each site that was chosen for a possible field evaluation study, not to heavily contaminated areas at the sites.

The composition of the sludges used in the study is shown in Tables III and IV. Sludge composition varied widely. Water was a major component of all the sludges since they were stored in open pits. Inorganic solids from soil were prominent in several sludges. The pH of several of the sludges was low, partially due to a high pentachlorophenol content (6). The oil carriers used to help the preservative penetrate the wood in the treating process caused the oil and grease content of the sludges to be high. The oil can cause difficulty in land treatment by reducing the access of oxygen to the preservatives, thereby reducing aerobic microbial activity. The oil carriers can also increase the mobility of preservative compounds in the soil.

The pentachlorophenol and PAH contents of the sludges varied widely. A pentachlorophenol treating solution usually contains about 70 to 80,000 ppm pentachlorophenol, and a creosote treating solution may contain 400 to 500,000 ppm PAHs. Therefore if the water was removed, some of the sludges would be similar in concentration to an actual treating solution (Table III).

The major PAHs in the sludges are shown in Table IV. The chemical composition of creosote sold for wood-treating purposes varies considerably, and creosote is usually described by its physical properties (distillation fractions) rather than by chemical content. The chemical variability of the sludges is due to their initial variability, their use and reuse in wood treating, and subsequent exposure to the environment for several years in sludge pits.

The major PAHs in sludges are two, three, four, five, and six ring compounds, with the two, three, and four ring compounds accounting for the greater portion. Both toxicity and resistance to transformation increase as the number of rings increase.

Transformation Studies. The total PAH breakdown was similar in soils from all four sites if similar loading concentrations were used (Tables V-VII). The individual PAHs can be divided into three groups according to the results of this research: those with apparent half lives of ten days or less, those with apparent half lives of one hundred days or less, and those with apparent half lives of more than one hundred days. Naphthalene, 2-methylnaphthalene, 1-methylnaphthalene, biphenyl, acenaphthalene, acenaphthene, dibenzofuran, and fluorene usually had apparent half lives of ten days or less. Phenanthrene, anthracene, carbazole, and fluoranthene usually had apparent half lives between ten and one hundred days. Pyrene, 1, 2-benzanthracene, chrysene, benzo-a-pyrene, and benzo-ghi-perylene usually had apparent half lives greater than one hundred days. In several cases these last five compounds showed essentially no transformation within the time frame of the experiment.

The transformation rates of individual PAHs were apparently related to molecular size and structure, as noted in previous studies (4). The zero to ten day half life group contained compounds with two aromatic rings, the ten to one hundred day half life group contained compounds with three aromatic rings, and the one hundred plus day half life group contained compounds with four or more aromatic rings. However, some of the larger, most recalcitrant compounds apparently were transformed readily in some soils (Note Tables VI-VII). This indicates that even the most persistent PAHs might be remediated with land treatment if further study could elucidate the precise conditions necessary for consistent results.

Carbazole, a compound containing a nitrogen bridge between two aromatic rings, varied greatly in persistence in different soils and loadings. This may be due to the nitrogen atom affecting water solubility and other properties of carbazole under varying local oxidation/reduction potentials and pH.

Table VI. Half Lives and 95% Confidence Limits of PAHs and PCP in Grenada Soil Loaded with Site Sludges at 0.33, 1.0, and 3.0% by Soil Dry Weight

Compounds	0.33% Loading			1.0% Loading			3.0% Loading		
	t 1/2 (days)	Lower limit	Upper limit	t 1/2 (days)	Lower limit	Upper limit	t 1/2 (days)	Lower limit	Upper limit
Naphthalene	1	1	1	1	1	1	ND[b]	ND	ND
2-Methylnaphthalene	1	1	1	1	1	1	ND	ND	ND
1-Methylnaphthalene	1	1	1	1	1	1	ND	ND	ND
Biphenyl	1	1	1	1	1	1	1	1	1
Acenaphthylene	4	2	23	1	1	1	ND	ND	ND
Acenaphthene	4	3	9	1	1	1	ND	ND	ND
Dibenzofuran	4	3	9	1	1	1	116	75	248
Fluorene	4	3	11	1	1	1	ND	ND	ND
Phenanthrene	5	3	18	2	9	17	7	2	NT
Anthracene	10	5	NT[a]	27	19	45	8	2	NT
Carbazole	3	2	5	1	1	1	ND	ND	ND
Fluoranthene	65	51	91	36	25	65	21	14	42
Pyrene	68	53	95	45	30	86	21	19	24
1,2 Benzanthracene	3466	169	NT	107	66	285	23	18	31
Chrysene	173	95	866	102	64	248	72	43	72
Benzo-a-pyrene	3466	433	NT	NT	NT	NT	NT	NT	NT
Benzo-ghi-perylene	770	126	NT	NT	NT	NT	ND	ND	ND
Pentachlorophenol	46	39	55	53	26	NT	21	14	37

[a]NT = no transformation observed.
[b]ND = not detected.

Table VII. Half Lives and 95% Confidence Limits of PAHs and PCP in Wiggins Soil Loaded with Site Sludges at 0.33, 1.0, and 3.0% by Soil Dry Weight

Compounds	0.33% Loading			1.0% Loading			3.0% Loading		
	t 1/2 (days)	Lower limit	Upper limit	t 1/2 (days)	Lower limit	Upper limit	t 1/2 (days)	Lower limit	Upper limit
Naphthalene	1	1	1	6	3	NT[a]	3	2	8
2-Methylnaphthalene	1	1	1	6	3	NT	3	2	8
1-Methylnaphthalene	1	1	1	3	2	6	4	2	22
Biphenyl	1	1	1	3	2	6	1	1	1
Acenaphthylene	5	1	3	3	2	6	1	1	1
Acenaphthene	3	2	6	41	5	NT	17	12	32
Dibenzofuran	3	2	6	58	24	NT	28	16	99
Fluorene	3	2	6	58	22	NT	23	14	53
Phenanthrene	4	3	7	58	2	NT	20	13	43
Anthracene	4	3	9	NT	NT	NT	50	30	139
Carbazole	4	3	10	NT	NT	NT	53	35	116
Fluoranthene	29	20	53	58	30	693	46	22	NT
Pyrene	6	3	30	NT	NT	NT	347	99	NT
1,2 Benzanthracene	43	7	NT	693	99	NT	139	63	NT
Chrysene	3	2	5	NT	NT	NT	7	99	NT
Benzo-a-pyrene	ND[b]	ND	ND	NT	NT	NT	4	2	173
Benzo-ghi-perylene	ND	ND	ND	1	1	1	ND	ND	ND
Pentachlorophenol	116	32	NT	91	29	NT	105	35	NT

[a]NT = no transformation observed.
[b]ND = not detected.

Acenaphthylene and acenaphthene, differing only in the presence or absence of a double bond (and two hydrogens), show the effect of small changes in structure. Acenaphthene had much longer average half life than acenaphthylene. Apparently, the double bond is easier to attack, although the single bond in acenaphthene also lowers the vapor pressure, which may affect the apparent half life by volatilization.

PCP transformation occurred in all the soils but was slow in Columbus soil (Table V), which was from a site not exposed to PCP treatment wastes. Grenada soil (Table VI) transformed PCP with half lives ranging from one to two months, a very practical range for land treatment operations. Meridian soil (Table VIII) also exhibited rapid transformation rates except at the highest loading rate. Wiggins soil (Table VII) transformed PCP with half lives of three to four months, still an appropriate range for land treatment operations since Wiggins is located in the deep south where soil temperatures are high enough for effective microbiological activity most of the year. Although the Columbus soil did exhibit some transformation of PCP, the low rates indicate that land treating PCP at that location would probably be impractical.

The different loading rates did not markedly affect the apparent half lives of the sludge constituents except in the Columbus soil. All the loading rates used in this study were at concentrations that previous field experience indicated would not inhibit transformation. However, the 3.0% loading in the Columbus soil gave the highest concentration of PAH's in soil of all the sites in the study, since the Columbus sludge contained the highest concentration of PAH's. This may account for the longer half lives in the 3.0% loading in the Columbus soil.

Migration Studies. No PAHs and PCP were found above background levels (from control cores) in the drainage water from the tested soil cores. Furthermore, there were no detectable increases in PAHs or PCP in soil below the zone of incorporation in the loaded cores. This is not unexpected, since the soil organic matter (from added chicken manure) and clay tightly binds these compounds. At the low concentrations of PAHs and PCP for which landfarming appears suitable (up to 30,000 ppm PAHs, 3,000 ppm PCP) migration would be very slow in most soils with high organic matter and clay. In most cases, transformation would have ample time to take place before migration would present problems.

General Discussion

The results of these experiments indicate that PAHs and PCP can be transformed in soil at rates practical for land treatment. Although the variability of the data is relatively large in some cases, the general trend is clear. Based on the treatability data from the soils tested to date, land treatment of creosote and PCP wood treating wastes appears to provide a viable management alternative for site remediation. The data variability does indicate the need for conducting site-specific treatability studies to discern the appropriate operation and management scenario for a given site.

Table VIII. Half Lives and 95% Confidence Limits of PAHs and PCP in Meridian Soil Loaded with Site Sludges at 0.33, 1.0, and 3.0% by Soil Dry Weight

Compounds	0.33% Loading			1.0% Loading			3.0% Loading		
	t 1/2 (days)	Lower limit	Upper limit	t 1/2 (days)	Lower limit	Upper limit	t 1/2 (days)	Lower limit	Upper limit
Naphthalene	1	1	1	6	2	NT[a]	1	1	1
2-Methylnaphthalene	1	1	1	7	3	NT	1	1	1
1-Methylnaphthalene	1	1	1	8	3	NT	1	1	1
Biphenyl	4	1	NT	8	27	NT	1	1	1
Acenaphthylene	1	1	NT	8	3	NT	1	1	1
Acenaphthene	1	1	1	7	25	NT	6	3	NT
Dibenzofuran	1	1	1	6	2	NT	10	3	NT
Fluorene	1	1	1	7	2	NT	8	3	NT
Phenanthrene	5	2	NT	38	16	NT	8	3	NT
Anthracene	4	2	NT	28	4	NT	6	3	NT
Carbazole	ND[b]	ND	ND	7	3	NT	1	1	1
Fluoranthene	41	19	NT	NT	NT	NT	90	37	NT
Pyrene	53	32	205	NT	NT	NT	NT	NT	NT
1,2 Benzanthracene	139	58	NT	15	5	NT	12	3	NT
Chrysene	NT	NT	NT	16	5	NT	11	3	NT
Benzo-a-pyrene	ND	ND	ND	NT	NT	NT	NT	NT	NT
Benzo-ghi-perylene	ND	ND	ND	NT	NT	NT	ND	ND	ND
Pentachlorophenol	43	34	60	72	30	462	NT	NT	NT

[a]NT = no transformation observed.
[b]ND = not detected.

Further study of treatment of PCP and the higher molecular weight PAHs is needed to determine the most advantageous environmental conditions and management techniques for more rapid transformation of these compounds. Many of these compounds were readily transformed in some cases. Therefore, further study may reveal reliable techniques for enhancing land treatment as a practically useful management alternative for these recalcitrant compounds. Since the environmental problems that the wood treating industry faces are almost unlimited and since the resources available to solve these problems are quite limited, land treatment is very attractive as a safe and economical solution.

Literature Cited

1. Soil Survey Manual. Agric. Handbook No. 18, USDA. U.S. Govt. Printing Office, Washington, DC. 1951.
2. Test Methods for Evaluating Solid Waste. 1B. SW-846. Third Edition. U.S. EPA. 1986.
3. Cerniglia, C. E. 1984. In: Adv. in Appl. Microbiol., A.I. Laskin, Ed. Vol. 30, Academic Press, New York, NY. Vol. 30, pp. 30-71.
4. Bulman, T. L., S. Lesage, P. J. A. Fowles, and M. D. Webber. The Persistence of Polynuclear Aromatic Hydrocarbons in Soil. PACE Report No. 85-2, Petroleum Assoc. for Conservation of the Canadian Environ., Ottawa, Ontario. 1985.
5. Means, J. C., G. S. Wood, J. J. Hassett, and W. L. Banwart. Environ. Sci. Technol. 1979. 14:1524-1528.
6. Crosby, D. G. 1981. Pure Appl. Chem. 53:1052-1080. 1981.

RECEIVED November 10, 1989

Chapter 16

Factors Influencing Mobility of Toxic Metals in Landfills Operated with Leachate Recycle

Joseph P. Gould[1], Wendall H. Cross[1], and Frederick G. Pohland[2]

[1]School of Civil Engineering, Georgia Institute of Technology, Atlanta, GA 30332–0355
[2]Department of Civil Engineering, University of Pittsburgh, Pittsburgh, PA 15261–2294

Ten simulated landfill columns containing shredded municipal refuse have been operated for over three years. Five have been operated in the conventional single-pass mode while the other five have been operated with leachate containment and recycle. To three of the columns have been added sludges containing an array of toxic metals (Cd, Cr, Ni, Hg, Pb, Zn) at "low", "moderate" and "high" levels. The concentrations of these metals in the leachates from these columns have been monitored along with other parameters of significance in control of metal mobility such as pH, sulfide, sulfate, chloride and ORP.

Indications have been obtained of the presence of a complex array of chemical and physical mechanisms acting to minimize the mobility of these metals in the leachates from these columns. These mechanisms confer upon the municipal solid waste a substantial assimilative capacity for these metals and offer the potential for the practical codisposal of measured levels of toxic metals in municipal landfills given adequate attention to monitoring and control in such systems.

Many industrial activities common both in the United States and elsewhere generate often complex solid residues of wastewater treatment which contain one or more toxic heavy metals. These metal sludges are subject to stringent disposal requirements which often entail considerable expense and effort to ensure that the metals do not escape into and damage the environment. Earlier practices governed by less stringent controls resulted in the disposal of such sludges along with municipal solid wastes. In addition, many toxic inorganic compounds are codisposed with domestic wastes as a consequence of household and other nonindustrial activities. Thus, although disposal of toxic metals in designated hazardous waste landfills is currently mandated,

0097–6156/90/0422–0267$07.25/0

codisposal of such wastes in nominally municipal landfills has been a not uncommon past practice and, under some conditions, continues at present. The research whose results are described herein was designed to evaluate the impact of codisposal of heavy metal sludges with municipal solid waste (MSW) and to investigate the mechanisms controlling mobilization and attenuation of these metals in leachates from the experimental simulated landfill systems.

Materials and Methods

Column Design. The ten simulated landfill columns were designed in pairs with the operational features illustrated in Figure 1. Each column was constructed of two 0.9 m diameter steel sections with a total height of 3.0 m. Five of the columns were designed to operate with single pass leachate flow while the other five were designed to permit leachate collection and recycle. All columns were provided with appropriate appurtenances and were sealed gas-tight after loading.

The single pass and recycle control columns were loaded with shredded MSW free of significant concentrations of the metals of interest only, whereas the remaining four pairs of columns received selected organic priority pollutants along with the MSW. Three pairs of these latter columns also received incremental loadings of toxic heavy metals in the form of alkaline metal finishing waste treatment sludge supplemented where necessary with additional heavy metals. Specifically, lead and mercury were added in the form of reagent grade divalent metal oxides (HgO and PbO). Prior to addition, the metal solids were intimately blended, mixed with measured quantities of sawdust to enhance contact with leachate and mechanically reduced to a finely divided uniform particulate material. The metal loading levels for the simulated landfill columns are presented in Table I.

Table I. Loading Levels for Simulated Landfill Columns

	Shredded Municipal Refuse, kg (dry) 267		
Heavy Metals, g	Low	Moderate	High
Cadmium	26.1	52.3	104.6
Chromium	45.3	90.6	181.2
Mercury	20.3	40.6	81.2
Nickel	39.7	79.4	158.8
Lead	104.9	209.8	419.8
Zinc	45.7	91.4	182.8

Column Designations

C & CR	Control Single Pass & Recycle
O & OR	Organic Priority Pollutants Only
OL & OLR	Organics + Low Metal Loading
OM & OMR	Organics + Moderate Metal Loading
OH & OHR	Organics + High Metal Loading

After loading and sealing had been completed, Atlanta tap water was added to bring each column to field capacity and initiate the immediate production of leachate for recycle and/or analysis. Thereafter, water was added at a rate of six liters per week to each of the columns. Water addition to the recycle columns continued until sufficient leachate had accumulated to accommodate recycle and analysis schedules.

Initiation of Methane Formation. The columns were operated for a prolonged period in the acid forming phase in order to fully ascertain the behavior of the systems under the conditions prevailing during that stage. Following this period, it was decided to initiate methane formation by the addition of supernatant from a municipal wastewater treatment plant anaerobic sludge digester. The addition of this sludge took place over a period of several weeks beginning around Day 600 of column operation.

Leachate Analysis. Once leachate was generated from each of the ten columns, a comprehensive analytical program was established for the common indicator parameters and the added priority pollutants. Of specific interest here are the analytical methods summarized in Table II. Samples used for metal analysis were first digested using nitric acid and hydrogen peroxide. The analysis of anions was first preceded by a clean-up on a cation exchange resin to remove interferences.

Table II. Analytical Methods (1)

Parameter	Method	Instrument
Lead	Flame or Graphite Furnace Atomic Absorption	Perkin Elmer 303 Perkin Elmer 703
Mercury	Flameless Atomic Absorption	Perkin Elmer 703
Chloride & Sulfate	Ion Chromatography	Dionex 2000i/SP
Sulfide, ORP, pH	Electrode Methods	--

Results and Discussion

Theoretical Considerations. The behavior of metals in these systems will be controlled by a complex array of chemical and physical mechanisms. Depending on the metal under consideration, these might include any or all of the following.

pH. The hydrogen ion concentration will directly influence metal behavior with, as is well known, metal solubilities generally decreasing with decreasing hydrogen ion concentration. In addition, the hydrogen ion concentration will indirectly influence metal solubility by its impact on such processes as the dissociation of an acid to yield a precipitant anion and reduction-oxidation (redox) reactions.

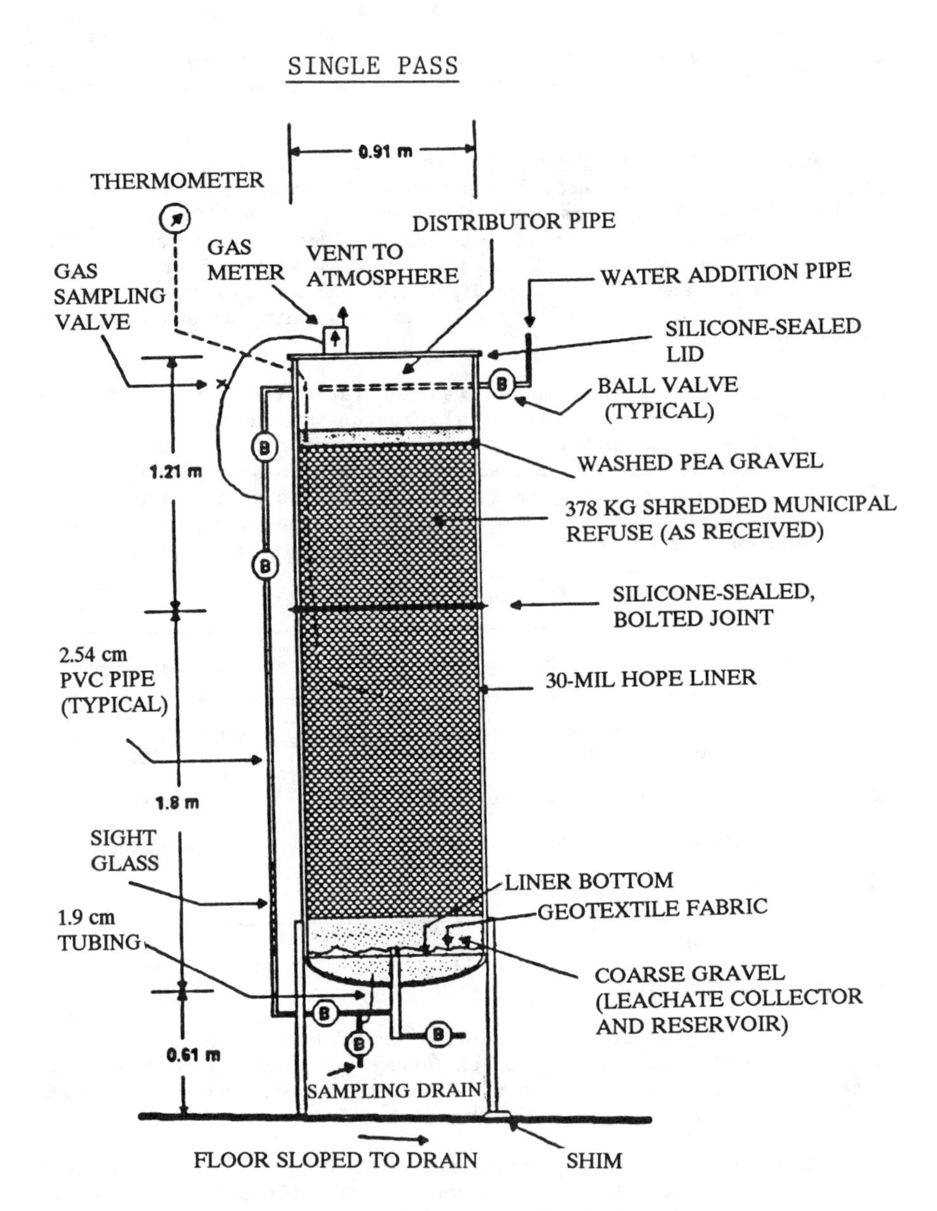

Figure 1. Schematic of column design.

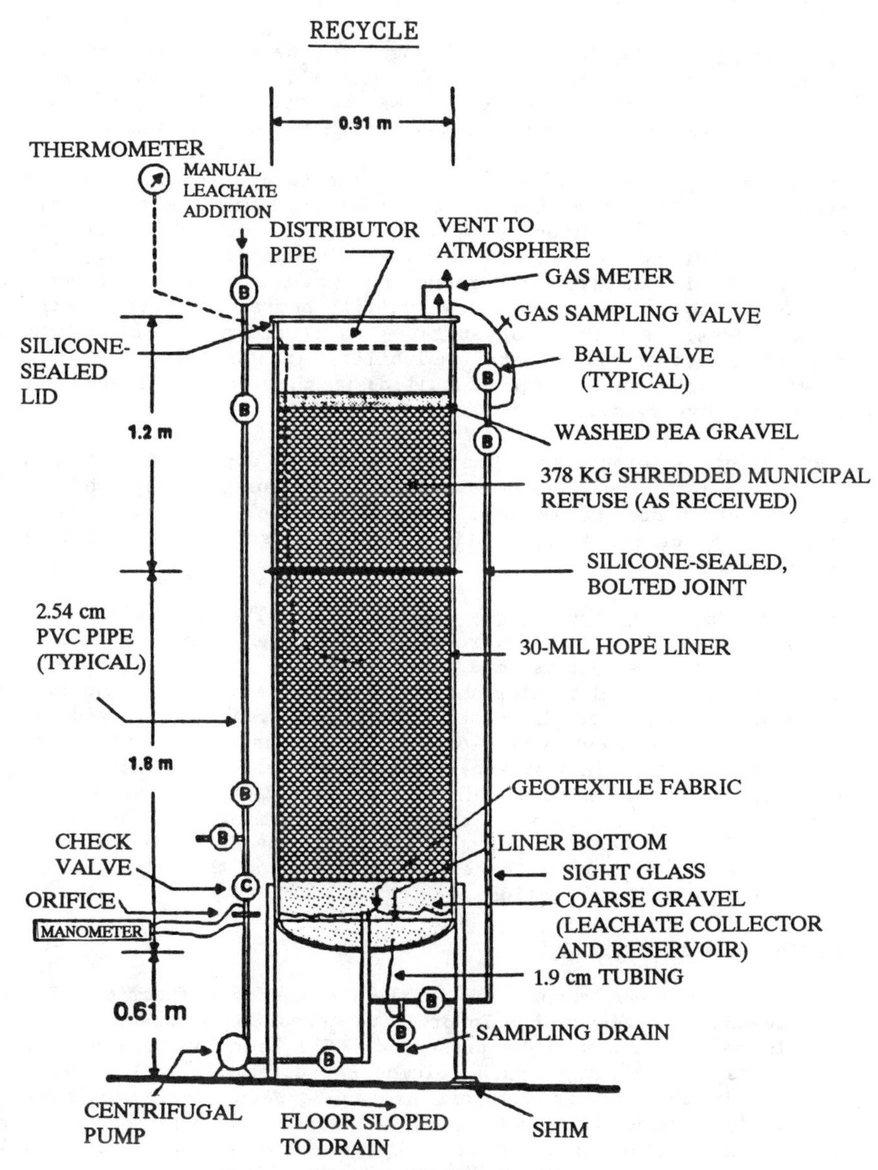

Figure 1. *Continued.*

Precipitation. Many inorganic and organic anions can precipitate metals as sparingly soluble salts. Some anions such as sulfide and hydroxide have broad precipitant capabilities while many others such as sulfate and chloride will form sparingly soluble salts with only a limited number of metals.

Complexation. The formation of complexes between heavy metals and ligands tends to increase metal solubility although there are conditions under which the opposite may be expected. Many organic and inorganic ligands will form complexes with toxic heavy metals.

Reduction/oxidation. Electrochemical processes can influence metal speciation and behavior both directly by modifying the nature of the metal itself and indirectly by conversion of other species in the landfill environment. Thus, the toxic non-metal, selenium, can be removed from landfills by reduction to the neutral element or conversion to the selenide ion which will be readily precipitated by ferrous ions. Reduction of sulfate is considered to be a pivotal process in controlling metal solubility in landfills in that it provides an abundant source of sulfide which is a potent precipitant for many heavy metals.

Solid/Solute Interactions. The interaction of dissolved species with the refuse solids provides many opportunities for processes which will act to attenuate metal mobility. These include adsorption, ion exchange, interaction with solid phase ligands and metal hold-up in interstitial water.

Interactions With the Sludge Mass. The alkaline metal sludges themselves represent microenvironments within the landfill masses which will be substantially less acid than the general refuse mass especially during the early acid-forming phase of landfill evolution. Mobility of dissolved metals through these environments can be expected to be significantly retarded as the local pH is increased by the reservoir of sludge alkalinity. In addition, interactions between the metal sludges and leachate anions such as sulfate and sulfide will tend to replace hydroxide in the sludge with these anions forming much less soluble salts. This process of encapsulation of the hydroxide solids by sulfide, etc., will have the effect of reducing the mobilization of metals from the sludge solids.

General Column Behavior

pH. The trends in leachate pH in the columns as a function of time are presented in Figure 2. In order to economize on space, only the recycle column data will be presented. The behavior of the various parameters in the single pass columns has differed little from that observed in the recycle columns. As can be seen, following a early period characterized by very acid pH values and high variability, leachate pH levels stabilized in the range of approximately 5.2 to 5.6, a range maintained by the buffering effect of several low molecular weight volatile organic acids (acetic through hexanoic). It was only with the onset of active methane formation and the consequent depletion of these acids in many of the columns that the leachate pH increased to values in the neutral range.

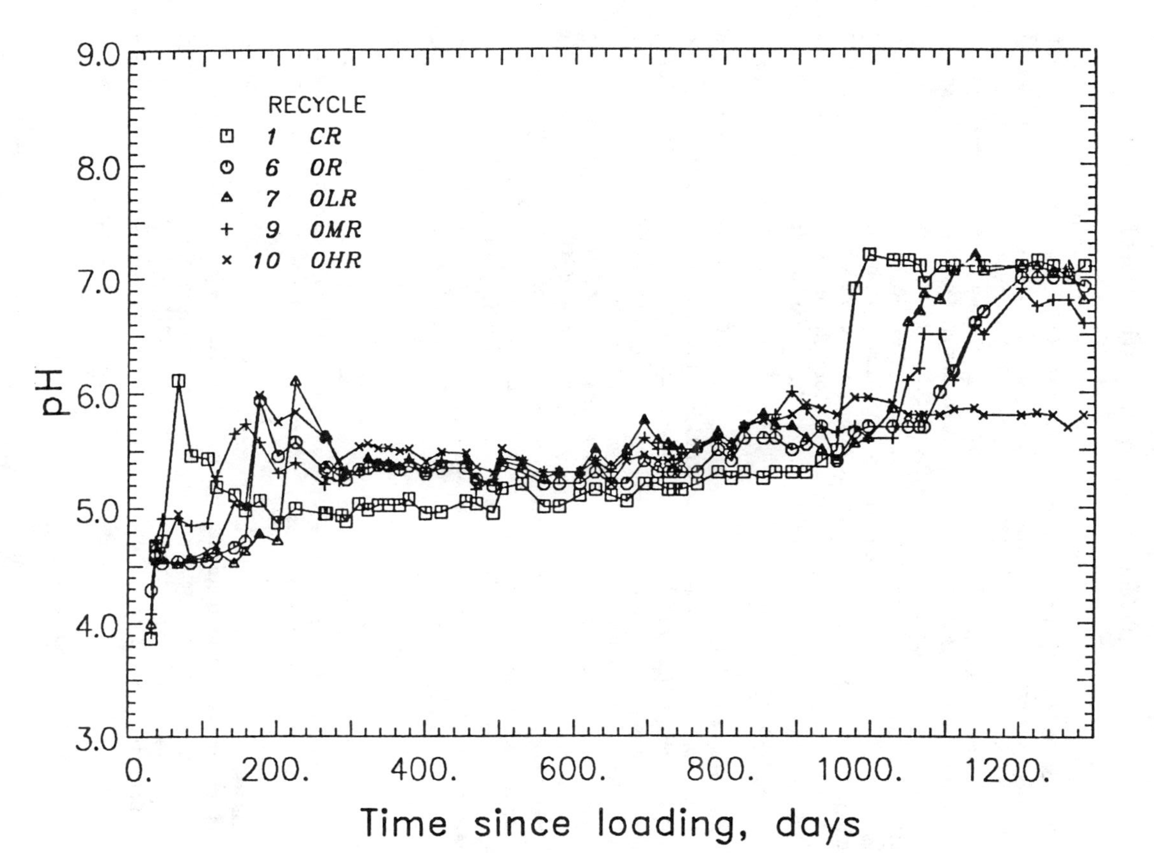

Figure 2. Leachate pH as a function of time.

ORP. Trends in oxidation-reduction potentials (ORP) in these columns are shown in Figure 3. As can be seen, the values measured in the leachates have been consistently equal to or more negative than zero millivolts. It should be noted that these values are surely substantially less negative than the values actually present and controlling within the columns. Clearly, the environment within the columns has been consistently and remains significantly reducing.

Sulfate and Chloride. These species will play a major role in the behavior of the codisposed heavy metals. Chloride is an important complexing agent for certain of the metals, while sulfate is a significant precipitant for lead and, as the precursor of sulfide, is vital in attenuation of several of the heavy metals. Chloride concentrations as a function of time are shown in Figure 4. Chloride can be viewed as a conservative substance or tracer in these systems, and thus displayed washout behavior as expected in the single pass columns. In the case of the recycle columns, chloride as expected maintained an essentially constant concentration until approximately Day 700 following which its concentrations have decreased steadily. This recent decrease in chloride levels has been ascribed to dilution resulting from the addition of anaerobic sludge.

Sulfate concentrations as a function of time are presented in Figure 5. As was the case for chloride, single pass column behavior was consistent with wash-out of a conservative tracer. In the recycle columns however, a prolonged period of steady sulfate concentrations up to approximately Day 700 of column operation was followed by a precipitous decline in sulfate concentrations to values approaching zero in many cases. This decrease, far too great in magnitude to be ascribed to dilution by anaerobic sludge, can be explained by the onset of rapid sulfide formation coincident with the initiation of methanogenesis around Day 700. Prior to this time, the sulfate concentration curves present no evidence for any significant sulfate reduction/sulfide formation.

Sulfide. Presented in Figure 6 are sulfide concentrations as a function of time in the leachates from these columns. Sulfide concentrations were consistently at or below the detection limit for the electrode method used (≈0.1 mg/L) until Day 900 for the recycle columns and Day 1,000 for the single pass columns. This is not inconsistent with the behavior of sulfate in the systems. The 200 to 300 day lag in appearance of sulfide after the start of the rapid decline in sulfate concentrations can be ascribed to the abundance of sinks for sulfide in these columns the major of which will be ferrous iron.

Behavior of Metals

Chromium. Chromium is, in many ways, the simplest of the toxic metals in these columns with respect to its chemical behavior. At the ORP values characteristic of the columns (Figure 3) chromium will exist in the +3 oxidation state and will be subject to precipitation by hydroxide only among the significant anions in the leachates. The

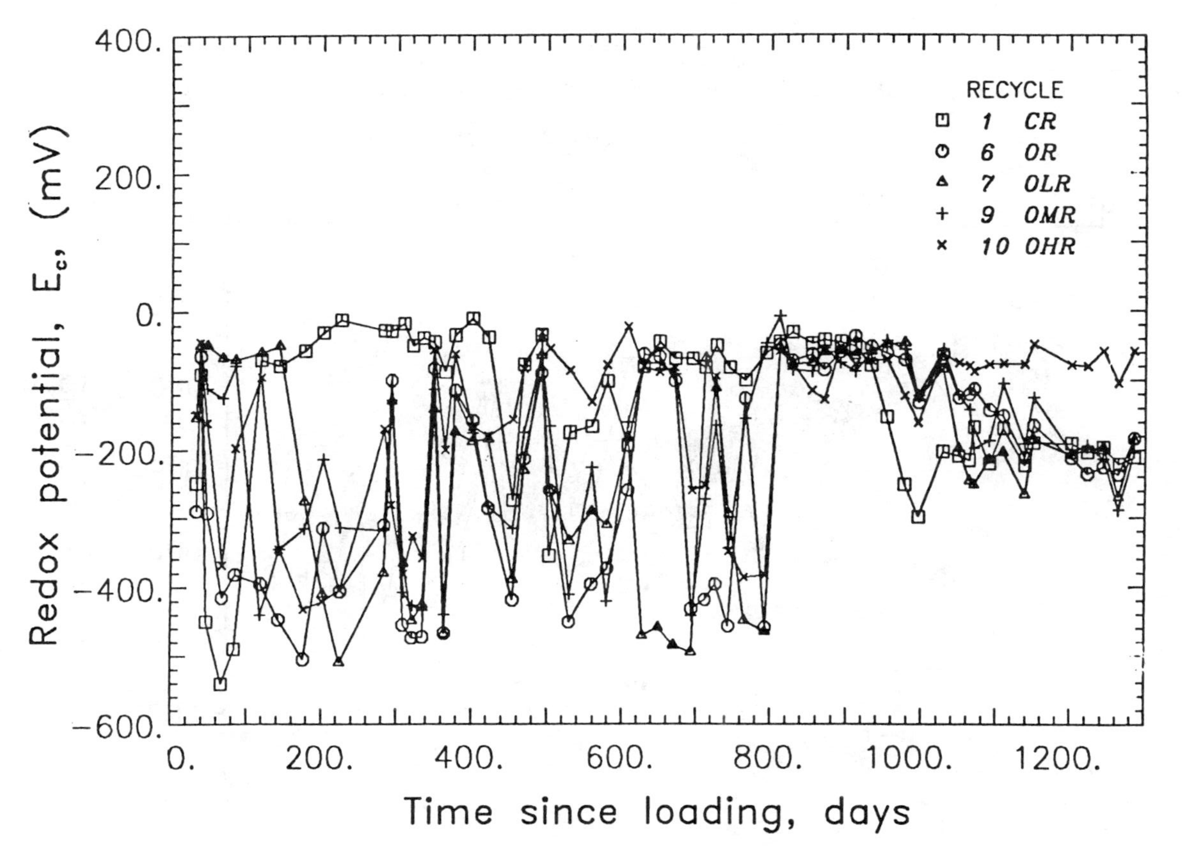

Figure 3. Leachate ORP as a function of time.

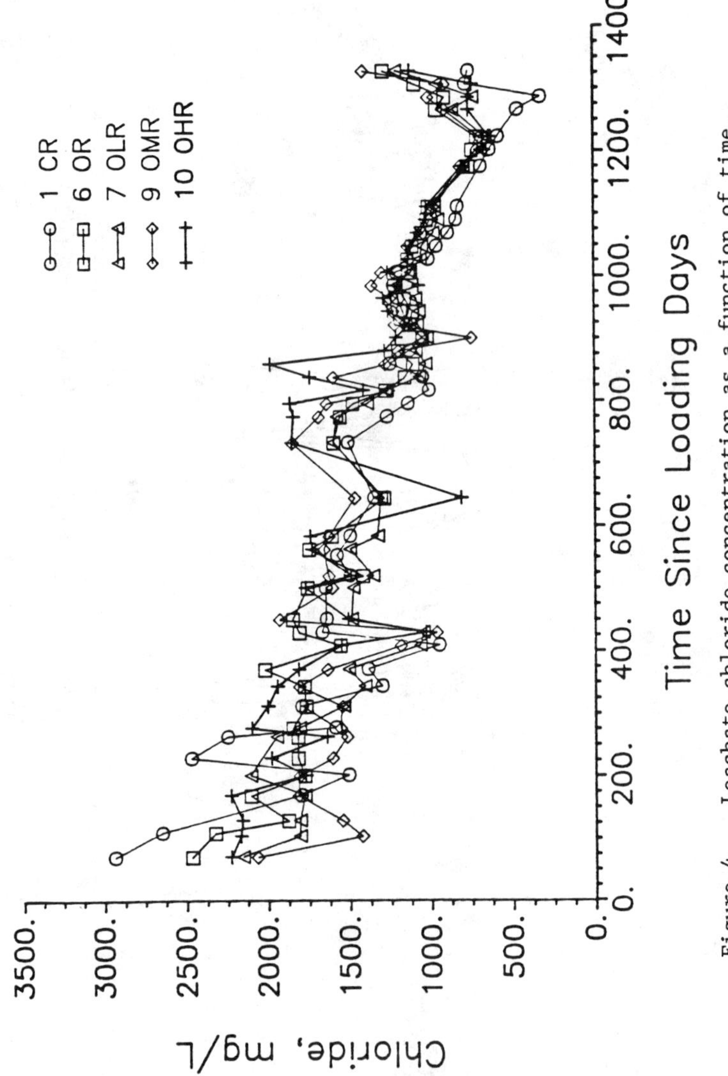

Figure 4. Leachate chloride concentration as a function of time.

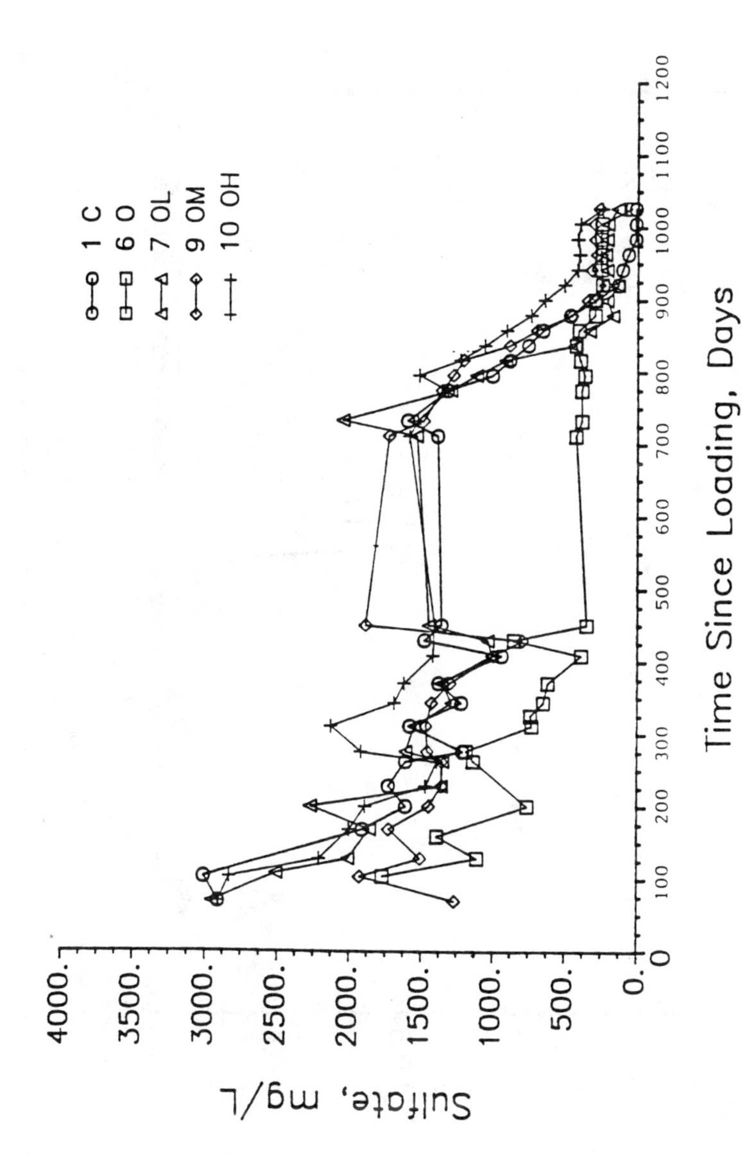

Figure 5. Leachate sulfate concentration as a function of time.

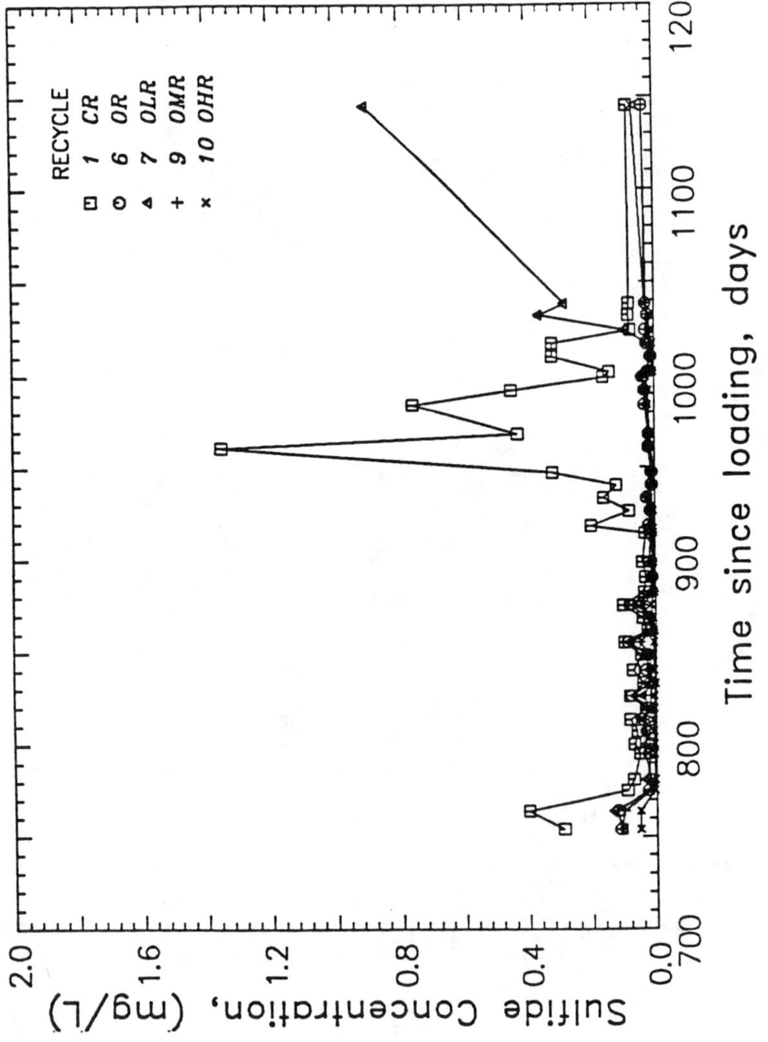

Figure 6. Leachate sulfide concentration as a function of time.

chromic ion (Cr^{3+}) is unusual in that it does not form a sparingly soluble sulfide. The hydroxide is, on the other hand, a very sparingly soluble salt (Equation 1) and, even at pH values as low as the 5.2 to 5.6 observed during the major portion of the acid phase of column development, precipitation as the hydroxide can be expected to control the behavior of chromium.

$$Cr^{3+} + 3OH^{-} \rightleftharpoons Cr(OH)_3(s) \text{-----} pK_{s0}=31.0 \qquad (1)$$

Examination of the concentration versus time data for chromium in these columns (Figure 7) reveals behavior totally consistent with the expectation of hydroxide control for chromium. During the early portion of column operation the chromium concentrations in the leachates were subject to substantial variations and occasionally attained rather high values. By Day 500, chromium concentrations in all columns had decreased to less than one mg/L after which no significant chromium breakthrough was observed. The rapid decrease in chromium levels in the leachates significantly predated the onset of rapid sulfide formation at about Day 700 and can only be ascribed to precipitation by hydroxide. The transition of the columns to the methane phase of operation with its corresponding increase in pH will have had the effect of further decreasing chromium solubility in these systems.

Nickel, Cadmium and Zinc. These three elements are characterized by having very similar chemistries and their behavior will be examined together. Evidence of the similarity of their reactions in these columns is clearly seen from their concentration versus time plots (Figures 8-10). The metals showed some evidence of pseudo-conservative behavior during the acid phase of operations with fairly high and consistent leachate concentrations apparent in the recycle columns while the single pass columns displayed behavior reminiscent of wash-out.

Chemically these elements are quite similar in their behavior existing only in the +2 oxidation states in the column environments and subject to precipitation only by hydroxide at pH values in excess of ≈ 8 or by sulfide at substantially lower pH values. Since the concentrations observed during the acid phase were much lower than would have been expected in the pH range characteristic of that phase, it is tempting to invoke sulfide as the controlling factor. Unfortunately, no evidence was obtained for the formation of sulfide during this phase of column operation. It should be noted that the solubilities of the sulfides of these metals are so low (Table III.) that sulfide control was possible even at concentrations well below the detection limit of the electrode method used and at the pH conditions characteristic of the acid phase leachates.

In the absence of sulfide control, other mechanisms which might act to attenuate nickel, cadmium and zinc in these systems might include adsorption, ion exchange and precipitation in the alkaline microenvironments of the metal sludge solids.

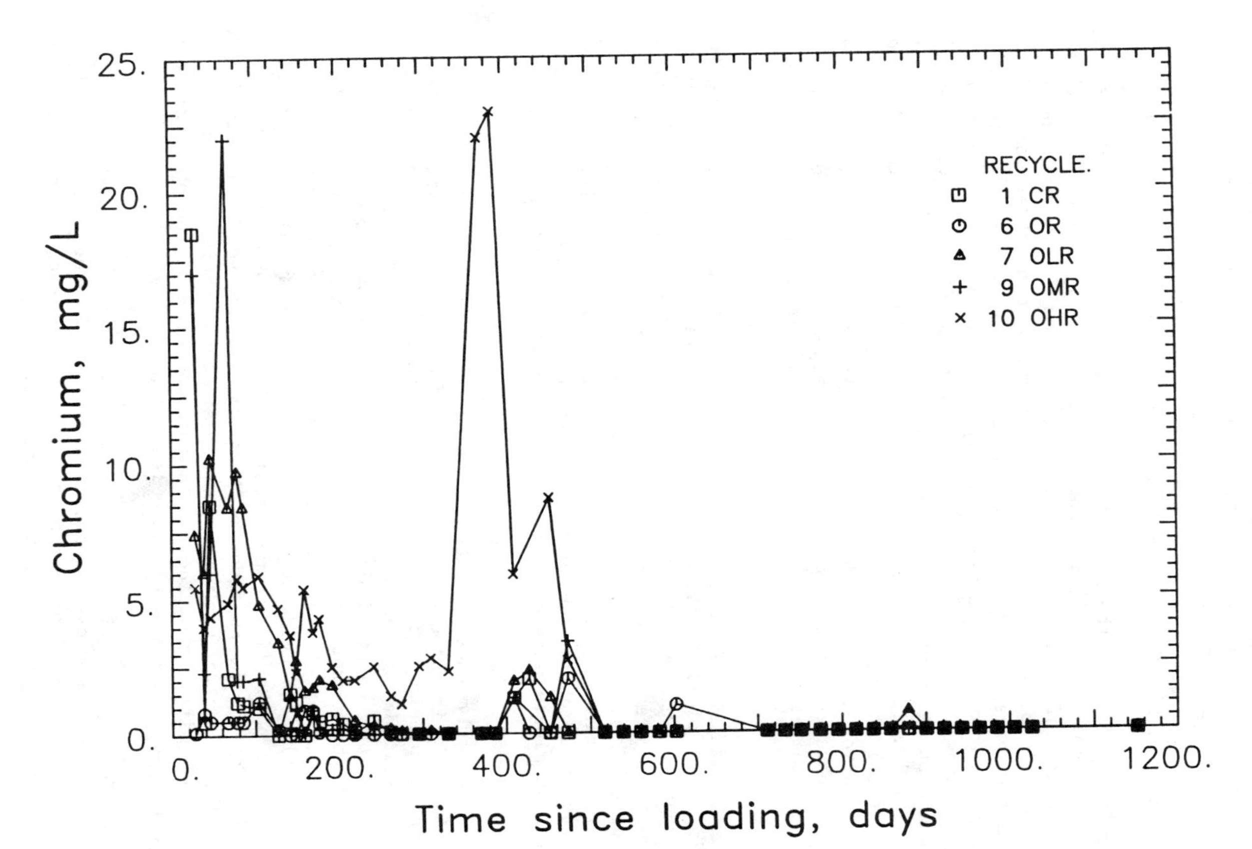

Figure 7. Leachate chromium concentration as a function of time.

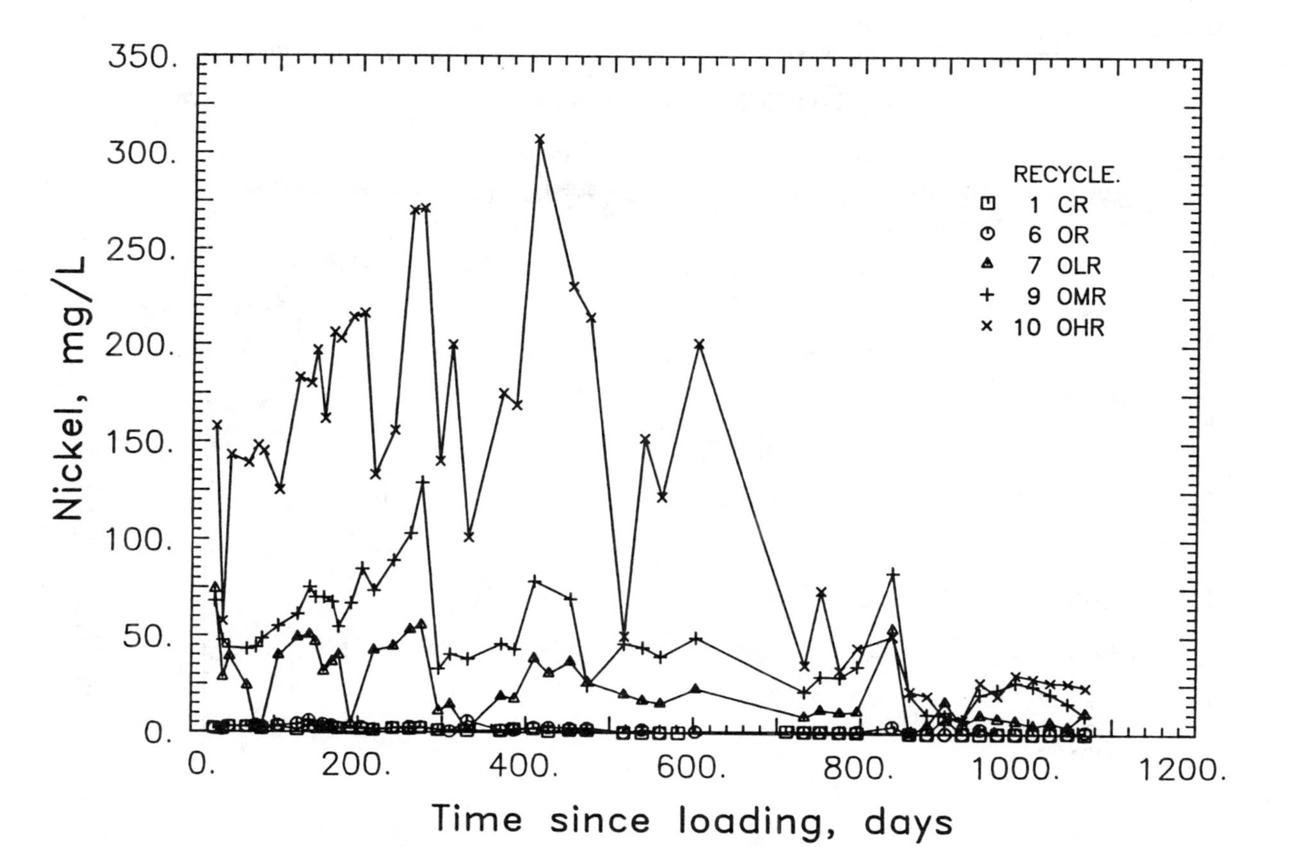

Figure 8. Leachate nickel concentration as a function of time.

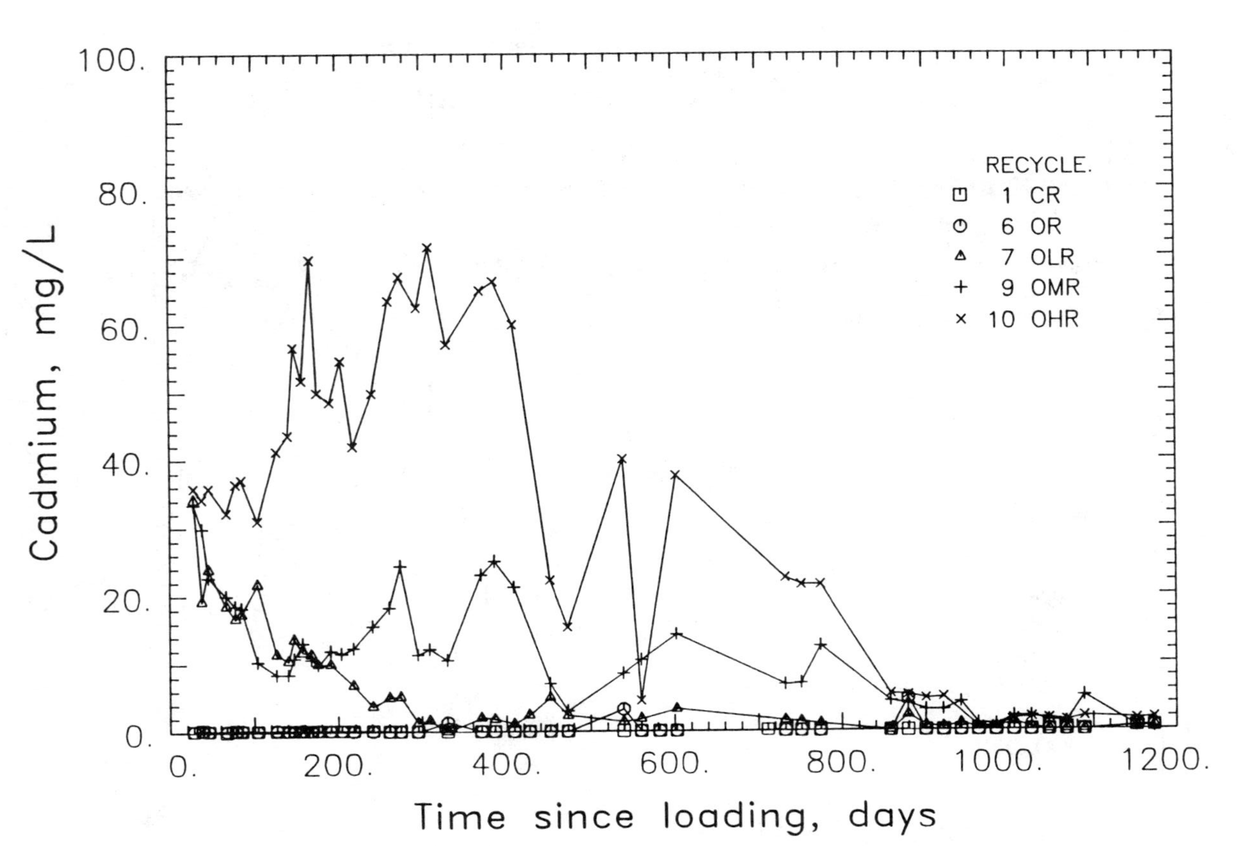

Figure 9. Leachate cadmium concentration as a function of time.

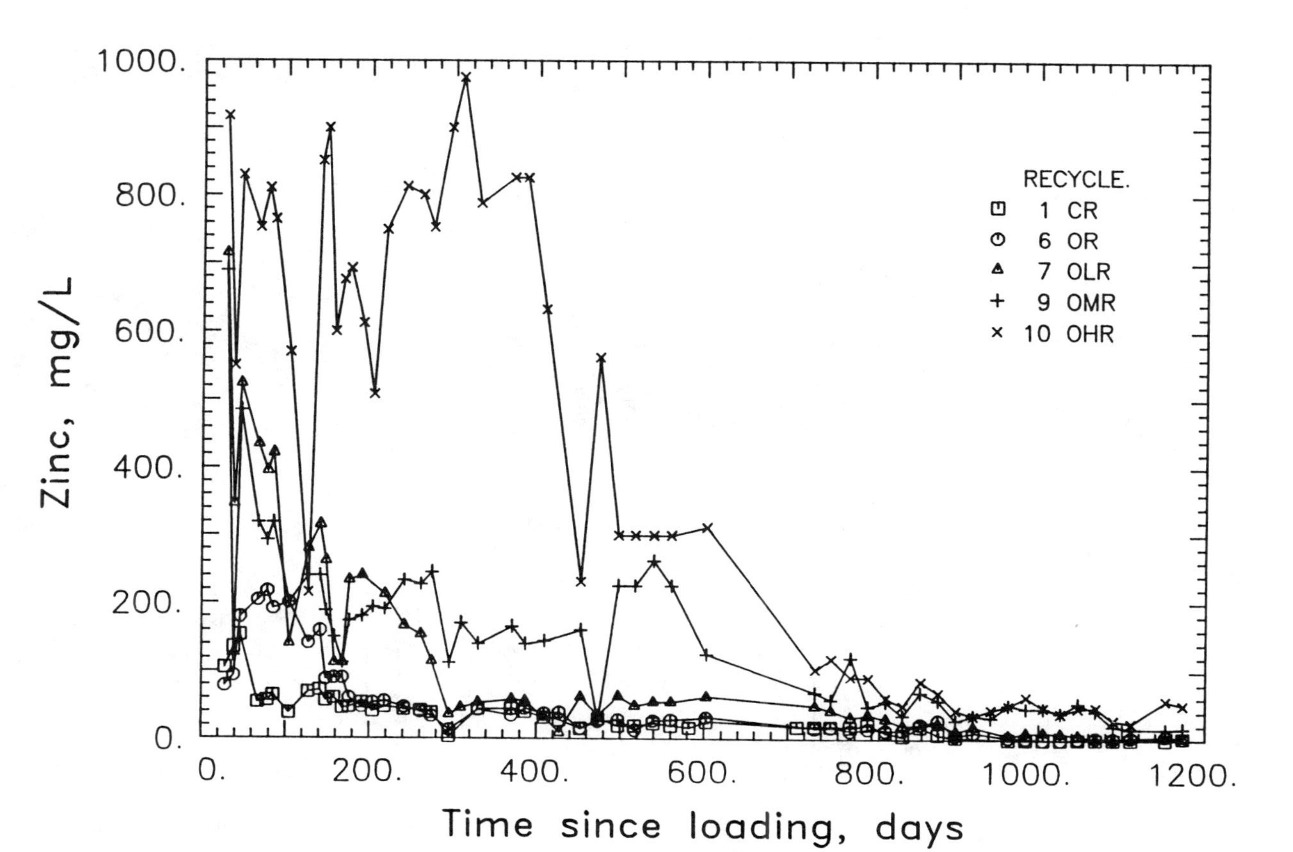

Figure 10. Leachate zinc concentration as a function of time.

Table III. Logs of Equilibrium Constants of Significance (2,3)

Solubility Products		Cumulative Formation Constants	
PbS	-26.6	$PbSO_4$	2.8
$PbSO_4$	-7.8	$Pb(SO_4)_2^{-2}$	3.6
$Pb(OH)_2$	-15.5	$PbCl^+$	1.3
$PbCl_2$	-4.7	$PbCl_2$	1.7
PbOHCl	-13.7	$PbCl_3^-$	1.9
$Pb_2Cl(OH)_3$	-36.1	$PbCl_4^{-2}$	1.5
HgS	-52.0	$HgCl_2$	13.9
$Hg(OH)_2$	-25.7	$HgCl_3$	14.8
		$HgCl_4^{-2}$	15.8
CdS	-27.2		
NiS	-23.8		
ZnS	-24.8		
Acidic Dissociation Constants			
H_2S	Log K_{a1} = -7.0		
	Log K_{a2} = -13.6		

The impact of the initiation of methane formation in these systems with the coincident onset of sulfate reduction/sulfide production is clearly evident (Figures 8-10). Beginning approximately at Day 650, the leachate concentrations of nickel, cadmium and zinc decreased rapidly. This decrease almost surely represented the effect of substantial precipitation of these metals as their sparingly soluble sulfide salts and possibly the impact of sludge particulate encapsulation by sulfide with consequent diminution of primary metal mobility.

Lead. The behavior of lead in these systems (Figure 11) is especially interesting in that it suggests that a transition from one chemical control mechanism to another can occur in response to the onset of methanogenesis in the refuse columns. Examination of Figure 11 indicates that the lead concentrations during the acid phase were on the order of 50 μM. If sulfide were to be assumed to by the primary control on lead solubility during this phase, the total sulfide concentration in the leachate ($C_{T,S}=[H_2S] + [HS^-] + [S^{2-}]$) would be on the order of 10^{-12}M to yield the observed lead concentration in the leachate at a pH of 5.5 based on lead solubility equilibria (Table III). This sulfide concentration is orders of magnitude lower than the limit of detection of the analytical method employed and thus cannot be documented in any reliable manner. Examination of other solubility and complex equilibria for lead in equilibrium with sulfate and chloride suggested that there were alternate possibilities for solubility control during the acid phase in these systems.

Figure 12 is a predominance area diagram for lead in equilibrium with chloride and sulfate and provides evidence that, in the absence of sulfide control, the major solid lead species to be expected in these systems was the sulfate. Based on complex equilibria (Table III), the distribution of soluble lead species in the leachate was computed and the total soluble lead concentration estimated (Table IV). Based on these computations, leachate lead concentrations on the order of 4 to 6 mg/L were predicted. These values are reasonably well in accordance with the observed lead concentrations. Again, the inability to measure sulfide at levels at which it would have been able to control lead solubility in these systems (Figure 13) dictates a degree of caution in the making of unequivocal statements regarding sulfate control of lead solubility during the acid phase. However, the results of this analysis are highly suggestive of the presence of sulfate control during this period.

The response of lead to the onset of active sulfate reduction/sulfide production was dramatic (Figure 11). By Day 750 in all columns, lead concentrations in the leachates had fallen below the analytical detection limit and have remained at this level since. There can be little question that, in the presence of measurable levels of sulfide even at pH levels below 6, lead will be essentially totally immobilized as the sulfide salt.

Mercury. The behavior of mercury in these refuse columns (Fig. 14) appears to have been dictated not primarily by solubility or complex equilibria (Table III) but by its unusual redox chemistry. Whether

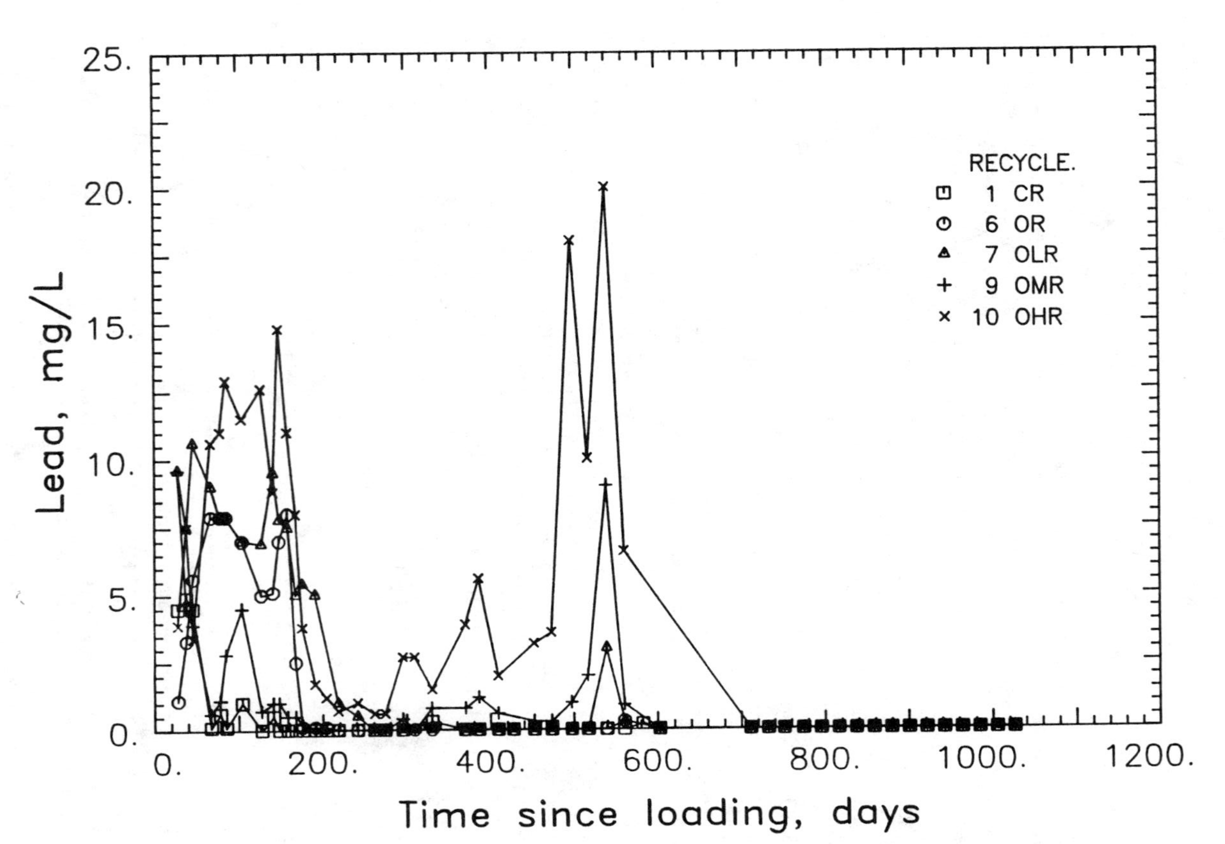

Figure 11. Leachate lead concentration as a function of time.

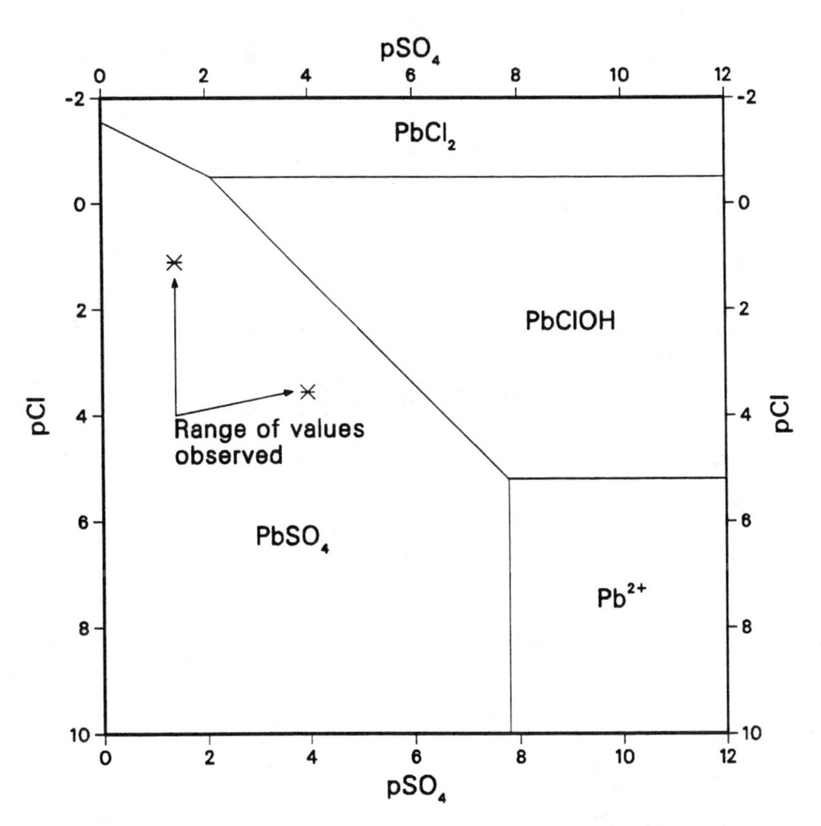

Figure 12. Predominance area diagram for the system $Pb^{2+}/SO_4^{2-}/Cl^-/OH^-$.

Table IV. Distribution of Lead Species in Typical Leachate*

Pb^{+2}	6%
$PbCl^+$	5%
$PbCl_2$	1%
$PbSO_4$	77%
$Pb(SO_4)_2^{-2}$	11%

$$[Pb]_T = [Pb^{+2}] + [PbCl^+] + [PbCl_2] + [PbSO_4] + [Pb(SO_4)_2^{-2}] = 16\ [P^{+2}]$$

*Chloride and sulfate at 1,500 mg/l

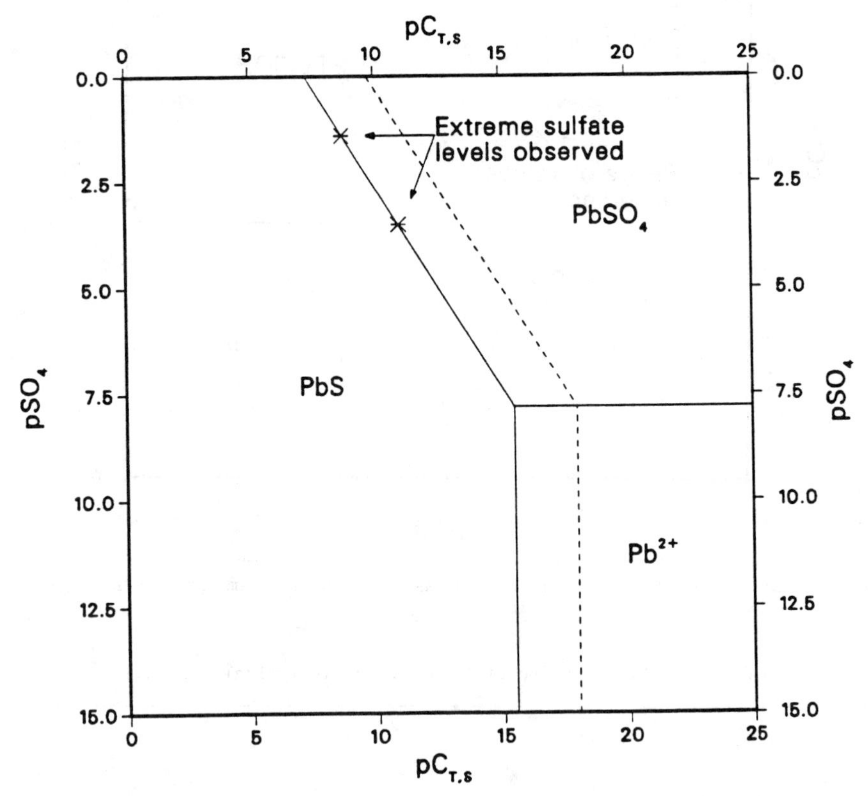

Figure 13. Predominance area diagram for the system $Pb^{2+}/S^{2-}/Cl^{-}/OH^{-}$.

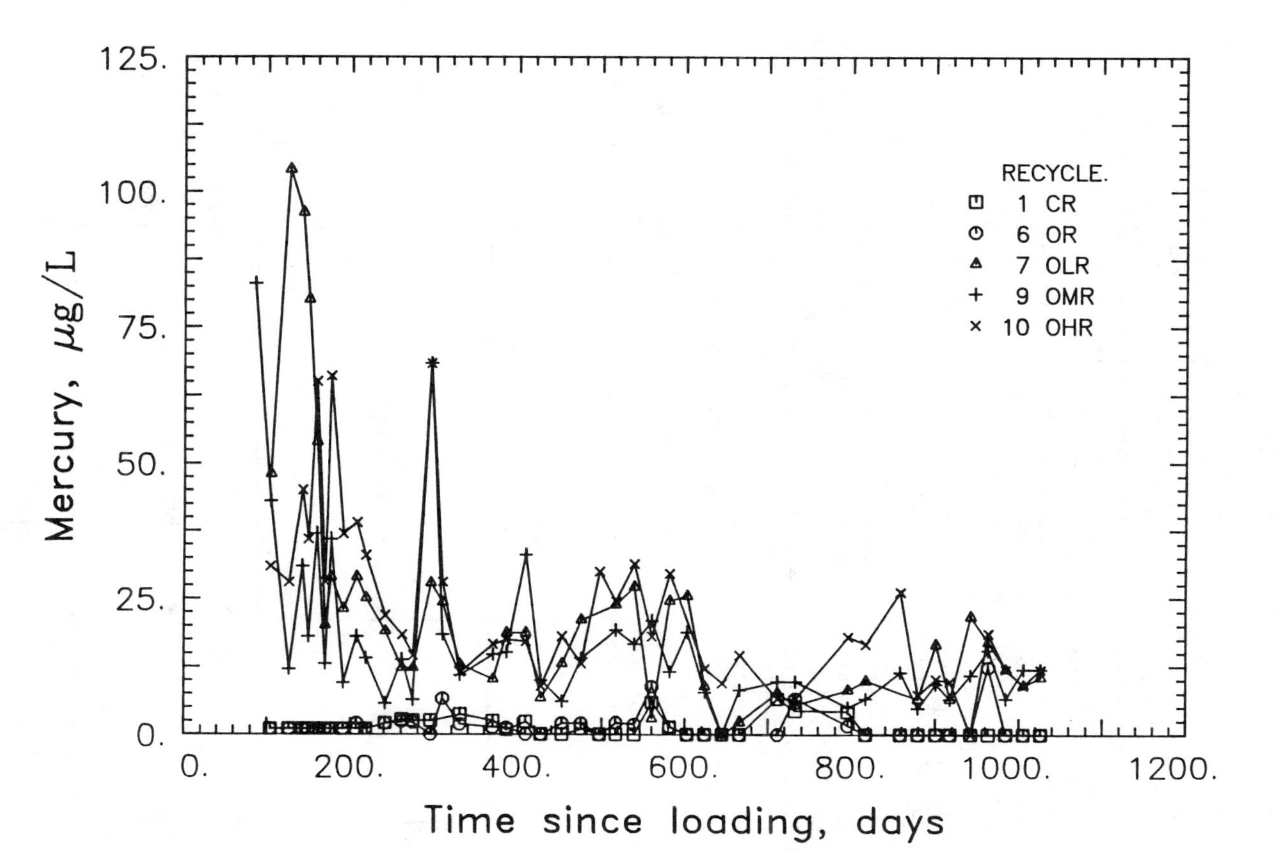

Figure 14. Leachate mercury concentration as a function of time.

present as chloromercuric complexes ($HgCl_2$, $HgCl_3^-$, $HgCl_4^{2-}$) or as solid mercuric sulfide, HgS, the reducing conditions in the columns render reduction of mercury to the metal highly likely.

Figure 15 is a pC-pε diagram for mercury with respect to chloride. As can be seen, column ORP values (Figure 3) are sufficiently reducing to permit reduction of the mercury to the metallic form. On examination of Figure 14, it can be seen that leachate mercury levels decreased rapidly early in column operation to values in the range of 10 to 50μg/L, a level which has been maintained essentially constantly since. This concentration is consistent with reported values for solubility of metallic mercury in water of 20-40μg/L (4). It is also noteworthy that the mercury concentrations have shown no response to the onset of sulfide generation in spite of the extremely low solubility of mercuric sulfide (Table III). It would appear that, at least to this point, sulfide has had no control with respect to the behavior of mercury in these columns.

The possibility that mercury is subject to reduction to the metal in landfills is significant in that it presents the possibility of transport from the columns by a vapor phase mechanism due to the high volatility of metallic mercury. Hughes (5) has cited this as an important mechanism for the transport of metallic mercury from landfills. Finally, it is interesting to note that the conditions in these columns which include low pH levels, reducing ORPs, high organic concentrations and substantial biological activity are precisely the conditions under which alkylation of mercury has been observed (5). While it has not been possible to detect alkylmercuric compounds to this point, it is reasonable to expect that they have been produced and

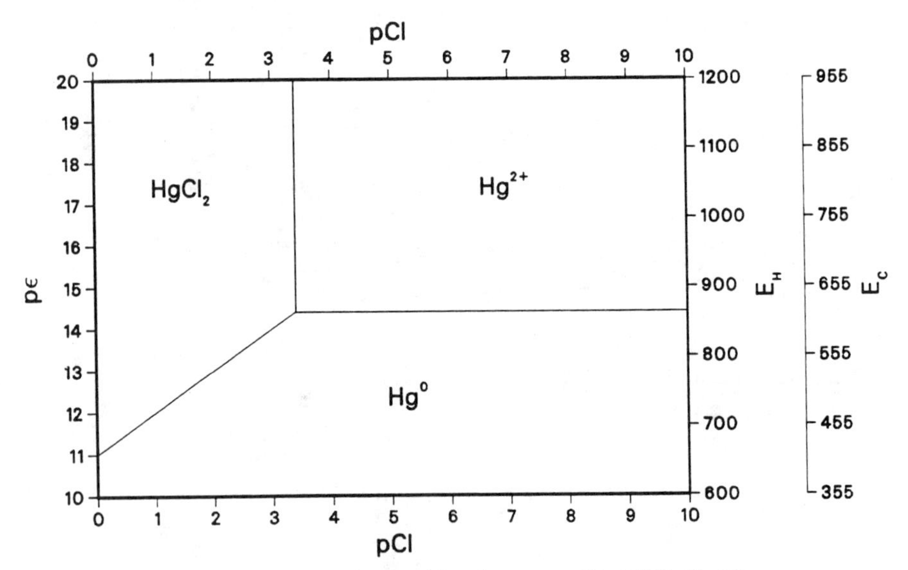

Figure 15. pCl-pε diagram for the system Hg°C/$HgCl_2$/Cl^-.

are present in these systems. The unusual toxicological properties of these compounds makes them of particular significance in assessing the health impact of codisposal of mercury containing wastes with MSW.

Conclusions

The behavior of an array of toxic metals codisposed with municipal solid wastes in simulated landfill columns was indicative of the presence of a broad range of attenuating mechanisms limiting the mobility of the metals in the leachates from the columns. Chemical mechanisms included direct reduction to the element (Hg), precipitation by reductively generated sulfide (Cd,Zn,Pb & Ni) and precipitation by sulfate in the case of lead during the acid phase. Acid phase attenuation of zinc, cadmium and nickel was suggestive of either the operation of physical sorptive mechanisms or the impact of localized regions of alkaline conditions within the columns associated with the codisposed metal sludges themselves. That all observed processes had the effect of reducing metal mobilities suggests strongly that municipal solid waste has a capacity for minimizing the mobility of toxic metals which might be used to great advantage in conjunction with a judicious program of codisposal of metal processing wastes residues with domestic refuse.

Acknowledgments

This research was supported by the U.S. Environmental Protection Agency under the terms of Cooperative Agreement CR-812158. The results and conclusions presented herein are not necessarily those of the U.S. Environmental Protection Agency.

Registry Numbers

Cadmium 7440-43-9; chromium 7440-47-3; lead 7439-92-1; mercury 7439-97-6; nickel 7440-02-0; zinc 7440-66-6.

Literature Cited

1. APHA.AWWA.WPCF (1985). Standard Methods for the Examination of Water and Wastewater. Sixteenth Ed. Amer. Public Health Assn, Washington, D.C.
2. Sillen, L.G. and Martell, A.E. (1964). Stability Constants of Metal-Ion Complexes. Special Publication No. 17 of the Chemical Society, London, 1,150 pp.
3. Sillen, L.G. and Martell, A.E. (1971). Stability Constants of Metal-Ion Complexes, Suppl. Special Publication No. 25 of the Chemical Society, London, 838 pp.
4. Friberg, L. and Vostal, J. (1972). Mercury in the Environment, Chapter 3. CRC Press, Cleveland, OH.
5. Hughes, W.L. (1957). A Physicochemical Rationale for the Biological Activity of Mercury and its Compounds. Ann. New York Acad. Sci., 66, 454.

RECEIVED January 12, 1989

Chapter 17

Sorptive Behavior of Selected Organic Pollutants Codisposed in a Municipal Landfill

Debra R. Reinhart[1,3], Joseph P. Gould[1], Wendall H. Cross[1], and Frederick G. Pohland[2]

[1]School of Civil Engineering, Georgia Institute of Technology, Atlanta, GA 30332
[2]Department of Civil Engineering, University of Pittsburgh, Pittsburgh, PA 15261–2294

The impact of organic compound characteristics, leachate components, and the solid waste matrix on the fate of selected organic compounds codisposed with municipal refuse was evaluated. Test compounds included dibromomethane, trichloroethene, 1,4-dichlorobenzene, 2-nitrophenol, nitrobenzene, 2,4-dichlorophenol, 1,2,4-trichlorobenzene, naphthalene, and gamma-hexachlorocyclohexane. Bench top tests were employed to characterize the sorption process. Equilibrium sorbed and solution concentrations for nine test compounds were linearly related, with the coefficient of linearity well correlated with the octanol/water partition coefficient. The sorption partition coefficients normalized to carbon content were comparable to coefficients measured for similar hydrophobic compounds on soils and sediments. The majority of refuse sorption of hydrophobic test compounds was found to occur on low energy surfaces such as lipids and plastics. The affinity of the refuse for organic compounds enhances the assimilative capacity of the landfill and affords opportunity for more complete degradation of these constituents.

[3]Current address: University of Central Florida, Box 25,000, Orlando, FL 32816

0097–6156/90/0422–0292$06.00/0

Hazardous materials are frequently disposed in municipal refuse landfills. The sources of these materials include household refuse, wastes from small quantity generators, and illegally disposed wastes. Hazardous materials are also found in older landfills which predate enactment of legislation regulating hazardous waste disposal. In addition, hazardous pollutants may be released within the landfill following the degradation of otherwise innocuous refuse. One of the consequences of hazardous material disposal can be the contamination of leachates and groundwater (1-3).

The landfill environment provides numerous pathways for transformation and transport of codisposed pollutants. Pollutant advection may occur as moisture (originally present in the landfill or subsequently added by precipitation, irrigation, or surface runoff) percolates through the landfill. The large surface area of the refuse matrix provides an opportunity for compound sorption and microbial attachment. Moreover, because of the extended reaction time available within a landfill, there is an opportunity for microbial acclimation to xenobiotic compounds. The predominately anaerobic state promotes reducing reactions, including, among others, reductive dehalogenation, reduction of substituent groups on aromatic compounds, and ring reduction. The diverse microbial consortia found in most landfills have the capability to degrade or transform a variety of complex organic compounds to simpler intermediates, humic-like substances, or ultimately to gaseous endproducts. Gas production, in turn, tends to transport volatile compounds out of the landfill.

Because of the complexity of the landfill environment, it is difficult to assess the separate impact of each attenuating mechanism, in situ. Laboratory studies were therefore conducted to facilitate the evaluation of individual transport and transformation phenomena. Presented in this paper, are results of equilibrium studies used to characterize the sorption process and elucidate those factors which influence the equilibrium distribution between liquid and solid phases.

Experimental Section

Glassware, Reagents, and Solvents. All glassware was washed and baked in an oven at 325 degrees C overnight. Organic solvents were HPLC or Optima grade and were used as received from Fisher Scientific. Organic-free water was prepared by passing distilled water through an ion exchange resin, activated carbon, and an ultraviolet unit. Chemicals used were high purity reagent grade trichloroethene (TCE), dibromomethane (DBM), 2-nitrophenol (NP), 1,2,4-trichlorobenzene (TCB), 2,4-dichlorophenol (DCP), hexachlorocyclohexane (lindane - LIN), and anthracene d-10 (Aldrich

Chemical Co.); 1,4-dichlorobenzene (DCB), nitrobenzene (NB), naphthalene (NAP), and anhydrous sodium sulfate (Fischer Scientific); and hexamethylbenzene (Pfaltz and Bauer, Inc.).

Extraction of Organic Compounds from Aqueous Samples. Sample cleanup and concentration methods for aqueous organic compound analysis followed U.S. Environmental Protection Agency procedures (4). A measured volume of sample (approximately 30 mL) was acidified to pH 2 with 1 + 1 sulfuric acid and placed in a separatory funnel. 100 microL of a 2-g/L solution of anthracene d-10, a surrogate standard spiking solution, were added. The sample was then sequentially extracted with three 30-mL volumes of methylene chloride. Extractions were combined in a 250-mL Erlenmeyer flask. Sodium sulfate, oven baked at 325 degrees C for four hours, was added to absorb water, contacted for approximately 10 min, then discarded.

Sample extracts were then transferred to a Kuderna-Danish evaporator consisting of a 10-mL concentrator, a 250-mL evaporative flask, and a 3-ball Snyder column. The apparatus was immersed in a 60 to 70 degree C water bath until the extract reached approximately 4 mL. The extract was concentrated to 1 mL by blowdown with dry nitrogen gas. The concentrated sample was placed in a 2-mL glass vial with a Teflon-lined crimp top and stored at 4 degrees C until analyzed.

10 microL of an internal standard (hexamethylbenzene, 2 g/L) were added to the 1-mL concentrated extract. A 3-microL sample was injected into a fused silica capillary column installed in a Hewlett Packard Gas Chromatograph, Model 5830A, equipped with an flame ionization detector. The capillary column was 25 m long, 0.23-mm inside diameter and coated with a 0.2-micron film of DB5-30N (J & W Scientific). Chromatographic conditions included an initial temperature of 140 degrees C held for 10 min and then increased to 240 degrees C at a rate of 10 degrees per min. Helium was used as a carrier gas at a flowrate of 30 mL/min. The injection of the sample was conducted in the splitless mode.

Refuse Analysis. Fresh municipal refuse was obtained from the Dekalb County Sanitation Department after shredding at the Buford Highway Pulverization Plant in Atlanta, GA. Refuse was dried at 103 degrees C, ground (geometric mean size, 2.0 mm), and stored at 4 degrees C in a sealed glass container until used. Refuse chemical characteristics are summarized in Table 1. Carbon, hydrogen, and nitrogen analyses were performed on an F & M CHN Analyzer, Model 185. Ash content was determined according to Standard Methods (5). Refuse density was determined by measuring the volume of organic-free water displaced by the addition of approximately 2 g of moisture-free refuse to 25 mL of organic-free water. Each determination was repeated 4 times.

Table I. Characterization of Refuse Material Used in Sorption Studies

Refuse Density, g/mL	0.69 ± 0.10
Ash, percent dry weight	31.3
Lipids, percent dry weight	4.6
Elemental Analysis (Ash-free basis)	
Nitrogen	2.2
Carbon	58.7
Hydrogen	8.6

Lipid content was determined according to the method described by Black (6). Preweighed samples of ground, dried refuse were placed in cellulose extraction thimbles. The thimbles were inserted into Sohxlet Extractors and the refuse extracted with 200 mL of ethyl ether for 24 hrs. The extract was transferred to a Kuderna-Danish apparatus and concentrated to about 20 mL in a 60 to 70 degree C water bath, then air dried overnight. The remaining material was then dried to a constant weight in an oven at 103 degrees C.

Isotherms for Volatile Compounds. A headspace bottlepoint technique was used to collect sorption data for TCE, DBM, DCB, and TCB. Preweighed aliquots of dried, ground refuse were placed in 160-mL serum bottles. Test solution was prepared by mixing organic-free water with the test compound for 48 hrs. Sodium azide was added at approximately 0.15 percent by weight to inhibit biological activity. The solution was buffered at a pH of 5.1 ± 0.1 by a 0.01 N acetate solution. 50 mL of test solution were added to each bottle. Each sorption reactor contained a different mass of sorbate and was prepared in triplicate. Serum bottles were agitated for 24 hrs at 23 degrees C. Following a quiescent period at 23 degrees C, 5 to 10 mL of headspace gas from the bottle headspace were removed using a gas tight glass syringe and analyzed for the test compounds using a Perkin-Elmer Sigma 1 Gas Chromatograph equipped with a 183-cm long, 2-mm inside diameter column packed with 80/100 Carbopack C coated with 0.1 percent SP-1000 (Supelco, Inc.). Using an experimentally determined Henry's Law Constant (the equilibrium ratio of gas and liquid concentrations), liquid concentration could be estimated from measured headspace concentrations (7). Control samples containing test solution and no sorbent were treated in an identical fashion to provide an initial sorbate concentration. From a mass balance, sorbed mass could be calculated.

Isotherms for Nonvolatile Compounds. Isotherms were developed using individual, completely mixed, batch reactors for each data point (bottlepoint technique) for DCP, NAP, LIN, NP, and NB. 130-mL French square bottles were loaded with preweighed aliquots of dried, ground refuse of varied amounts and filled with test solution prepared as above. Triplicate test samples and control samples without refuse were prepared. After agitation for 24 hrs at 23 degrees C, bottles were allowed to stand for 3 hrs and the liquid contents decanted. Supernatant was centrifuged for 20 min at 2000 RPM in a Beckman centrifuge, Model J2-21. The centrate was analyzed for test compounds and a mass balance calculation provided sorbed mass.

Desorption Tests. Desorption tests were performed in a fashion similar to the procedures described for nonvolatile compounds. Following sorption and decanting, the decanted sorbate solution was replaced with organic-free water and equilibrated again at 23 degrees C for 24 hrs. Each sorption reactor used for the desorption isotherm was desorbed sequentially twice. After decanting, supernatant was analyzed for the sorbate.

Time to Equilibrium Studies. Time to equilibrium studies were carried out in a 4 L completely mixed reactor. The reactor was sealed during the test to minimize volatilization. Selected test compounds were mixed with 3 L of organic-free water for 24 hrs. The solution was buffered to a pH of 5.1 with a 0.01 N acetate solution. Sodium azide at 0.15 percent by weight was added. Ten g of ground refuse were added to the test solution and continuously stirred as a slurry. Periodically, mixing was interrupted, the slurry was allowed to settle for 5 min, and an aliquot of supernatant was removed. The supernatant was centrifuged, extracted, and analyzed for test compounds. From time to equilibrium study results, sorption was observed to be substantially complete within 120 min.

Time to equilibrium studies for desorption were performed in a manner similar to the sorption equilibrium studies. 15 g of ground refuse were added to 3L of an equilibrated aqueous solution of LIN and mixed for 24 hrs. After a quiescent period, 1 L was decanted from the mixture, analyzed for LIN and replaced with 1 L of organic-free water. The solution was again mixed. Over the next 22 hrs, mixing was interrupted at preselected intervals and an aliquot of supernatant was removed and centrifuged. The centrate sample was then extracted, concentrated, and analyzed for LIN. Desorption was substantially complete within 6 to 8 hours.

Results and Discussion

Sorption Isotherms. Separate sorption isotherms were developed for each test compound. All isotherm data points confirmed strong linearity between sorbed and solute concentrations, as seen in Figure 1, where a typical sorption isotherm is presented. From a least squares regression analysis, the coefficient of linearity, or sorption partition coefficient, K_p (the equilibrium ratio between sorbed and solute concentration) for each compound was calculated, and is provided in Table 2 along with concentration ranges, number of data points, and correlation coefficients. Correlation coefficients ranged from 0.88 to 0.99. The headspace bottlepoint technique consistently produced linear isotherms with correlation coefficients of 0.99, while the conventional bottlepoint technique had lower correlation coefficients.

As recorded in Table 2, the partition coefficients covered a range of three orders of magnitude. Since all measurements were made using the same sorbent, at fixed solution pH, temperature, and ionic strength, variations in K_p can be attributed to differences in molecular characteristics of the test

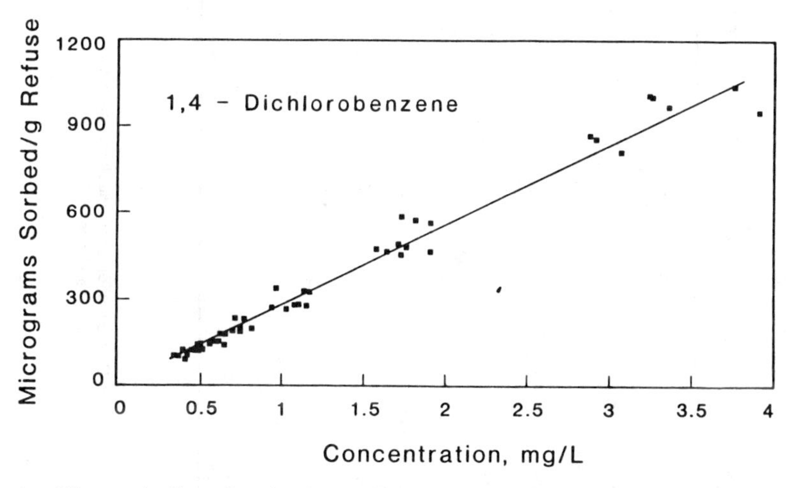

Figure 1. Sorption isotherm for 1,4-dichlorobenzene on refuse.

Table 2. Linear Regression Analysis of Refuse Sorption Isotherm Data for Test Organic Compounds

Compound	Partition Coefficient, K_p[a] mg/g	Regression Coefficient, r	Concentration Range mg/L	Number of Data Points
Dibromomethane	7.26	0.99	6.0 - 106	17
Trichloroethene	36.6	0.99	0.35 - 9.4	26
Nitrobenzene	17.4	0.97	9.7 - 203	13
2-Nitrophenol	29.5	0.94	6.0 - 177	18
1,4-Dichlorobenzene	285	0.99	0.33 - 3.83	62
2,4-Dichlorophenol	126	0.98	10.1 - 135	11
1,2,4-Trichlorobenzene	1350	0.99	0.015 - 0.180	11
Lindane	853	0.88	0.39 - 3.93	35
Naphthalene	216	0.91	0.06 - 0.76	19

[a] $K_p = \frac{S}{C}$ where: S = Sorption concentration, mg/g and C = Solute concentration, g/L.

compounds. The octanol/water partition coefficients, K_{ow}, the ratio of equilibrium concentrations of the solute in a two phase system of water and octanol (provided in Table 3), were found to be strongly correlated with the experimental partition coefficients, with the linear dependence described by Equation 1:

$$\log K_p = 0.81 \log K_{ow} - 0.21 \quad (1)$$

A correlation coefficient, r, equal to 0.98 was calculated. K_{ow} is often used as a measure of lipophilicity and has been found in other studies to be strongly correlated to sorption coefficients for soils and sediments (8-10).

Table 3. Refuse Sorption Coefficients and Octanol/Water Partition Coefficients for Test Organic Compounds

Compound	Log Partition Coefficient, log mL/g	Log Octanol/ Water Partition Coefficient
Dibromomethane	0.86	1.45
Trichloroethene	1.56	2.29
2-Nitrophenol	1.24	1.85
Nitrobenzene	1.47	1.76
1,4-Dichlorobenzene	2.45	3.38
2,4-Dichlorophenol	2.10	2.75
1,2,4-Trichlorobenzene	3.13	4.04
Lindane	2.93	3.72
Naphthalene	2.33	3.37

Many researchers (11-15) have found that the sorption of a single compound on a series of soils or sediments can be described by a coefficient, K_{oc}, derived by dividing K_p by the organic carbon fraction of the sorbent, f_{oc}. The successful use of K_{oc} implies that sorption of these compounds was largely a function of compound hydrophobicity and not of the nature of the organic matter itself.

K_p for each test compound sorbed on refuse was divided by the refuse carbon fraction (0.40). Plotted in Figure 2 is the line of best fit for K_{oc} versus K_{ow} along with the 95 percent confidence interval. Also shown are K_{oc} data from several investigators (8) for a large number of hydrocarbons, chlorinated hydrocarbons, chlorotriazines, and carbamates. Refuse data lie within the band of soil/sediment data confirming the significance of hydrophobic interactions between hydrophobic compounds and refuse, soil, and sediment organic surfaces.

The K_p for DCB was measured on two additional refuse materials. The first was obtained approximately two years after the refuse described above was obtained, from the same source, and processed in a similar manner. A K_p of 272 mL/g was measured, which is within the coefficient of variation for the K_p calculated for DCB replicate isotherms (285 mL/g $\pm$ 9.2 percent). The second material was stable refuse obtained from a 5-year old lysimeter which had been operated with leachate recycle until the methane fermentation phase had been completed (16). The K_p measured for this material was 176 mL/g, 62 percent of that measured on fresh refuse. Reduction in the sorption affinity of the stable refuse for DCB is probably due to the lower carbon content of the stable refuse (0.28) as compared to the fresh refuse (0.40). K_p normalized to carbon for DCB on fresh refuse was 705 mL/g and 631 mL/g on the stable refuse.

Sorption of DCB on refuse was examined in the presence of leachate obtained from an experimental landfill column operating in the acidogenic phase of landfill stabilization (characterized in Table 4). Leachate constituents depressed DCB sorption slightly, although the measured K_p of 258 mL/g was within the range of K_p values determined during replicate analysis of DCB sorption by refuse from an acetate buffer/azide solution. Since leachate components tend to be hydrophilic, having preferentially partitioned into the aqueous phase as water percolated through the refuse, these materials probably do not compete for hydrophobic sites on the refuse. However, leachate constituents may act to solubilize hydrophobic organic compounds resulting in slightly lower affinity for the refuse.

<u>Sorption on Refuse Components.</u> The affinity for sorption of three compounds (DCB, TCB, and NAP) by various refuse components (lipids, plastics, wood, garden materials, metal, fabric, and paper) was measured.

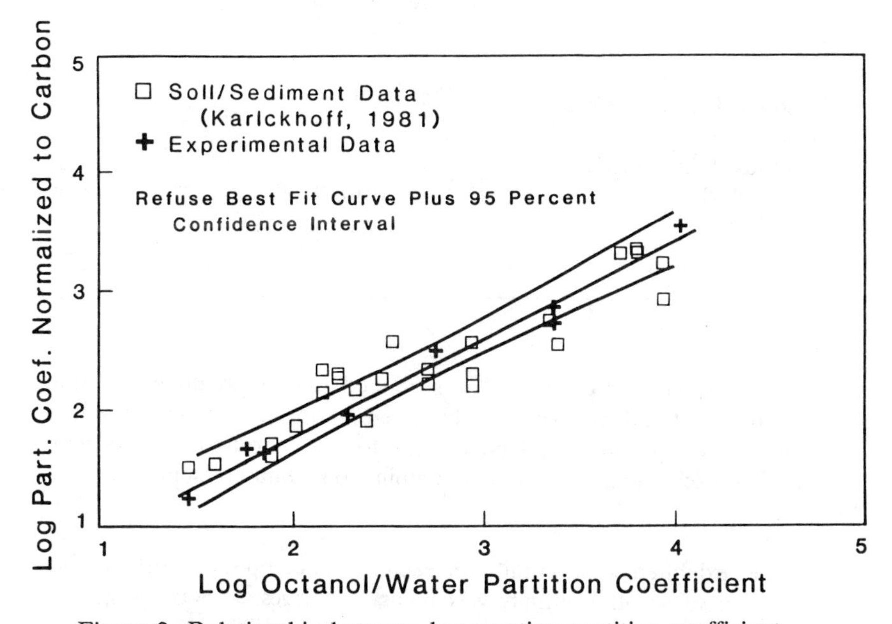

Figure 2. Relationship between log sorption partition coefficient normalized to carbon content and log octanol/water partition coefficient for hydrophophic compounds on refuse, soil, and sediment.

Table 4. Characterization of Leachate Used in Sorption Studies

Chemical Oxygen Demand, mg/L	22,500
Biochemical Oxygen Demand, mg/L	17,800
pH	5.1
Conductivity, mmhos/cm	9.4
Volatile Acids, mg/L as Acetic Acid	10,780

Individual materials were ground or sectioned into approximately 5-mm sized particles. Lipid material was obtained by the extraction procedure described in the Experimental Section. One to three sorption reactors were prepared for each compound/sorbent combination. Data are summarized in Table 5.

All surfaces exhibited some affinity for each of the compounds tested. The general order of sorption affinity was lipids > plastics > wood = garden materials > paper > fabric > metal. Jones, et al. (17) measured similar affinities for trichloroethene on several refuse components including polyethylene, paper, and a textile material. Trichloroethene and toluene were poorly sorbed on cellulose and oxidized humins, while lipids and lignins were strong sorbents, according to an investigation by Gabarini and Lion (18).

Figure 3 shows partition coefficients calculated for each refuse component relative to the partition coefficient for the total refuse. Refuse components are plotted in order of increasing surface energy. Surface free energy can be defined in terms of molecular forces interacting at solid/polar liquid

Table 5. Determination of Sorption Capacity of Various Refuse Components for Dichlorobenzene, Trichlorobenzene and Naphthalene

Material	Typical Percent of Refuse by Weight[a]	1,4-Dichlorobenzene		1,2,4-Trichlorobenzene		Naphthalene	
		Partition Coefficient, L/mg	Concentration, mg/L	Partition Coefficient, L/mg	Concentration, mg/L	Partition Coefficient, L/mg	Concentration, mg/L
Plastic	3	115	0.33	735	0.42	17	0.61
Garden Materials	12	78	0.38	694	0.43	45	0.56
Wood	2	70	0.39	260	0.80	46	0.10
Paper	40	31	0.44	120	1.10	32	0.65
Tin Can	6	20	0.45	39	1.39	2	0.65
Fabric	2	37	0.47	17	1.50	21	0.61
Soil	4	129	0.55	113	0.28	27	0.14
Ethyl Ether Extractable	4.6[b]	2144	0.84	----	----	--	----
Total Refuse	---	186	0.29	1070	0.32	57	0.52

[a]From Tchobanoglous, et al. (1977)
[b]Extraction of municipal refuse.

interfaces. Liquid molecules located next to a low energy solid are attracted more strongly into the bulk of the liquid than by the solid. Tension exists in the interface and the solid is poorly wetted. Liquid molecules are more strongly attracted by a high energy solid than by other liquid molecules, resulting in effective wetting of the solid. Highly hydrogenated surfaces such as fats, oils, and waxes tend to have low surface energies and are poorly wetted. As the degree of chlorination, oxygenation, and nitrogenation increases, wettability increases (19). Sorption affinity of hydrophobic compounds appeared to be inversely related to the surface energy of the solid perhaps because water at the interface of a low energy surface can by more easily displaced by a hydrophobic compound than at a high energy surface. Therefore, it is likely that the majority of the sorption of the hydrophobic compounds by refuse occurred at low energy surfaces such as fats, oils, waxes, microbial cell walls, hydrocarbon chains on humic materials, lignin, plastics, and leather.

Desorption. The reversibility of sorption may have significant importance to the environmental fate of organic compounds. Nonlabile fractions of sorbed compound may be immobile and unavailable for microbial degradation, and furthermore, slow desorption may lead to prolonged presence of compounds in the aqueous phase at low concentrations. Therefore, a desorption isotherm was developed. The LIN desorption data points are presented in Figure 4 along with sorption isotherm data. According to a Student t test (20), sorption and desorption data are statistically equivalent at the 95 percent confident level. Sorption is assumed, therefore, to be reversible and single valued.

Desorption of eight compounds from two refuse samples was evaluated 120 days after placing the organic compounds on the refuse. Samples were extracted with methylene chloride to measure total compound loadings, summarized in Table 6. Preweighed aliquots were then placed in French square bottles and mixed with organic-free water buffered with an acetate/azide solution to a pH of 5.1. After 24 hours of agitation at 23 degrees C, samples were decanted, the supernatant was centrifuged, and the centrate was extracted and analyzed for the organic test compounds. From a mass balance, sorbed concentrations were calculated. Desorption data are summarized in Table 6.

Long-term desorption tests, unlike the LIN desorption isotherm, revealed desorption hysteresis in almost every case. With the exception of DCP, K_p for desorption tests was significantly greater than for sorption. Of the eight compounds tested, DCP had the highest aqueous solubility (4500 mg/L as compared to 50 mg/L or less for the remaining compounds) and the lowest affinity for the solid phase.

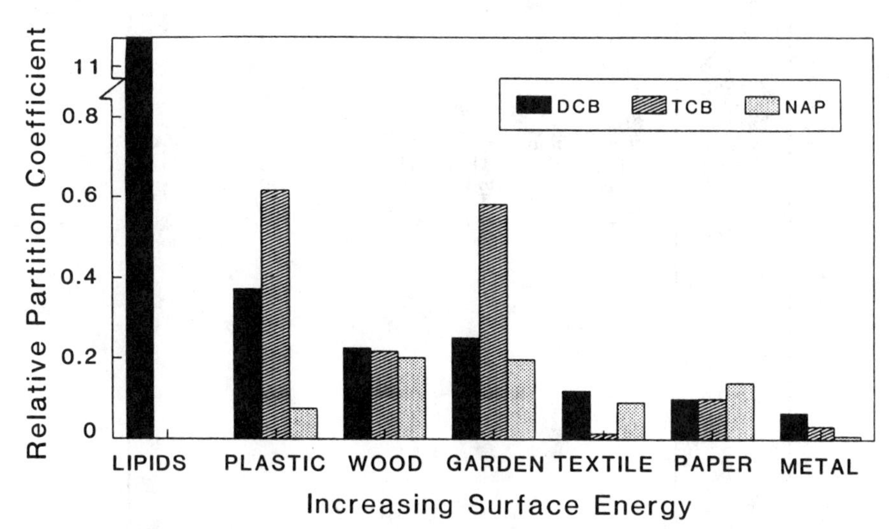

Figure 3. Sorption of 1,4-dichlorobenzene (DCB), 1,2,4-trichlorobenzene (TCB), and naphthalene (NAP) by various components of municipal refuse.

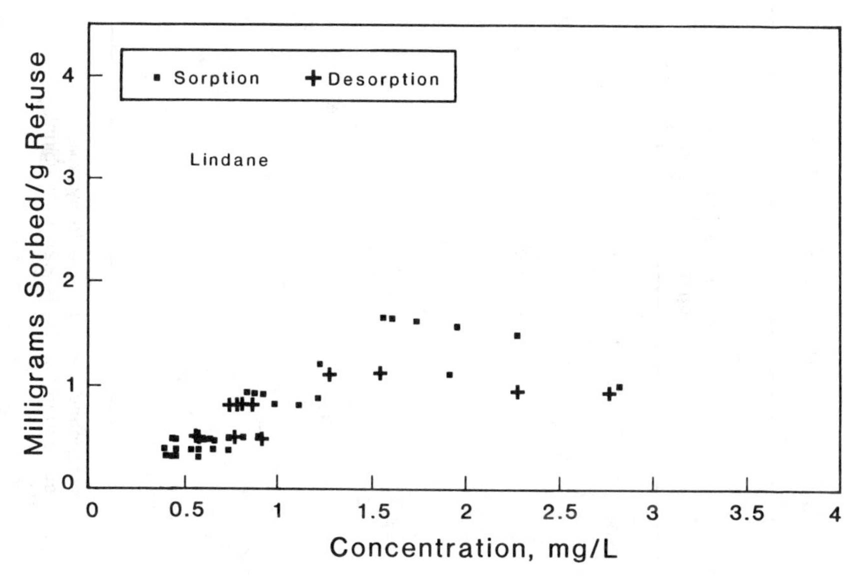

Figure 4. Presentation of sorption and desorption isotherm data for lindane on refuse.

Table 6. Long-Term Refuse Desorption Study after 120 Days of Sorption

Compound	Compound Loading, mg/g	Partition Coefficient from Isotherm, mL/g	Partition Coefficient Desorption after 120d, mL/g ± Standard Deviation
Dichlorobenzene	5.44[a] 3.24[b]	285	1762 ± 456 1065 ± 521
Dichlorophenol	0.12[a] 0.57[b]	126	142 ± 98 108 ± 61
Trichlorobenzene	8.77[a] 8.64[b]	1350	11648 ± 3217 7157 ± 3909
Lindane	7.67[a] 1.90[b]	853	3447 ± 436 3599 ± 885
Naphthalene	5.44[a] 5.80[b]	216	1129 ± 291 655 ± 190
Hexachlorobenzene	10.10[a]	---	23,986
Dieldrin	3.87[a]	---	8,160
Bis(2-ethylhexyl)phthalate	22.70[a]	---	17,955

[a]Total spiked organic compound loading is 6.41% by weight.
[b]Total spiked organic compound loading is 2.28% by weight.

For refuse Sample 1, K_p for desorption for each compound except DCP, was 4 to 8 times greater than for sorption. For refuse Sample 2, K_p for desorption was 3 to 5 times greater than K_p for sorption. This ratio increased with increasing K_p (or affinity for the refuse). K_p for Sample 1 for desorption exceeded Sample 2 for four of the five compounds where sufficient data were available for a comparison (LIN is the exception). Sample 1 had a test organic compound loading of 6.4 g/100 g while Sample 2 had a loading of only 2.3 g/100 g.

From long term desorption test results, desorption hysteresis appeared to increase with the length of sorbed time, compound hydrophobicity, and the total quantity of sorbed hydrophobic compounds. Hysteresis may be a product of one or more phenomena. Sorbed concentration on refuse column samples were greater than those of isotherm tests. Desorption tests may have been conducted in a region of nonlinearity. Alternatively, a time-dependent second sorption step may have existed where the sorbed compound diffused deeper into the solid phase. Desorption then required the compound to diffuse back out of the preferred hydrophobic environment into the aqueous phase. There is strong support in the literature for the existence of a nonlabile fraction which increases with time of sorption on soils and sediments (21-23).

Since the combined loading of the test compounds on refuse samples actually increased organic carbon content by 2.3 to 6.4 percent, the nature of the sorbent may have been altered. The refuse may have become more attractive to the hydrophobic compounds, resulting in a change in the equilibrium position of the solid/aqueous phase partitioning in favor of the solid phase. Such behavior was also noted by Jackson, et al. (24) where partition coefficients for dioxin on soils were positively correlated to the quantity of previously sorbed nonpolar organic compounds.

Conclusions

The organic rich landfill solid phase showed strong affinity for relatively hydrophobic organic compounds, resulting in low aqueous phase concentrations. Such retention and consequential delayed migration enhance the in situ opportunity and assimilative capacity of a landfill and afford greater opportunity for more complete biological and chemical degradation of the otherwise recalcitrant constituents. In addition, the following could be concluded:

(1) Based on an examination of the sorptive affinity of various refuse materials for test compounds, an increasing affinity was associated with decreasing solid phase surface energy and increasing surface wettability. Therefore, the majority of refuse sorption would be expected to occur on low energy organic surfaces such as fats, oils, waxes, microbial cell walls, hydrocarbon side chains of humic-like substances, lignin, plastic, and leather.

(2) Since the organic content of the refuse largely determines sorptive affinity for a given organic compound, as the landfill stabilization process proceeds and the organic fraction declines, some reduction in affinity is expected.

(3) Since sorption of hydrophobic compounds on refuse occurs rapidly due to a physical partitioning, and desorption is slower, leaching of hydrophobic compounds may be kinetically limited, particularly in the context of liquid flow time in a porous refuse medium.

(4) Since a strong linear correlation was observed between the sorption partition coefficient and the octanol/water partition coefficient, a molecular descriptor of compound hydrophobicity, it is possible to predict the approximate affinity of refuse for a variety of organic compounds.

Registry Numbers: Dibromomethane, 74-95-3; trichloroethene, 79-01-6; 1,4-dichlorobenzene, 106-46-7; 1,2,4-trichlorobenzene, 120-82-1; 2-nitrophenol, 88-75-5; 2,4-dichlorophenol, 120-83-2, gamma-hexachlorocycloyhexane, 58-89-9; nitrobenzene, 98-95-3; naphthalene, 91-20-3.

Acknowledgments

Funding for this study was provided by the U. S. Environmental Protection Agency, EPA Cooperative Agreement CR-812158-01.

Literature Cited

1. Brown, K. W. and K. C. Donnelly, "An Estimation of the Risk Associated with the Organic Constituents of Hazardous and Municipal Waste Landfill Leachates," Journal of Hazardous Wastes and Hazardous Materials, 5(1), 3 (1988).
2. Schultz, B. and P. Kjeldsen, "Screening of Organic Matter in Leachate from Sanitary Landfills Using Gas Chromatography Combined with Mass Spectrometry," Water Research, 20(8), 965 (1986).
3. Chian, E. S. K. and F. B. DeWalle, "Characterization of Soluble Organic Matter in Leachate," Environmental Science and Technology, 11(2), 159 (1977).
4. U.S. Environmental Protection Agency, "Addendum to Sampling and Analysis Procedures for Screening of Industrial Effluents for Priority Pollutants," EMSL, Cincinnati, OH (1979).
5. Standard Methods for the Examination of Water and Wastewater, 16th Edition, American Public Health Association, Washington, D.C. (1985).
6. Black, C. A., Methods of Soil Analysis Part 2: Chemical and Microbiological Properties, American Society of Agronomy, Madison, WI (1965).

7. Reinhart, D. R., "Fate of Selected Organic Pollutants during Landfill Codisposal with Municipal Refuse," Ph.D. Thesis, Georgia Institute of Technology, Atlanta, GA (1989).
8. Karickhoff, S. W., "Semi-Empirical Estimation of Sorption of Hydrophobic Pollutants on Natural Sediments and Soils," Chemosphere, 10(8), 833 (1981).
9. Schwarzenbach, R. P. and J. Westall, "Transport of Nonpolar Organic Compounds from Surface Water to Groundwater - Laboratory Sorption Studies," Environmental Science and Technology, 15(8), 1360 (1981).
10. Abdul, A. S. and Thomas L. Gibson, "Equilibrium Batch Experiments with Six Polyaromatic Hydrocarbons and Two Aquifer Materials," Journal of Haqardouw Waste and Hazardous Materials, 3(2), 125 (1986).
11. Lambert, S. M., "Functional Relationship Between Sorption in Soil and Chemical Structure," Journal of Agriculture and Food Chemistry, 15(4), 52 (1967).
12. Karickhoff, S. W., et al., Sorption of Hydrophobic Pollution on Natural Sediments," Water Research, 13(4), 241 (1979).
13. Means, J. C., et al., "Sorption of Polycyclic Aromatic Hydrocarbons by Sediments and Soils," Environmental Science and Technology, 12(12), 1524, (1982).
14. Chiou, C. T. et al., "A Physical Concept of Soil-Water Equilibrium for Nonionic Organic Compounds," Science, 206, 831 (1979).
15. Briggs, G. C., "Theoretical and Experimental Relationships Between Soil Adsorption, Octanol-Water Partition Coefficients, Water Solubility, Bioconcentration Factor, and the Parachor," Journal of Agricultural Food Chemistry, 29(5), 1050 (1981).
16. Ghosh, S. B., "Factors Affecting the Mobility of Selected Radionuclides Codisposed with Municipal Refuse within Landfills," M.S.C.E. Thesis, Georgia Institute of Technology, Atlanta, GA (1987).
17. Jones, C. J., et al., "Adsorption of Some Toxic Substances by Waste Components," Journal of Hazardous Materials, 2, 219 (1977).
18. Gabarini, D. R. and L. W. Lion, "Influence of the Nature of Soil Organics on Sorption of Toluene and Trichloroethene," Environmental Science and Technology, 20(12), 1263 (1986).
19. Shafrin, E. G. and W. A. Zisman, "Constitutive Relations in the Wetting of Low Energy Surfaces and the Theory of the Retraction Method of Preparing Monolayers," Journal of Colloid Chemistry, 64, 519 (1960).
20. Zar, J. H., Biostatistical Analysis, Prentice-Hall, Inc., Englewood Cliffs, NJ (1974).
21. Karickhoff, S. W., "Organic Pollutant Sorption in Aquatic Systems," Journal of Hydraulic Engineering, 110(6), 707 (1984).

22. Pignatello, J. J., "Slow Desorption of Halogenated Aliphatic Hydrocarbons in Soils," Presented before the Division of Environmental Chemistry, American Chemistry Society, Los Angeles, CA (1988).
23. Oliver, B. G., "Desorption of Chlorinated Hydrocarbons from Spiked and Anthropogenically Contaminated Sediments," Chemosphere, 14(8), 1087 (1985).
24. Jackson, D. R., et al., "Leaching Potential of 2,3,7,8-TCDD in Contaminated Soils," Proceedings of 11th Annual Research Symposium on Land Disposal of Hazardous Waste, April 1985.

RECEIVED November 10, 1989

Chapter 18

Recovery of Organic Compounds from Sewage Sludge by Proton Transfer

Mamadou S. Diallo[1,3], James H. Johnson, Jr.[1,4], Ramesh Chawla[1], Joseph N. Cannon[1], and Frank E. Senftle[2]

[1]School of Engineering, Howard University, Washington, DC 20059
[2]Physics Laboratory, U.S. Geological Survey, Reston, VA 22092

This paper examines the feasibility of a new process that recovers valuable organic materials from sewage sludge. Sludge organic compounds, which have unshared electron pairs or which are unsaturated, protonate and dissolve in strong acid media such as concentrated sulfuric acid. These dissolved compounds can be precipitated and recovered unchanged from the concentrated acid solution by diluting the solution containing the dissolved organic compounds with water.

Experimental results show that up to 75% of the solids in sludge can be dissolved during protonation and up to 25% of the initial solids can be precipitated upon dilution with water. The recovered product has a high organic content (90-98%) and represents approximately 30% of the organic compounds in the initial sludge. NMR spectroscopic analysis of an extract of the recovered product shows it to be primarily composed of aromatic organic compounds.

This study suggests that this new process has the potential to produce a high organic content product that could be used as a fuel or a source of organic chemicals. However, more research is needed to turn this process into a viable sludge reuse technique.

[3]Current address: Environmental and Water Resources Engineering, University of Michigan, Ann Arbor, MI 48109–2125
[4]Address correspondence to this author.

0097–6156/90/0422–0311$06.00/0

The disposal of sewage sludge is becoming a major problem for municipalities in the United States and throughout the world. In 1984, cities in the United States generated 6.5 million dry tons of sludge (1). The production of sludge is expected to more than double by the year 2000 to an estimated 15 million dry tons as the urban population increases and as more municipalities comply with the Clean Water Act requirements (1). The conventional techniques for disposal or reuse of sewage sludge are land application, landfilling, incineration and ocean disposal (1). Because these techniques are rarely cost effective and are becoming more and more difficult to implement due to public concern over their long-term impacts on the environment, there is an urgent need for cost effective and environmentally acceptable sludge management alternatives. The necessity for a viable solution to the problem has triggered a number of studies aimed at developing new sludge reuse technologies that can recover valuable products from municipal sludge (2-7). These resource recovery studies have been primarily concerned with the conversion of sludge to a fuel oil (2-5) and the extraction of precious metals from incinerated sludge ash (6, 7).

Bayer and Kutubuddin (2) and Bridle and Campbell (3) were able to obtain a high heating value oil (21-34 MJ/kg) by heating dry sludge to 300-350°C in a reducing environment for about 30 minutes. Molton and Fassbender (4) and Molton et al. (5) demonstrated the feasibility of a continuous unit for the conversion of primary sewage sludge to a liquid fuel and reported a 73% recovery of energy in a Sludge-to-Oil-Reactor System (STORS). Although the conversion of sewage sludge to a fuel oil has been shown in the laboratory, the effectiveness of fuel production as a viable sludge reuse technique is yet to be established.

The extraction of valuable metals from sewage sludge is another resource recovery technique that has received considerable attention. Fassel (6) and Gabler (7) investigated the recovery of metals such as aluminum, chromium, copper, silver, lead and tin from incinerated sludge ash and were able to develop extraction schemes that could recover most of these metals. Recovery of precious metals is currently being used as an alternative reuse technique by the cities of Palo Alto, California and Rochester, New York which recover gold and silver, respectively, from their incinerated sludge ash. (Farell, J., Municipal Environmental Research Laboratory, U.S. EPA, Cincinnati, OH, Private Communication, 1988). Because the current extraction schemes are only suitable for the recovery of metals from incinerated sludge ash (7), recovery of precious metals from sludge offers only a partial answer to sludge reuse, that is, it does not address the recovery of sludge organic compounds which comprise 40 to 80% of the solids in sewage sludge (8).

This paper describes a new sludge reuse process that addresses the recovery of both its organic and inorganic fractions. This process is based on the dissolution of organic compounds in concentrated sulfuric acid solutions and their precipitation and recovery from the acid solutions upon dilution with water. Preliminary work showing the possibility of using this process to recover organic compounds from sewage sludge was conducted by Hackney (9) and Senftle et al. (10). This paper reports the results of an experimental study for the preliminary assessment of a new sludge reuse technique based on this process.

Theoretical Background and Conceptual Process Flowsheet

It has long been known that a wide variety of organic compounds, with the exception of saturated hydrocarbons and some aromatic hydrocarbons and their halogenated derivatives, can be protonated at temperatures from 0 to 35°C and thereby rendered soluble in concentrated sulfuric acid solutions (11-15). Protonation is a proton transfer reaction between an organic base and an acid molecule or ion that is driven by the activity of the proton in strong acid media. Proton transfer reactions are easy to promote and unlike many chemical reactions, bonding by proton transfer only involves 1s orbitals and does not necessitate a major reorganization of electronic valence shells (12, 15). Because of this, protonated compounds do not undergo the major structural changes that occur in many chemical reactions and thus can be recovered almost unchanged in most cases (11, 13).

Recovery of protonated species is brought about by simply diluting the solution containing the protonated species with water (11, 13). Dilution causes a decrease in the proton activity, that is, the strength of the proton transfer bond decreases thereby causing deprotonation and precipitation of the previously protonated (dissolved) species. Because protonation is a pre-equilibrium step for many acid catalyzed reactions, organic compounds can also undergo more complex reactions in concentrated sulfuric acid solutions such as hydrolysis, isomerization, decarboxylation, electrophilic substitutions, etc. (13). Reaction conditions (acid strength, temperature, reaction time, dilution ratio, etc.) can, however, be selected to mainly promote the protonation step as opposed to other types of reactions, and therefore allow maximum recovery of the organic compounds dissolved in the concentrated acid solution.

A conceptual flowsheet of a sludge reuse plant based on the dissolution and recovery processes described above is shown in Figure 1. In this version of the process, wet sludge (20%) is fed to a protonation reactor (mixer) where it is continuously mixed with the protonating solvent, in this case, a concentrated sulfuric acid solution. The acid slurry from the protonation reactor is pumped to a centrifuge where the undissolved solid residue is separated from the acid solution containing the dissolved organic compounds and subsequently stored. The acid slurry is then pumped to a deprotonation reactor (mixer) where the dissolved organic compounds are precipitated upon mixing with water. The acid slurry containing the precipitated organic compounds is sent to a settling tank. The overflow from the tank is sent to an acid recovery system, whereas the underflow is pumped to a second centrifuge where the organic compounds are recovered from the acid slurry containing the precipitated organic compounds. The supernatant liquid from the centrifuge is pumped to the acid recovery system and the centrifuge cake is sent to a dryer. The dried product is recovered and stored. The process shown in Figure 1 only utilizes standard unit operations equipment provided with protection against corrosion. Except for the acid recovery system, which is not fully described at the present time, all process equipment shown in Figure 1 can be purchased from any major supplier of process equipment.

Experimental Protocol

A series of dissolution and recovery experiments were conducted in this study. The main objectives of these experiments were to establish optimum operating conditions (reaction time, acid strength and reactant ratio) and to determine the fractions of dissolved solids and organic materials recovered for a preliminary assessment of the feasibility of the process shown in Figure 1.

Materials. An anaerobically digested sludge (centrifuge cake) from the Blue Plains Wastewater Treatment Plant in Washington, D.C. was used in this study. A sludge sample of 80% moisture and 75% (per dry weight) organic content was used in the dissolution experiments. The moisture content of this sample was determined by measuring the loss of weight of a 5g aliquot sample dried for 24 hours in an air circulation oven, whereas its organic matter content was determined by measuring the weight loss of an identical sample which was dried and ignited for 65 minutes at 550°C in a muffle furnace. This method of measuring the organic matter content of

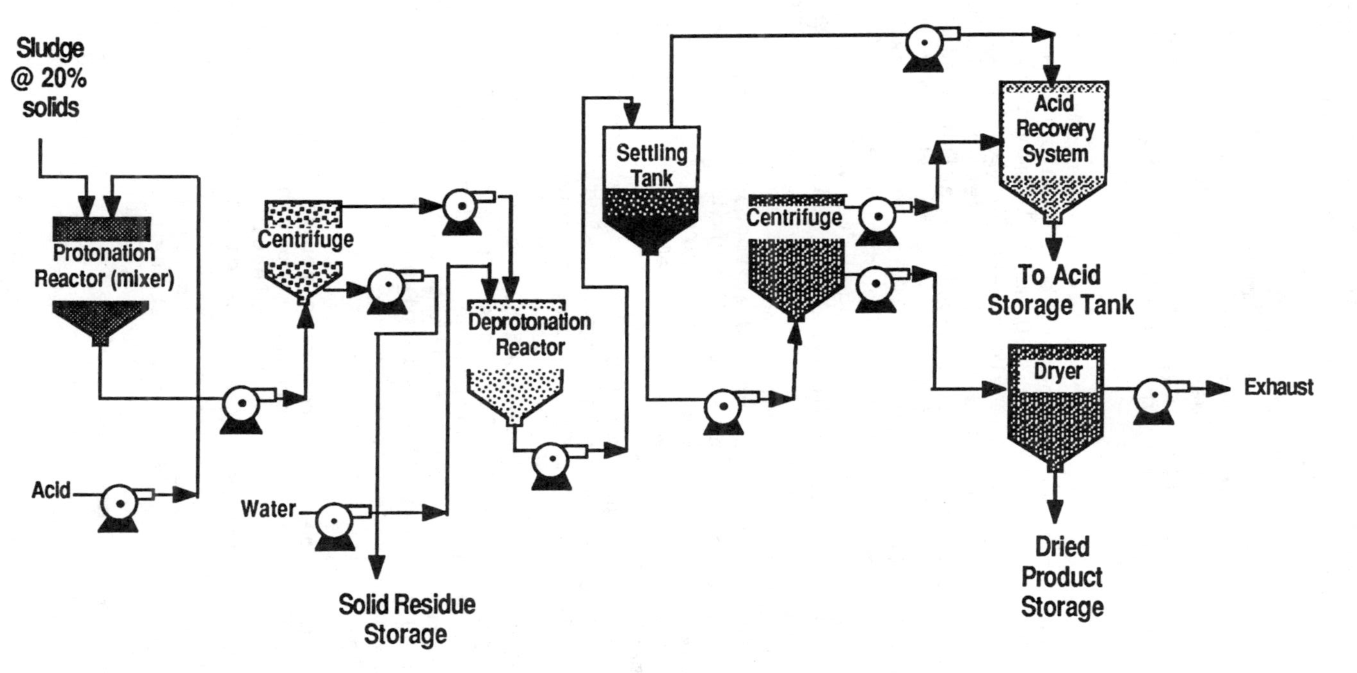

Figure 1. Process flowsheet. (Reproduced from Ref. 17.)

a sample is a standard procedure that is widely used in wastewater analysis (16). Because preliminary experiments showed that dried sludge was easier to handle than wet sludge, sludge samples were dried for 24 hours at 105°C and ground in a mortar pestle prior to use.

Reagent grade sulfuric acid (96% H_2SO_4 by weight) from the Mallinckrodt Company and acid solutions of 64 and 41% H_2SO_4 by weight were used in the dissolution experiments. These solutions were prepared by diluting appropriate amounts of reagent grade acid with deionized water.

Dissolution Experiments. Dissolution experiments were carried out at room temperature in batch reaction vessels which consisted of 250 mL Erlenmeyer flasks. To each flask, 3g of dried sludge and 60 mL of sulfuric acid solution were added. The flasks were immediately capped, placed on a Dubnoff metabolic shaking incubator from the Precision Scientific Company and shaken for 24 hours at approximately 60 cycles/min. After shaking, the contents of the flasks were collected in 60 mL polypropylene tubes and centrifuged for 20 min at 2500 rpm in an IEC centrifuge (Model HN). Following centrifugation, the supernatant solutions containing the dissolved solids were collected in 100 mL beakers, covered with parafilm and stored at room temperature for later use. The tubes containing the undissolved residue were filled with deionized water, handshaken for 2 min and centrifuged for 10 min to remove the remaining supernatant solutions. After centrifugation, the contents of the tubes were collected and the residues were recovered by vacuum filtration using 5.5 cm Whatman 934 AH glass fiber filters. The residue samples were dried using the procedure previously described in the "Materials" section and weighed to the nearest hundredth of milligram on a Mettler laboratory balance (Model H54AR) to determine the fractions of undissolved solids. The dried residues were then ignited for 65 min at 550°C and weighed to obtain the undissolved organic fractions. The percentages of dissolved solids and organics were determined by difference.

Recovery Experiments. Precipitation and recovery of dissolved organics were brought about by diluting the acid solutions containing the dissolved organic compounds with deionized water. The dilution experiments were carried out in batch reaction vessels which consisted of 1000 mL beakers. Because preliminary experiments showed that the precipitated product was easier to recover when a dilution ratio of 1 mL of solution to 15 mL of water was used, 60 mL of dissolved organic solution and 900 mL of deionized water were added to each beaker. The solutions containing the dissolved organic compounds were slowly added to the beakers to minimize the effects of heat of mixing on the precipitated products. These products were recovered by filtration and weighed to determine the fractions of the initial solids recovered after deprotonation. The organic contents of the recovered products were assayed using the same procedures described previously to assay the undissolved organic fractions.

Results and Discussion

Effects of Reaction Time and Acid Strength on Sludge Dissolution. Dissolution experiments were carried out to determine the percentage of sludge dissolved as a function of reaction time for the three acid solutions. Results of these experiments, shown in Figure 2, suggest that approximately 75% of the initial dry sludge is dissolved after 24 hours in the more concentrated acid solutions (96 and 64% H_2SO_4 by weight) as opposed to less than 70% in the more dilute solution (41% H_2SO_4 by weight). Figure 2 also shows that the fraction of sludge dissolved increases with time for the more dilute acid solution, whereas for the more concentrated solutions this fraction reaches a maximum at approximately 24 hours and then gradually decreases as reaction time increases. Because the extent of protonation decreases with acid strength (13), the observed increase of the percentage of solids dissolved with time for the more dilute acid solution might have been due to increased contributions from reactions other than protonation. The data given in Table I, which show lower concentrations of organic materials in the products recovered from the more dilute acid solution, seem to corroborate this explanation.

Effect of Reactant Ratio on Sludge Dissolution. The effect of reactant ratio on the fraction of dry sludge dissolved for an acid solution of 64% H_2SO_4 by weight and a reaction time of 24 hours was studied by mixing varying volumes of acid solutions with a fixed amount (3g) of dry sludge. The results of these studies are shown in Figure 3 and indicate that the "fraction of sludge dissolved as a function of reactant ratio" exhibits two maxima. These maxima approximately occur at acid volumes of 100 and 200 mL. This pattern is similar to that reported by Senftle et al. (10) and is expected to change with acid strength. Senftle et al. attributed the shape of a curve similar to that shown in Figure 3 to a competitive protonation/deprotonation equilibrium process based on the autoprotolysis equilibrium of concentrated sulfuric acid (Senftle, F.E., Miller, K.E.; Buckley, J.Y. J. Sol. Chem. In Preparation).

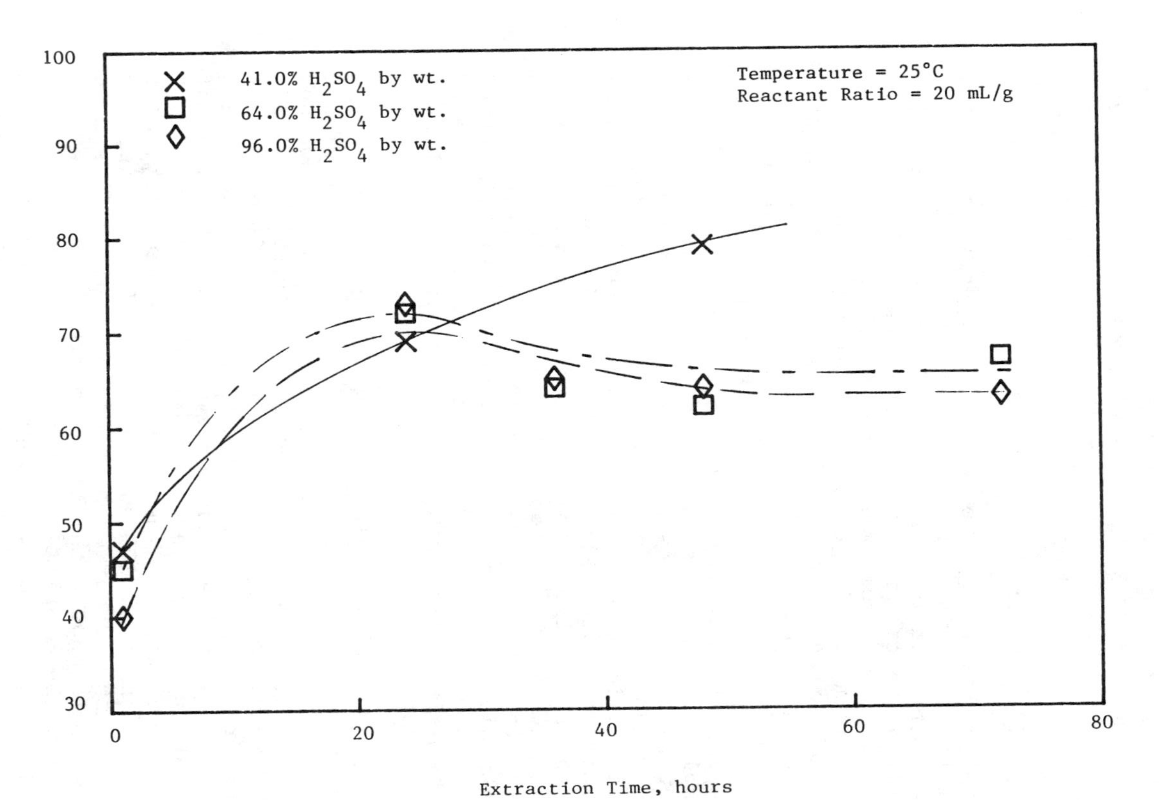

Figure 2. Fraction of sludge dissolved as a function of reaction time and acid strength. (Reproduced from Ref. 17.)

Table I. Total Solids and Organics Recovered as a Function of Acid Strength and Reaction Time[1]

Acid wt%	Time Hours	Solids wt%[2]	Organics wt%[3]	Quality wt%[4]
96.0	1.0	8.0	10.4	97.8
	24.0	18.6	24.3	98.0
	36.0	18.8	24.4	97.4
	48.0	22.8	29.7	97.8
64.0	1.0	5.4	6.9	96.0
	24.0	11.0	14.3	97.8
	36.0	12.9	16.6	96.4
	48.0	14.2	19.4	97.5
41.0	1.0	3.4	4.2	91.4
	24.0	9.4	12.0	95.5
	48.0	10.4	12.7	91.0

[1]Data were obtained under the following conditions
Reaction Temperature: 25°C
Reactant Ratio: 20 mL of acid/gm of dry sludge
[2]Fraction of solids recovered after deprotonation
[3]Fraction of organic materials recovered after deprotonation
[4]Fraction of organic materials in the solids recovered
SOURCE: Reprinted from ref. 17.

Effects of Acid Strength and Reaction Time on Product Recovery. The effects of reaction time and acid strength on the fractions of recovered solids and their organic contents were also studied. Results of these studies are given in Table I and show that up to 25% of the organics could be recovered. Maximum recovery of organic material is obtained using the more concentrated acid solution (96%). Although the organic content of the recovered products is high in all cases (over 90%), that of the products recovered from the more concentrated acid solution is somewhat higher (over 96%) in all cases. This seems to suggest that the percentage of recovered product and its quality (organic content) increase with acid strength.

Recovery of Organic Compounds From Wet Sludge. In all previous experiments, dried sludge was used because it was more convenient to handle than wet sludge. However, drying sludge before its treatment by sulfuric acid is not cost effective. Therefore, a series of experiments were conducted to evaluate the extraction and recovery of organic materials from wet sludge. The results of these experiments are shown in Table II and suggest that a significant fraction of wet sludge, up to 75% can be dissolved. Table II also shows that product recoveries comparable to those shown in Table I were obtained in these experiments. This seems to indicate that wet

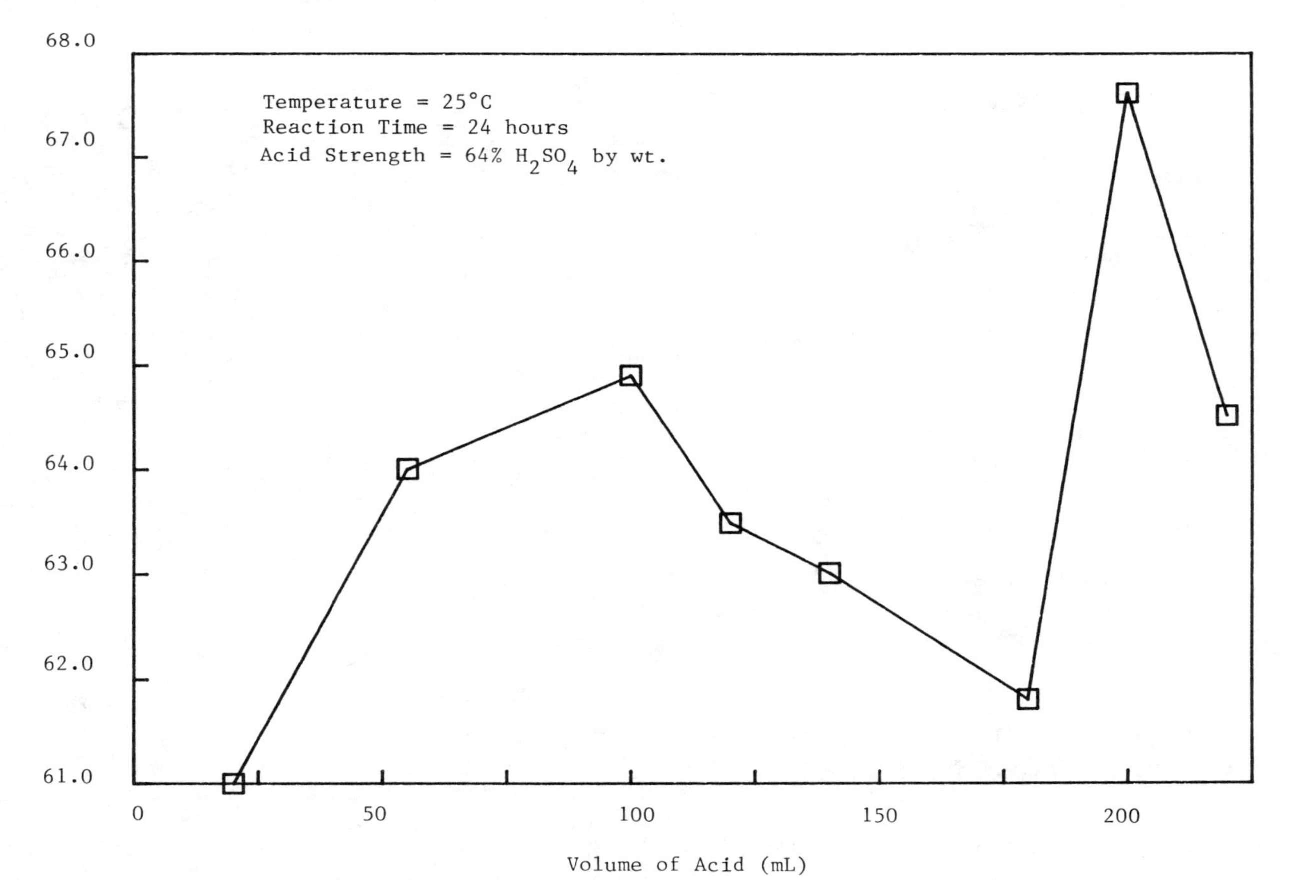

Figure 3. Fraction of solids dissolved as a function of acid volume.

sludge (80% moisture) could be used as raw materials for a sludge reuse plant based on the flowsheet shown in Figure 1. The utilization of sewage sludge of moisture content higher than 80% as raw material for the process shown in Figure 1 was deemed to be not very practical and because of this, sludge samples of higher moisture contents were not tested in this study.

Table II. Effect of Sludge Moisture on percent of Solids Dissolved and Recovered[1]

Moisture wt %	Acid wt%	Ratio[2] mL/g	Solids[3] wt %	Products[4] wt %
80.0	96.0	11.0	75.2	23.0
80.0	64.0	15.0	74.9	20.5
0.0	64.0	15.0	69.5	14.0

[1]Reaction Temperature: 25°C
[2]Acid-to-Sludge Ratio
[3]Fraction of dissolved solids
[4]Fraction of solids recovered after deprotonation

Characterization of the Recovered Products. Proton nuclear magnetic resonance (NMR) spectroscopy was used to characterize the main structural groups associated with the recovered product. Samples for the NMR studies were prepared by dissolving approximately 0.25g of recovered solids with 50 mL of methylene chloride in a 250 mL Erlenmeyer flask for 16 hours under continuous stirring. Following dissolution, the contents of the flask were filtered and the filtrate was collected and dried at 100°C until all the liquid was evaporated. After evaporation, the resulting residue was redissolved in chloroform-d (Merck) and analyzed using a NT-200 Nicolet/GE superconducting NMR spectrometer operating at 200 MHz for protons with an average of 16 scans per sample. The resulting spectra, shown in Figures 4 and 5, illustrate that most of the peaks are observed in the downfield chemical shift region (7.2 - 8.7 ppm) which is indicative of aromatics and heteroaromatics. These findings suggest that the recovered products contain primarily aromatic compounds and are consistent with H/C ratios of similar products reported by Hackney (9).

The composition and fuel values of the recovered products were not determined in this study. Results of carbon, oxygen, nitrogen and sulfur analyses reported by Hackney (9), however, indicate that the heating values of products recovered from sludge using a similar process vary from 15.8 to 32.9 MJ/kg, whereas their sulfur contents range from 1.28 to 10.6% (on a dry weight basis).

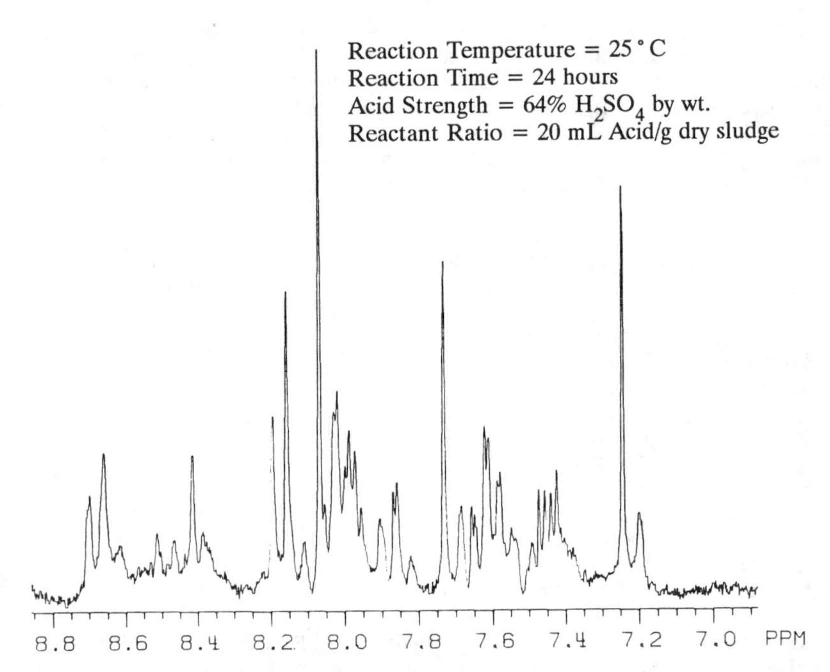

Figure 4. NMR spectrum of recovered product extract - Downfield Chemical Shift region.

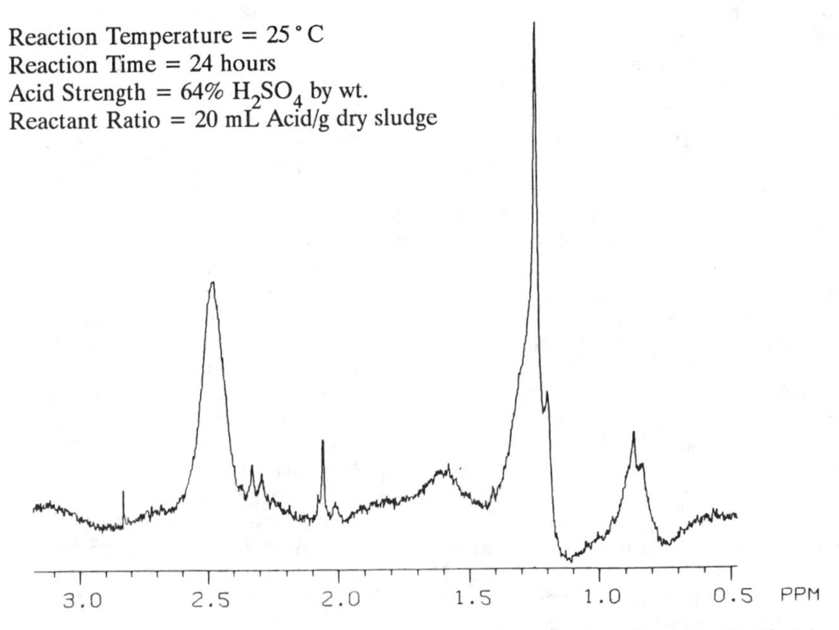

Figure 5. NMR spectrum of a recovered product extract - Upfield Chemical Shift Region.

Characterization of the Undissolved Sludge Residue. The organic fraction of the undissolved sludge residue was characterized using a procedure identical to that previously used to analyze the recovered product. NMR spectra of the undissolved residues are given in Figures 6 and 7 and show a number of peaks in the downfield chemical shift region (7.2 -8.4 ppm) which suggest that all the aromatic compounds from the sludge were not dissolved and recovered. Some peaks are also observed in the upfield chemical shift region (1.9 - 2.2 ppm) which is indicative of aliphatic compounds, some of which may be recoverable. Although the NMR measurements suggest the presence of protonable organic compounds that may be recoverable, the amounts of these recoverable organics cannot be inferred from the spectra. Further investigation is needed to determine the amounts of these recoverable compounds and the conditions for their recovery.

The inorganic fractions of the undissolved solid residues were not analyzed in this study. Results of a spectrographic analysis of undissolved solid residues obtained from similar sludge samples using a similar process reported by Hackney (9), however, indicate that the undissolved residues contain a number of minor elements, e.g., aluminum, copper, titanium and silver.

Preliminary Assessment of Process Feasibility. In the introductory part of this article, the need for cost effective and environmentally acceptable sludge management alternatives was stated. Recovery of organic compounds by proton transfer might provide a new and effective sludge reuse technique. This process can generate an organic rich solid product and a by-product that is rich in valuable inorganic compounds. Because of its high organic content (over 90% on a weight basis), the recovered product could be potentially used as a solid fuel or as a source of aromatic organic compounds. A by-product from this process, which is the undissolved solid residue, could also be potentially used in a multitude of applications. Possible applications include uses such as construction materials (structural fill applications and concrete applications), chemical feedstocks (mineral filler for asphalt, plastics and paints) and metal sources. Although the recovery of organic compounds from sewage sludge by proton transfer has been technically demonstrated, at least two impediments to commercial implementation of the process still exist.

The first impediment is the relatively low yield of the process. As indicated by the experimental results, less than a third of the initial product organics was recovered and approximately 75% of the initial product remained in the solid residue. Although the recovered

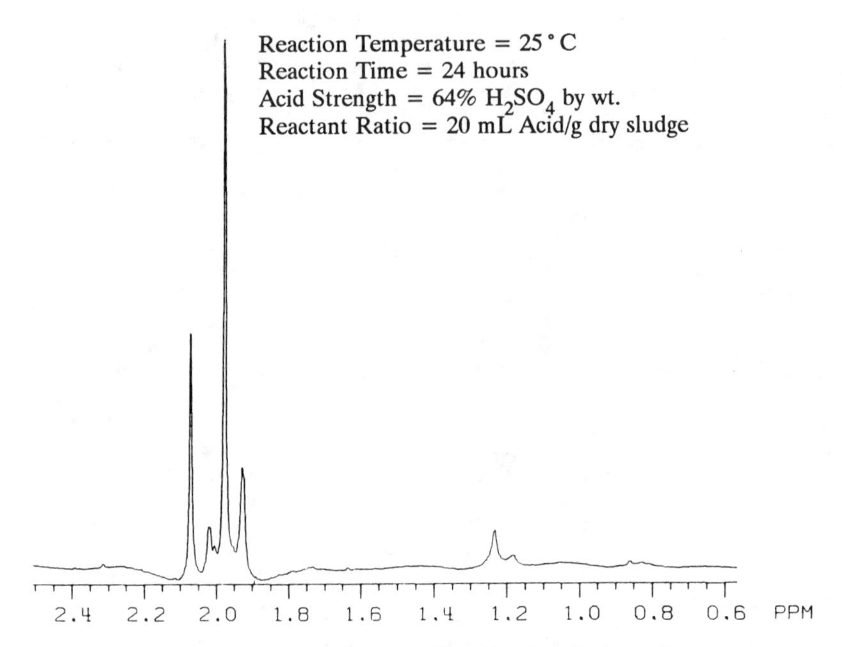

Figure 6. NMR spectrum of an undissolved solid residue extract - Upfield Chemical Shift Region.

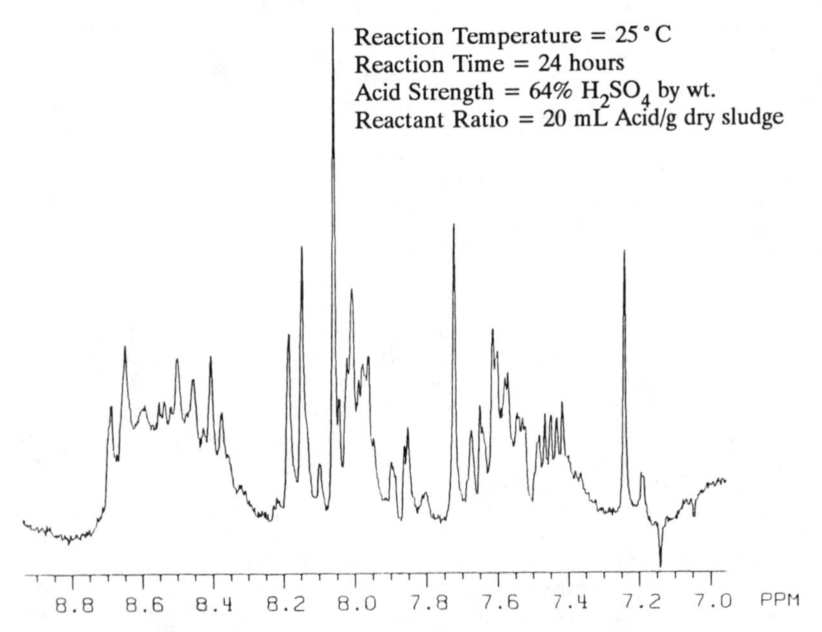

Figure 7. NMR spectrum of an undissolved solid residue extract - Downfield Chemical Shift Region.

material has a high organic content, it represents only 20 to 30% of the amount of organic compounds in the initial solid. Yield improvement strategies presently under investigation include the utilization of other protonating solvents and the optimization of the protonation/deprotonation processes to extract and recover more organic materials.

The second impediment comes from the necessity to develop a suitable acid recovery system. Johnson et al. (17) conducted a preliminary cost analysis of a sludge reuse plant based on the flowsheet shown in Figure 1. They found the proposed process to be competitive with current sludge disposal techniques (incineration and composting) for wastewater treatment plants serving municipalities of a million or more people if 95% of the spent acid is recovered and recycled. Consequently, the feasibility of the proposed technique depends on the availability of a suitable acid recovery system.

Because concentrated acid solutions, e.g. sulfuric acid solutions, have been used as catalysts and reaction media in a multitude of industrial processes (production of fertilizer, production of iron and steel, preparation of pigments and iron, etc.), a number of processes have been proposed to recover acid from waste streams containing spent acid (18-20). Most of these processes have been designed to recover acid from waste streams of specified compositions, e.g. spent acid solutions containing metal sulfates (18), chlorinated and nitrated benzoic acid derivatives (19), hydrocarbons (20), and organic impurities (21). The utilization of these already implemented processes might lack the flexibility to recover acid from a waste stream generated by the process proposed in this study (Figure 1). Therefore more sutdies are needed to develop a specific acid recovery system for this process.

A new approach developed by Eyal et al. (22) might provide a better way of recovering acid from complex waste streams containing organic and inorganic compounds. This new approach is based on a liquid-liquid extraction process that utilizes "a new family of extractants" which consist of amine and an organic acid-containing extractant otherwise known as "couple extractant". According to Eyal et al. (22), the utilization of these extractants can theoretically enable a selective and near complete recovery of acid from any acid-containing waste stream. Because the properties of these couple extractants "can be adjusted to treatment of various waste streams", the utilization of a properly formulated couple extractant could help recover most of the acid from a waste stream generated by a process similar to that shown in Figure 1. However, no detailed application of this new approach to recover acid for our proposed process has been made and a suitable acid recovery system for this process is yet to be developed.

Although the feasibility of this new sludge reuse technique primarily depends upon the removal of the impediments previously mentioned, the proposed process has the potential to become an economical competitive and an environmentally acceptable sludge reuse technique.

Summary and Conclusions

This research was a preliminary assessment of the feasibility of a new sludge reuse process that recovers valuable organics from municipal sludge by protonation of sludge organic compounds in concentrated sulfuric acid followed by dilution with water.

Bench-scale laboratory experiments conducted using aliquots of an anaerobically digested sludge sample from the Blue plains Wastewater Treatment Plant in Washington, D.C. showed that up to 75% of the solids in sludge could be dissolved and up to 25% of the initial sludge could be precipitated upon dilution with water and recovered. The precipitated product has a high organic content (90-98%) and represents up to 30% of the initial organics. NMR spectroscopic analysis of a recovered product extract showed the recovered product to contain primarily aromatic and heteroaromatic compounds and thus could be used as source of organic chemicals or a fuel.

Acknowledgments

This research was partially funded by the Pacific Northwest Laboratory under DOE Contract DE-AC06-76RL 1830. Thanks are extended to Mrs. Lily Wan and Mr. Kenneth Henley of the School of Engineering, Howard University for their help in the dissolution and recovery experiments and Dr. Anne H. Turner of the Howard University Department of Chemistry for technical assistance in the NMR experiments.

Literature Cited

1. Use and Disposal of Municipal Wastewater Sludge: 1984; U.S. EPA. Technology Transfer. Center for Environmental Research Information. Cincinnati, OH, 1984, EPA 625/10-84-003.
2. Bayer, E., Kutubuddin, M. In Recovery of Energy Materials from Residues and Waste; Thome Kozmiensky, K.J.; Karl, J.E., Eds.; Freitag-Verlag Umwilttech, Berlin, West Germany, 1982, p. 314.
3. Bridle, T.; Campbell, H.W. Proc. 35th West Can. Water Savage Conf., 1983, p. 195.
4. Molton, P.M.; Fassbender, A.G. Proc. U.S. EPA, Int. Conf. on Conversion of Municipal Sludge, Hartford, CT, 1983, p1.

5. Molton, P.M.; Fassbender, A.G.; Brown, M., Proc. Nat. Conf. on Municipal Sludge Manager, Orlando, FL, 1983.
6. Fassell, W.M. Proc. Nat. Conf. on Municipal Sludge Management, Pittsburgh, PA, 1974, p. 195.
7. Gabler, R.C. Incinerated Municipal Sewage Sludge as a Potential Secondary Resource for Metals and Phosphorous, U.S. Bureau of Mines, 1979, Report Investigation No. 8390.
8. Sommers, L.E.; Nelson, D.W.; Kirleis, A.W.; Strachan, S.D.; Inman, J.C.; Boyd, S.A.; Gravel, J.G.; Behel, A.D. Characterization of Sewage Sludge and Sewage Sludge Soil Systems, U.S. EPA, Municipal Environmental Research Laboratory, Cincinnati, OH, 1984, EPA 600/2-84-046.
9. Hackney, P.I., M.E. Thesis, Howard University, Washington, DC, July 1988.
10. Senftle, F.E.; Hackney, P.I.; Jackson, F.E.; Parker, A.; Miller, K.; Brown, Z. Low Ash Solid Fuel via Acid Extraction of Organic Matter from Sewage Sludge. U.S. Geological Survey, Reston, VA, 1988, Open File Report No. 881239.
11. Gillespie, R.J.; Robinson, E.A. In Advances in Inorganic Chemistry and Radiochemistry; Emeleus, H.J.; Sharpe, A.G., Eds.; Vol. 1, Academic Press, New York, 1959, p. 385.
12. Bell, R.P. The Proton in Chemistry; Cornell University Press, Ithaca, NY, 1959.
13. Liler, M. Reaction Mechanisms in Sulfuric Acid and Other Strong Acid Solutions; Academic Press, New York, 1971.
14. Stewart, R. The Proton: Applications to Organic Chemistry; Academic Press, New York, 1985.
15. Olah, G.A.; Prakash, G.K.S.; Sommer, J. Superacids; John Wiley & Sons, New York, 1985.
16. Standard Methods for the Examination of Water and Wastewater, Fifteenth Edition, American Public Health Association, Washington, DC, 1980, p. 95.
17. Johnson, J.H., Jr.; Chawla, R.C.; Cannon, J.N.; Diallo, M.S.; Wan, L.; Henley, K.G. Proc. DOE Conf. on Separations Innovative Concepts, Crystal City, VA, 1988, p7.1.
18. Peterson, J.C. U.S. Patent 3 900 955, 1975.
19. Boettler, A.J.; Dyer, A.M. U.S. Patent 3 749 648, 1973.
20. Lewis, R.M.; Weston, C. U.S. Patent 3 652 708, 1972.
21. Estes, S.L. U.S. Patent 3 506 399, 1970.
22. Eyal, A.; Baniel, A.; Mizrahi, J. In Emerging Technologies in Hazardous Waste Management; American Chemical Society; Washington, DC, 1990; p 214.

RECEIVED November 10, 1989

Chapter 19

Energy Waste Stabilization Technology for Use in Artificial Reef Construction

Chih-Shin Shieh[1], Iver W. Duedall[1], Edward H. Kalajian[2], and Frank J. Roethal[3]

[1]Department of Chemical and Environmental Engineering, Florida Institute of Technology, Melbourne, FL 32901
[2]Department of Civil Engineering, Florida Institute of Technology, Melbourne, FL 32901
[3]Waste Management Institute, Marine Sciences Research Center, State University of New York, Stony Brook, NY 11794

Stabilization methodology has been applied in recent years to a variety of energy wastes, such as coal ash, FGD sludge, oil ash, and incineration ash. The primary purpose of the research has been to develop and evaluate stabilized waste materials that can be used for the construction of artificial reefs. The methodology involves mixing granular solid wastes with chemical additives, fabrication of the reef block, followed by curing at a constant temperature for a period of time, so that a solid block is produced. On-going monitoring programs over a number of years have been designed to demonstrate the environmental feasibility of the stabilization technology. To date, the reef blocks fabricated using particulate energy wastes have maintained their physical and chemical integrity in the marine environment without causing adverse effects on marine organisms.

Rapid growth in modern civilizations has greatly increased the use of energy in the past decades. The annual per capita rate of energy consumption could be as high as 300,000 kWh by the year 2020 (1). Generation of energy comes mainly from the combustion of oil, gas, coal, and municipal solid wastes (MSW). The wastes produced from these energy sources are in gaseous or solid forms. The production and disposal of energy wastes will become an increasingly serious problem during the coming decades.

Solid wastes produced from the combustion of oil, coal, and municipal solid waste are oil ash, coal ash and flue-gas desulfurization (FGD) sludge, and incineration

0097–6156/90/0422–0328$06.00/0

ash, respectively. In the United States the total production of coal wastes, including fly ash, bottom ash, and FGD sludge, will be 500 million tons in the year 2000 (2); the production of MSW incineration ash would be 19 million tons by the year 2000 (2); the production of oil ash would be 1500 tons per year per oil-burned power plant (3). Trace metals and contaminants of environmental concern are enriched in coal wastes (4), oil ash (5-8), and MSW incineration ash (9-13). These trace metals and contaminants of environmental concern are arsenic (As), selenium (Se), cadmium (Cd), lead (Pb), zinc (Zn), dioxins, furans, and polycyclic aromatic hydrocarbons (PAHs). Because of the enrichment of these constituents in the solid energy wastes, a safe and environmentally sound management system is required. Currently in the U.S., landfilling is the major option for the disposal of the solid energy wastes, except in land-limited metropolitan regions which adjoin the coastal zone. Disposal of these contaminant-enriched solid wastes in landfills may cause environmental problems in areas where shallow aquifers are more susceptible to contamination. Florida is a state with shallow aquifers and there is a move in the state of Florida to greatly reduce the landfilling of solid wastes.

In this paper, the technology of stabilization is presented as an alternative to the disposal of solid energy wastes. Particulate energy wastes (i.e., coal wastes, oil ash, and MSW incineration ash) can be transformed into solid block forms by the stabilization process. The goal is to utilize the stabilized energy waste blocks in a beneficial manner, such as construction of artificial reefs in the ocean. Stabilized energy waste blocks have been tested in the laboratory for their suitability for reef construction, leaching behavior in aqueous solutions, and effects on marine organisms. Following laboratory evaluations, monitoring studies, including engineering investigations, chemical examinations, biological colonization, and bioaccumulation of metals, have been carried out to demonstrate the environmental feasibility of using the stabilized energy waste blocks as reef materials. The results of these studies conducted on different types of stabilized energy waste blocks are summarized and directions for future development of stabilization technology are suggested.

Stabilization Methodology

The process of waste stabilization to reef block construction begins with the mixing of particulate waste materials with water and chemical additives, such as lime ($Ca(OH)_2$) and cement, followed by fabrication of the solid waste blocks using conventional concrete block technology. A series of mix designs are first generated in order to screen for an optimum mix. After fabrication, blocks are cured at a constant temperature for a period of time so that a solid product is produced. A final mix is selected based on the development of desirable compressive strengths of the stabilized products. The stabilized energy waste blocks must be strong in order to withstand the effects of mechanical manufacture, transportation, and emplacement as a reef.

Coal fly ash, which is a pozzolan, is normally used as the matrix for the stabilization of FGD sludge and oil ash, both of which have no pozzolanic characteristics. Some MSW incineration ash, such as bottom ash, may also has pozzolanic property and therefore it can be stabilized by mixing the ash with chemical additives. Use of lime and cement in the stabilization process allows the cementation reaction to occur when the ash matrix is mixed with water.

Several factors must be considered in the stabilization methodology to achieve a strong stabilized product; these are waste-additives ratio, particle size distribution of the waste materials, water content of the mix, and curing condition. It is always desirable to maximize the content of waste material to be stabilized in order to use as much wastes as possible, while minimization of the chemical additives is desired from the economic standpoint. The optimum waste-additive ratios of each mix will be dependent upon the properties of the waste, water content of the mix, and curing. Particle size distribution of the wastes becomes a critical factor in determining the effectiveness of stabilization. Better stabilization will be achieved if the waste materials have a well-graded particle size distribution. Optimum water content in a mix provides the most efficient cementation reaction. However, excess water will result in reduction of the strength of the stabilized product (14). Curing of the mix at high temperature may accelerate the stabilization process. However, an optimum curing condition must be found for each mix design. The production of an optimum mix is, therefore, the result of a unique combination of: nature of the waste, waste-additive ratio, moisture content, and curing conditions.

The Stabilized Solid Energy Wastes

The components of mix designs will vary with the type of waste used in the mix and the intended use of the stabilized products. For a reef construction, according to Mentzios (15), the stabilized energy waste blocks should have the following characteristics: a minimum strength of 2000 kPa; remain stable on the sea floor, a minimized surface erosion, a minimized leaching of metals, suitability as settling base for biological colonization, and economically feasible. Table I lists the components in mix designs for the stabilized coal waste, oil ash, and incineration ash. These mixes have been used for a number of field demonstrations and will be discussed later in this paper.

Table I. Mix Components (%) of the Stabilized Coal Waste, Oil ash, and Incineration Ash

Component	Coal Waste[1]	Oil Ash[2]	Incineration Ash[3]
Coal fly ash	70.5	39.7	-
Oil Ash	-	39.7	-
FGD sludge	20.5	-	-
MSW Incineration Ash	-	-	85
Cement	6.0	15.9	15
$Ca(OH)_2$	3.0	4.7	-

1. Data from Duedall et al. (16).
2. Data from Mazurek (17).
3. Data from Roethel et al. (11).

Physical and Engineering Properties. Physical and engineering properties of the stabilized energy wastes blocks normally investigated are compressive strength, bulk density, porosity, and permeability. Studies of these properties provide an estimation of how the physical integrity of the blocks can be maintained in marine environments, how the blocks interact with seawater, and how the blocks remain stable on the sea floor. Determination of physical properties is performed using methods of the American Society for Testing and Materials (ASTM). For compressive strength ANSI/ASTM C39-72, "Standard Method for Compressive Strength Testing of Cylindrical Concrete Specimens" (18) is used; for porosity ASTM method C642-75, "Standard Test Method for Specific Gravity, Absorption and Voids in Hardened Concrete" (19) is used; for permeability the Darcy falling head method (20) and a constant-head method (21) are used.

Table II lists the physical and engineering properties of the stabilized energy waste blocks used in reef construction. The compressive strength of these blocks are in the range from 4500 kPa to 7700 kPa. Values are less than that of concrete block but far in excess of the minimum unconfined compressive strength criteria of 2000 kPa necessary for mechanical transportation and placement into the ocean as reef materials (22). The latest developments in mix designs for oil ash and MSW incineration ash indicates that the use of new mix additives will generate an unconfined compressive strength of the stabilized blocks that is comparative to concrete blocks (15, 23).

Table II. Physical and Engineering Properties of the Stabilized Energy Waste Blocks

	Compressive Strength (kPa)	Bulk Density ($g\ cm^{-3}$)	Porosity (%)	Permeability ($x10^{-7}$ $cm\ s^{-1}$)
Coal Waste[2]	6300	2.18	37	1.0
Oil Ash[3]	4500	0.16	50.	-
Incineration Ash[4]	7700	2.32	42	0.2
Concrete[2]	14500	2.40	13	0.1

1. 1 kPa = 0.145 psi
2. Data from Duedall et al. (16)
3. Data from Mazurek (17)
4. Data from Roethel et al. (11)

The bulk density of the stabilized energy waste blocks is normally lower than that of the concrete blocks. The value of bulk density is inversely proportional to that of porosity; a stabilized block has a higher bulk density, and hence, a lower porosity than concrete blocks.

The permeability of the stabilized energy waste blocks is about an order of magnitude higher than that of concrete block; a higher permeability for the stabilized blocks may result in more interactions of the block components with seawater.

Chemical Properties. Elemental composition of the stabilized energy waste blocks is determined by the composition of the matrix components (i.e., waste materials and chemical additives). Elemental analysis of the stabilized blocks is carried out by analyzing hydrofluoric/boric acid digests of the ground block materials using an atomic absorption spectrophotometer and methods reported by Silberman and Fisher (24). Table III shows the major and minor elements in some stabilized energy waste blocks. Vanadium (V) is especially enriched in the stabilized oil ash block due to the enrichment of V in oil ash (6, 8, 28). Sodium and K are enriched in the stabilized incineration ash blocks. Zinc, Pb, Cu, and Mn are significantly enriched in incineration ash blocks; this is the result of the enrichment of these metals in the ash materials. The elemental composition of the incineration ash is dependent upon the source of the waste being burned and the operational conditions of the plant (29).

Table III. Elemental Composition of the Stabilized Energy Waste Blocks

	Coal Waste[1]	Oil Ash[2]	Incineration Ash[3]
Ca (%)	10.0	11.0	15.70
Al (%)	8.2	6.1	4.71
Si (%)	16.0	17.5	16.76
Mg (%)	0.7	1.6	1.13
Na (%)	0.2	-	2.39
K (%)	1.1	-	0.82
Fe (%)	15.0	3.2	7.25
V (%)	-	1.42	-
Zn ($\mu g\ g^{-1}$)	160	313	3760
Pb ($\mu g\ g^{-1}$)	33	248	3580
Cr ($\mu g\ g^{-1}$)	96	-	178
Cu ($\mu g\ g^{-1}$)	63	146	1260
Mn ($\mu g\ g^{-1}$)	210	185	1020
Cd ($\mu g\ g^{-1}$)	0.8	1.38	23.6
Ni ($\mu g\ g^{-1}$)	84	-	-
As ($\mu g\ g^{-1}$)	68	-	-

1. Data from Labotka et al. (25).
2. Data from Shieh et al. (26).
3. Data from Breslin et al. (27).

Studies of metal leaching in aquatic environments are always initially conducted at the bench level because the stabilized energy waste blocks are enriched in trace metals of environmental concern and therefore it is important to determine metal mobility to satisfy compliance requirements. Leaching of trace metals from stabilized energy waste blocks can be investigated using U.S. EPA EP Leachate Test (30). Table IV indicates that the concentration of metals leached from the stabilized energy waste blocks are less than the level for drinking water and the toxic criteria.

Table IV. Metal Concentrations (mg L^{-1}) in EP Leachate of Stabilized MSW incineration Waste Blocks[1]

	Stabilized Blocks	Drinking Water	Toxic Criteria
As	<0.001	0.05	5.0
Cd	<0.001	0.1	1.0
Cr	0.17	0.05	5.0
Cu	0.14	1.0	-
Fe	0.004	0.3	-
Hg	<0.002	0.002	0.2
Mn	0.001	0.3	-
Pb	<0.001	0.05	5.0
Zn	0.047	5.0	-

1. Data modified from Roethel et al. (11)

The physical-chemical behavior of the stabilized energy waste blocks in seawater has been investigated using tank leaching studies (8, 31-33). The objectives of these studies have been to estimate the chemical stability of the blocks in seawater and to determine diffusion process associated with metals in the solid block matrix. Calcium is frequently selected as the tracer to accomplish these objectives because it is a major element responsible for the formation of cementitious compounds in the stabilized products. Figure 1 shows the trend in calcium release from the stabilized blocks into surrounding seawater. Based on the results of leaching studies, a diffusion model was developed by Duedall et al. (31) and modified by van der Sloot et al. (33) to predict the fate of metals in the stabilized product. The diffusion model predicts that leaching of elements from the stabilized blocks should follow the theoretical diffusion line with a slope of -1/2 when logarithm of leaching rate is plotted against logarithm of time. Deviations from the -1/2 slope line may be due to the rapid surface dissolution of Ca^{2+}, changes in block density, changes in tortuous path for ions, and other physical-chemical reactions occurring inside the blocks. The results shown in Figure 1 indicate that the stabilized incineration ash blocks have the most desirable leaching characteristics in seawater.

Artificial Reef in the Ocean

There have been a number of field demonstration studies using stabilized energy waste as artificial reefs. In 1977, a small reef comprised of stabilized coal fly ash and FGD sludge was placed in Conscience Bay, an arm off Long Island Sound (34). In 1979, fifteen thousand blocks

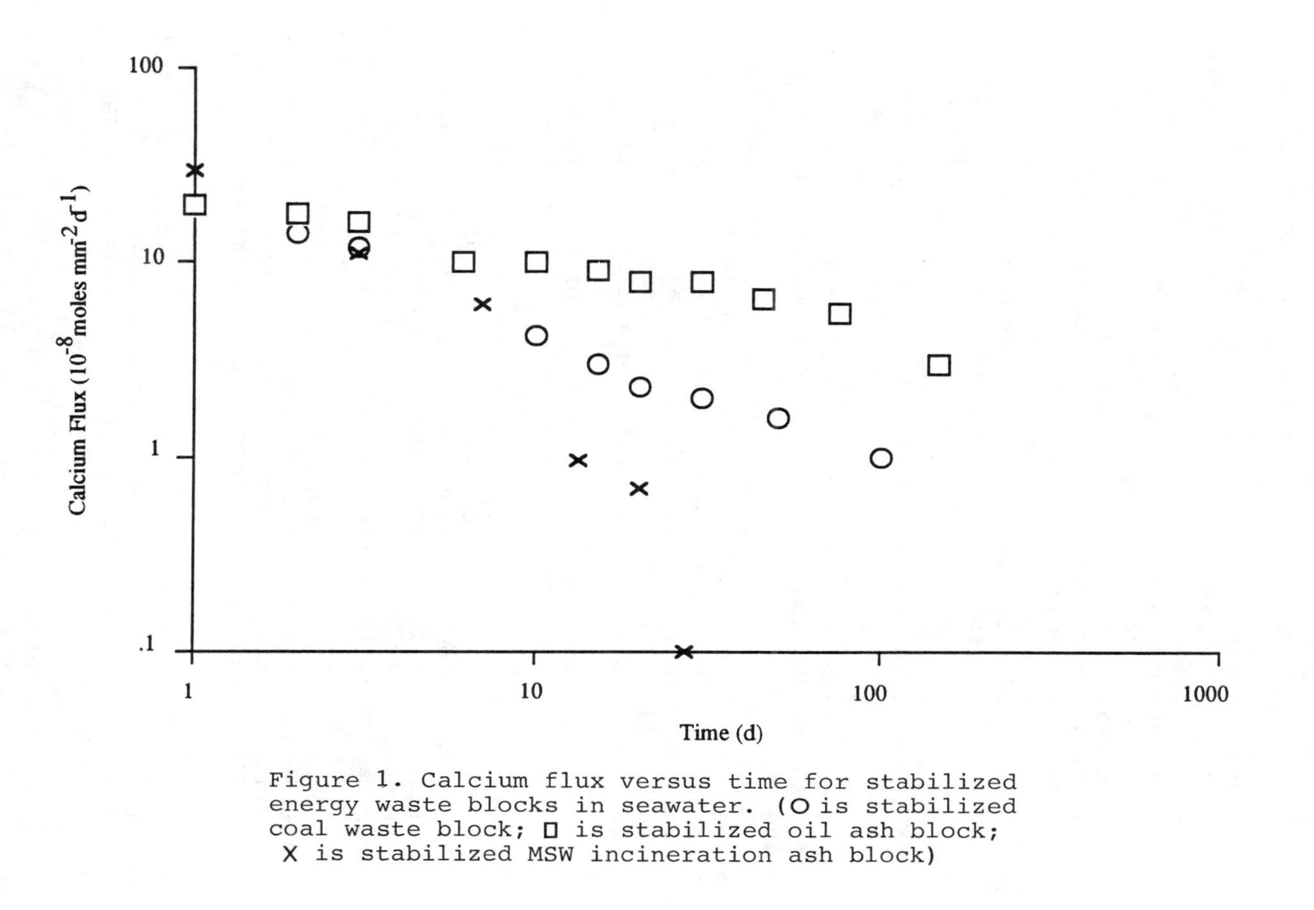

Figure 1. Calcium flux versus time for stabilized energy waste blocks in seawater. (O is stabilized coal waste block; □ is stabilized oil ash block; X is stabilized MSW incineration ash block)

(450 t) of stabilized coal waste blocks were emplaced into the ocean to form an artificial reef in the Atlantic Ocean south of Long Island, New York (1). On 7 April 1987, an experimental artificial reef comprised of stabilized oil ash blocks was placed into the ocean 2 km off the Vero Beach, Florida (35). Also, during April 1987, an experimental artificial reef made from stabilized MSW incineration ash was placed in 8 m of water in Conscience Bay, Long Island Sound, New York (27). Other stabilized ash reefs exist off the west coast of Florida, Delaware Bay and in the waters off Taiwan and Japan.

A field monitoring program is normally carried out to demonstrate the environmental feasibility and acceptability of using the stabilized energy waste blocks for the construction of artificial reefs in the ocean. Engineering studies are designed to determine the physical stability of the blocks in marine environments; chemical studies are designed to investigate the interaction of the stabilized blocks with seawater over the period of submersion; biological studies are designed to evaluate the suitability of the waste blocks for attracting and colonizing marine organisms; bioaccumulation studies are designed to determine the chemical effect of the stabilized blocks on marine organisms.

The results of monitoring studies have revealed that the physical integrity of the blocks are maintained in the ocean over the period of submersion. Variations of compressive strength of the reef blocks as a function of submersion time for a number of studies are shown in Figure 2 indicating that strengths of the blocks are maintained.

Variation in elemental composition of the reef blocks (Table V) has been examined by studying the elemental concentrations at different layers within the blocks. Portions of the surface layer (less than 1 cm), mid-depth, and central parts of the blocks have been collected and analyzed for selected components by $HF-H_3BO_3$ digestion techniques using flame and flameless atomic absorption spectrophotometry (26). A decreasing (less than 10%) Ca concentration at the surface layer of the blocks has been found. In contrast to the release of Ca, enrichment of Mg in the blocks has been observed. Brucite ($Mg(OH)_2$) formation and replacement of Mg for Ca may account for the observed Mg enrichment in stabilized energy waste blocks. The mid-depth and central parts of the blocks remain intact.

Metals of environmental concern appear to be effectively retained within stabilized energy blocks. Formation of cementitious compounds combined with the high alkalinity environment inside of the stabilized energy waste blocks provide a favorable environment within the blocks for the retention of metals. It has also been observed that interaction of the stabilized energy waste blocks with seawater occurred mainly on the surface layer of the blocks (26).

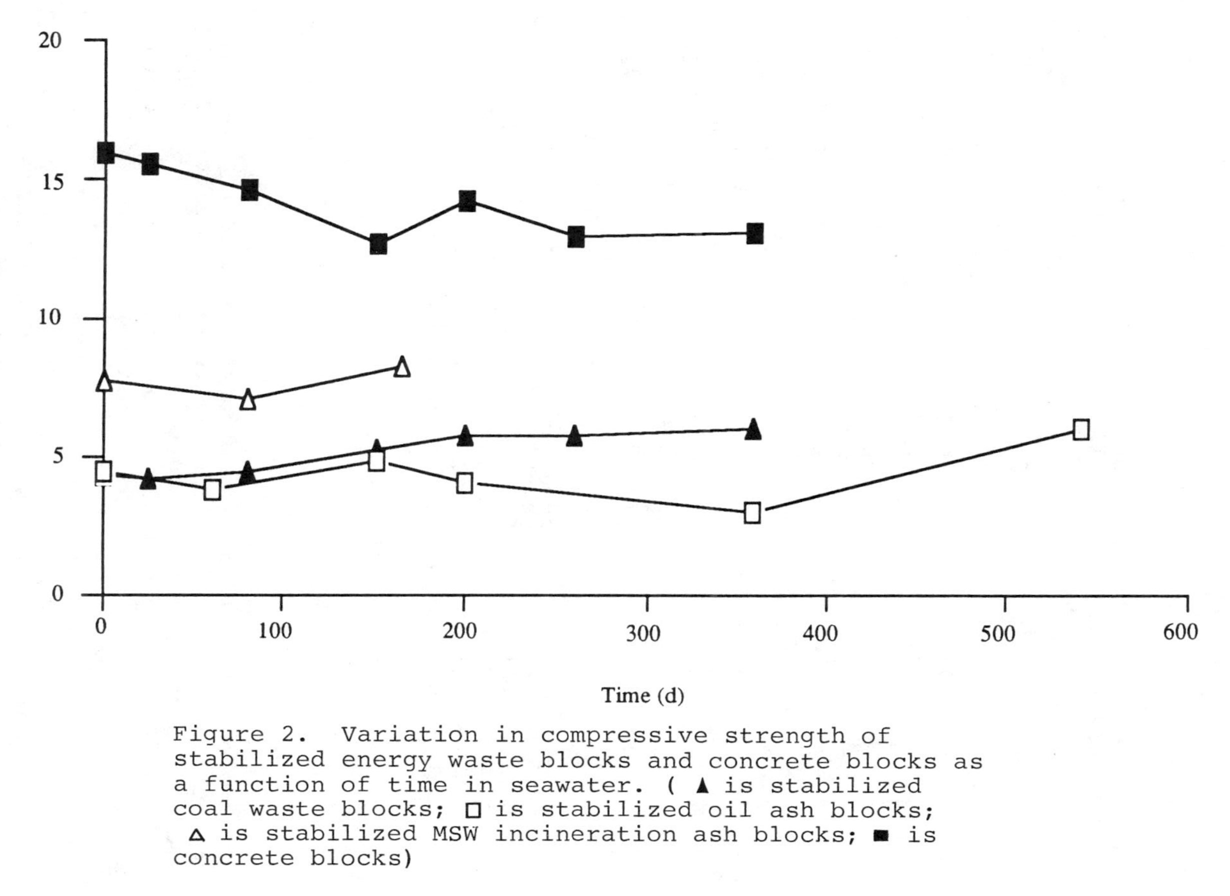

Figure 2. Variation in compressive strength of stabilized energy waste blocks and concrete blocks as a function of time in seawater. (▲ is stabilized coal waste blocks; □ is stabilized oil ash blocks; △ is stabilized MSW incineration ash blocks; ■ is concrete blocks)

Table V. Percentage Variation (%) in Elemental Concentration of Stabilized Energy Waste Blocks

	Coal Waste[1] (238 d)	Oil Ash[2] (240 d)	Incineration Ash[3] (165 d)
Ca	N.S.[4]	N.S.	N.S.
Mg	+50	+40	+25
Na	+122	N.A.[4]	N.S.
K	-20	N.A.	-16
Al	N.S.	N.A.	N.S.
Si	N.S.	N.A.	N.S.
Fe	+16	N.S.	N.S.
Pb	N.S.	N.S.	N.S.
Zn	N.S.	N.S.	N.S.
Cr	N.S.	N.S.	N.S.
Cu	N.S.	N.S.	N.S.
Mn	N.S.	N.S.	N.S.
Cd	+33	N.S.	N.S.
V	N.A.	N.S.	N.A.
Ni	+28	N.S.	N.A.

1. Data modified from Labotka et al. (25).
2. Data modified from Shieh et al. (26).
3. Data modified from Breslin et al. (27).
4. N.A. represents not analyzed; N.S. represents no significant variations (< 10% variations).

Colonization of marine organisms occurs rapidly on the surface of the stabilized waste blocks (Figure 3). Fouling organisms attach themselves onto the surface of the stabilized energy waste block within the first month after reef emplacement. No significant differences in settlement density have been found between the reefs composed of stabilized energy waste blocks and those composed of concrete blocks (22, 36). Recruitment of fish is also rapid. Previous studies show that there is no statistically significant difference in abundance of fish between reef types (22, 36).

Metal concentrations of the organisms attached to the surface of the reef blocks have been analyzed to assess

Figure 3. Colonization of organisms on the surface of the stabilized energy waste reef blocks.

the extent of bioaccumulation. Results show (Table VI) that most metal concentrations in the organisms colonizing the stabilized energy waste blocks are low compared with the values for those colonized on the concrete reef blocks. However, Ni and V are exceptions, with higher concentrations in organisms from stabilized oil ash reefs compared with those from concrete reefs. Elevated Ni and V values could result from either leaching of oil ash particles present in the gut of an organism or in situ bioaccumulation (37). At present, limited data are available to determine the biological effects of the elevated Ni and V in organisms associated with oil ash blocks. However, no apparent differences in mortality rates have been observed between stabilized oil ash reef and the control reef (36).

Table VI. Percentage Variation (%) in Elemental Composition of Marine Organisms Attached onto Stabilized Waste and Concrete Blocks

	Percentage variation (%)[1]		
	Coal Waste[2]	Oil Ash[3]	Incineration Ash[4]
Pb	N.S.	N.A.	N.S.
Zn	N.S.	-22	N.S.
Cr	N.S.	N.A.	N.S.
Cu	N.S.	N.S.	N.S.
Mn	N.A.	N.A.	N.S.
Cd	N.S.	N.A.	N.S.
Ag	N.A.	N.A.	N.A.
Hg	N.A.	N.A.	N.A.
Se	-32	N.A.	N.A.
V	N.A.	+233	N.A.
Ni	N.A.	+94	N.A.

1. Percentage variation (%) = $(MC_{wb} - MC_{cb})/MC_{cb}$, where MC_{wb} represents metal concentration in the organisms attached onto the stabilized waste blocks; MC_{cb} represents metal concentration in the organisms attached onto the concrete blocks. N.A. represents not analyzed; N.S. represents no significant variation (< 10% variations).
2. Organisms are mainly hydroids, Roethel (34).
3. Organisms are mainly barnacles, Metz and Trefry (37).
4. Organisms are mainly hydroids, Breslin et al. (27).

Future Directions in Stabilization Technology

The technology for stabilization of solid energy wastes for artificial reefs has been evolving in a positive direction. Future directions in the stabilization technology should involve consideration of cost, practicality of the method, public health consideration, and improvement in fundamental understanding of physical and chemical processes occurring within the stabilized products. Studies such as development of on-site fabrication technology, design of new shapes for a reef unit, applications of the technology to other uses, and continuation of basic research are required.

Development of On-Site Fabrication Technology. A technology for on-site stabilization needs to be developed to reduce the costs for transportation, manufacture, and emplacement. The on-site stabilization of the wastes will reduce the relocation of the wastes and, thus, the stress on human health and environmental quality. A complete process, including weighing of particulate waste materials and chemical additives, mixing of the matrix, fabrication of the mixes, and curing of the products will be carried out inside the unit.

Design of New Shapes for a Reef Unit. Blocks (20 cm x 20 cm x 40 cm) have been used as units for the construction of artificial reef in the past several years to demonstrate the feasibility of the stabilization technology. New shapes for an artificial reef unit are now being developed by Zawada (38) based on the consideration of the efficiency of reef functions, the stability of the reef units, and the accessibility of reef emplacement.

Other Applications in Marine Environments. Stabilized energy wastes should be considered for other purposes, such as seashore protection and dock construction. New mix designs with different strength, sizes, and shapes will be required in order to approach these applications.

Fundamental Research. It is well known that a series of cementation reactions are the primary processes responsible for the stabilization of waste materials. However, several aspects of the reaction are incompletely understood. Previous studies have shown that stabilization reduces the mobility of trace metals inside the stabilized waste blocks. But it is uncertain if the minimization of metal leaching is due to physical enclosement, chemical binding, or a combination of both processes. For a complete understanding of the significance of the stabilization process, future studies should be carried out to examine the relationships between: (i) porosity and permeability, (ii) permeability and metal leaching, (iii) compressive strength and metal leaching, and (iv) abrasion resistance and compressive strength.

Conclusions

The technology of stabilization has been applied to particulate energy wastes to form solid blocks for artificial reefs which appear to be environmentally acceptable. The stabilized energy waste blocks have maintained their physical integrity over the period of submersion in the ocean. No adverse effect on the marine environment and organisms has been found. Future studies should focus on the development of on-site treatment technology, design of new shapes, and applications of the stabilization technology for other purposes.

Acknowledgments

The Coal Waste Artificial Reef Program (C-WARP) was supported by the Electric Power Research Institute; the Stabilized Oil Ash Reef (SOAR) program is being supported by Florida Power & Light Co.; and the stabilized incineration residues reef program is being supported by the New York State Legislature.

Literature Cited

1. Duedall, I. W.; Kester, D. R.; Park, P. K.; Ketchum, B. H. In Wastes in the Ocean, Vol. 4, Energy Wastes in the Ocean; Duedall, I. W.; Kester, D. R.; Park, P. K.; Ketchum, B. H., Eds.; Wiley-Interscience: New York, 1985; pp 3-42.
2. NRC. Monitoring Particulate Wastes in the Oceans, The Panel on Particulate Waste in the Oceans; Marine Board, NRC-NAS: Washington, D.C., 1988; p 176 + appendix.
3. Harris, W. R.; Raabe, O. G.; Silberman, D. R. American Chemical Society, Abstract, Las Vegas, 1982, pp 78-9.
4. Gehrs, C. W.; Shriner, D. S.; Herbes, S. E.; Salmon, E. J.; Perry, H. In Chemistry of Coal Utilization; Elliott, M. A., Ed.; Wiley-Interscience: New York, 1981; pp 2159-223.
5. Natusch, D. F. S. Env. Health Perspect. 1978, 22, 79-90.
6. Henry, W. H.; Knapp, K. T. Env. Sci. Tech. 1980, 14, 450-56.
7. Krishnan, R. E.; Hellwig, V. C. Env. Progr., 1982, 1, 290-95.
8. Breslin, V. T. Ph.D. Thesis, Florida Institute of Technology, Melbourne, 1986.

9. Greenberg, R. R.; Zoller, W. H.; Gordon, G. E. Env. Sci. Tech. 1978, 12, 566-73.
10. Law, S. L.; Gordon, G. E. Env. Sci. Tech. 1979, 13, 432-38.
11. Roethel, F. J.; Schaeperkoetter, V; Gregg, R.; Park, K. The Fixation of Incineration Ash: Physical and Leachate Properties; Waste Management Institute, Marine Sciences Research Center, State University of New York at Stony Brook, Interim Report to New York State Legislative Commission on the Water Resource Needs of Long Island, New York, 1986, p 200 + appendix.
12. Denison, R. A. Governmental Refuse Collection and Disposal Association 26th Annual International Solid Waste Exposition, 1988, p 16.
13. Van der Sloot, H.; J. Wijkstra. In Physical and Chemical Processes: Transport and Transformation; Baungartner, D. J.; Duedall, I. W.' Eds.; Robert E. Krieger: Florida, 1989; in press.
14. Shieh, C. S. M.S. Thesis, State University of New York, Stony Brook, 1984.
15. Mentzios, A. M.S. Thesis, Florida Institute of Technology, Melbourne, 1989.
16. Duedall, I. W.; Roethel, F. J.; Seligman, J. D.; O'Connors, H. B.; Parker, J. H.; Woodhead, P. M. J.; Dayal, R.; Chezar, B; Roberts, B. K.; Mullen, H. In Ocean Dumping of Industrial Wastes, Ketchum, B. H.; Kester, D. R.; Park, P. K., Eds.; Plenum: New York, 1981; pp 315-46.
17. Mazurek, D. F. M.S. Thesis, Florida Institute of Technology, Melbourne, 1984.
18. American Society for Testing and Materials. In Annual Book of ASTM Standards: Part 14; ASTM: Philadelphia, 1979; pp 24-7.
19. American Society for Testing and Materials. In Annual Book of ASTM Standards: Part 14; ASTM: Philadelphia, 1979; pp 380-82.
20. Verbeck, G. J. Pore Structure. ASTM Special Technical Publication, Philadelphia, 1968, 169a, 211-19.
21. Lambe, T. W. Soil Testing for Engineering; Wiley-Interscience: New York, 1951, p 165.
22. Woodhead, P. M. J.; Parker, J. H.; Carleton, H. R.; Duedall, I. W. Coal Waste Artificial Reef Program, Phase 4B; Interim Report. EPRI Report CS-3726, Electric Power Research Institute: Palo Alto, California; 1984.
23. Waste Management Institute. The Fixation of Incineration Residues: Phase II, Interim Report; Prepared for New York State Legislative Commission on the Water Resource Needs of Long Island. Waste Management Institute, Marine Sciences Research Center, State University of New York, Stony Brook; 1987; p 221.

24. Silberman, D.; Fisher, G. L. Anal. Chem. 1979, 106, 299-307.
25. Labotka, A. L.; Duedall, I. W.; Harder, P. J.; Schlotter, N. J. In Wastes in the Ocean, Vol. 4, Energy Wastes in the Ocean; Duedall, I. W.; Kester, D. R.; Park, P. K.; Ketchum, B. H., Eds.; Wiley-Interscience: New York, 1985; pp 717-39.
26. Shieh, C. S.; Duedall, I. W.; Kalajian, E. H.; Wilcox, J. R. The Env. Prof. 1989, 11, 185-88.
27. Breslin, V. T; Roethel, F. J.; Schaeperkoetter, V. P. Mar. Poll. Bull. 1988, 19, 628-632.
28. Fisher, G. L.; McNeill, K. L.; Prentice, B. A.; A. R. McFarland. Environ. Health Perspectives, 1983, 51, 181-186.
29. Sawell, S. E.; Constable, T. W. 10th Canadian Waste Mgt. Conf., Winnipeg; 1988, p 20.
30. Hazardous Waste, Proposed Guidelines and Regulations and Proposal on Identification and Listing; Federal Register, 1978, 43, No. 243, Part IV, Section 250.13, p 58955.
31. Duedall, I. W.; Buyer, J. S.; Heaton, M. G.; Oakley, S.; Okubo, A.; Dayal, R.; Tatro, M.; Roethel, F. J.; Wilke, R. J; Woodhead, P. M. J. In Wastes in the Ocean, Vol. 1, Industrial and Sewage Wastes in the Ocean; Duedall, I. W.; Ketchum, B. H.; Park, P. K.; Kester, D. R., Eds.; Wiley-Interscience: New York, 1983; pp 375-95.
32. Edwards, T.; I. W. Duedall. In Wastes in the Ocean, Vol. 4, Energy Wastes in the Ocean; Duedall, I. W.; Kester, D. R.; Park, P. K.; Ketchum, B. H., Eds.; Wiley-Interscience: New York, 1985; pp 741-55.
33. Van der Sloot, H.; Wijkstra, J.; van Stigt, C. A.; Hoede, D. In Wastes In the Ocean, Vol. 4, Energy Wastes in the Ocean; Duedall, I. W.; Kester, D. R.; Park, P. K.; Ketchum, B. H., Eds.; Wiley-Interscience: New York, 1985; pp 467-97.
34. Roethel, F. J. Ph.D. Thesis, State University of New York, Stony Brook, 1981.
35. Kalajian, E. H.; Duedall, I. W.; Shieh, C. S.; Wilcox, J. R. 6th Sym. Coast. Ocean Mgt/ASCE, 1989, pp 1102-15.
36. Nelson, W. G.; Navratil, P. M.; Savercool, D. M.; Vose, F. E. Mar. Poll. Bull. 1988, 19, 623-27.
37. Metz, S.; Trefry, J. H. Mar. Poll. Bull. 1988, 19, 633-36.
38. Zawada, D. M.S. Thesis, Florida Institute of Technology, Melbourne, 1989, In Preparation.

RECEIVED November 10, 1989

Chapter 20

Treatment and Disposal Options for a Heavy Metals Waste Containing Soluble Technetium-99

W. D. Bostick, J. L. Shoemaker, P. E. Osborne, and B. Evans-Brown

Oak Ridge Gaseous Diffusion Plant, Martin Marietta Energy Systems, Inc., P.O. Box 2003, Oak Ridge, TN 37831–7272

Various equipment decontamination and uranium recovery operations at the Portsmouth Gaseous Diffusion Plant generate a "raffinate" waste stream characterized by toxic heavy metals, high concentrations of nitric acid, and low levels of radioactive nuclides (^{235}U and ^{99}Tc). Dilution and adjustment of solution pH to a value of 8.2 to 8.5 precipitates heavy metals that can be hydrolyzed. The precipitant is concentrated by paper filtration to yield a filter cake heavy metals sludge (HMS) and HMS filtrate. The HMS fraction may be incorporated into cement-based grout containing ground blast furnace slag to reduce the mobility of its toxic and radioactive components. Sorption of soluble mercury, pertechnetate, and nitrate anions from the HMS filtrate was tested using organic resins and inorganic sorbents. Removal of Hg and ^{99}Tc by iron filings is efficient and economical, generating a small volume of spent sorbent amenable to co-disposal with HMS in a grout waste form, but is slow. For more rapid sorption, poly-4-vinylpyridine resin is very effective for the removal of soluble ^{99}Tc with little uptake of interfering anions at near-neutral influent pH values.

Technetium-99 (^{99}Tc) undergoes radioactive decay by low-energy (0.29 MeV) beta emission, with a half-life of 213,000 years and a specific activity of 0.0171 Ci/g. It is formed with a high yield (~6%) in the thermal neutron fission of ^{235}U. Typical low-enrichment light-water nuclear power reactors operated at optimum fuel exposure may yield 27.5 kg (467 Ci) of ^{99}Tc per 1000 MW-year of electric energy produced (1). Technetium from the nuclear fuel cycle may enter the environment as a result of disposal of aqueous wastes or during the separation and recovery of spent nuclear fuels (2).

The predominant form of Tc under oxic conditions is the pertechnetate anion (TcO_4^-) (3), which is highly soluble in water and readily mobile in the environment (1,2). Certain reduced forms of Tc [e.g., hydrolyzed Tc(IV) oxides (4)] have limited solubility. Technetium-99 is of particular concern because of its persistence and relative mobility (2,5).

0097–6156/90/0422–0345$06.75/0

Technetium entered the U.S. gaseous diffusion uranium enrichment complexes as a volatile fluorination product impurity in reprocessed uranium from plutonium-production reactors that was fed to the Paducah Gaseous Diffusion Plant (PGDP) (**6**). This UF_6 product, with ^{99}Tc impurity, was subsequently fed to the Oak Ridge and Portsmouth enrichment cascades. Most of the technetium remains within the cascade adsorbed on surfaces. However, some ^{99}Tc is periodically removed from these surfaces by wet-chemical decontamination for personnel protection when equipment is removed for maintenance or repair. At the X-705 facility of the Portsmouth (Ohio) Gaseous Diffusion Plant (PORTS), decontamination is achieved with nitric acid, which solubilizes uranium residue and ^{99}Tc. Uranium is recovered by solvent extraction, with the great majority of the technetium remaining in the aqueous raffinate waste.

Composition of the raffinate waste stream varies, but it is generally characterized by toxic heavy metals, high concentrations of nitric acid, and relatively low levels of radioactivity (Table I). The treatment of the raffinate stream generally consists of (1) dilution with an equal volume of water and pH adjustment to about 8.5 to precipitate the hydrolyzable heavy metals, (2) filtration of the precipitation slurry to yield a wet filter cake designated as heavy metals sludge (HMS) and an HMS filtrate, (3) processing of the HMS filtrate with a strong-base anion exchange resin to remove the soluble pertechetnate ion, (4) biodenitrification, and (5) sewage disposal (**7**).

This operation lowers the concentration of most heavy metals to levels below regulatory concern in the HMS filtrate (Table I). However, some nonhydrolyzable anionic species—such as pertechnetate, nitrate, and complexed mercury ions—are not removed by this treatment and remain in the HMS filtrate and HMS sludge as a result of incomplete dewatering (**8,9**).

Nitrate ion concentration in the HMS filtrate can be greatly reduced by biodenitrification, but toxic metals and radionuclides must first be removed to avoid formation of a voluminous contaminated biosludge that would not qualify for economical disposal by simple landfill. Anion exchange with a type 1 (quaternary amine) strong-base resin (Dowex SRB-OH) has been used at PORTS to remove residual technetium activity. However, this resin has demonstrated a modest technetium loading capacity because of competition by other anions, thus generating a relatively large volume of spent sorbent for storage or disposal. In addition, an amine-type resin in the presence of sorbed oxidant (e.g., nitrate ion) is thermodynamically unstable and may present a potential hazard upon prolonged storage of spent resin (**10**).

Inexpensive inorganic sorbents or reductants (ferrous sulfide and iron metal filings) are very efficient in removing technetium and certain other metal ions of regulatory concern, while generating a relatively small bulk of spent sorbent that is amenable to disposal in a grout waste form. However, the reduction of Tc(VII) to Tc(IV) is slow (batch equilibration times of several hours are recommended) (**8**). For more rapid contaminant removal (e.g., in treatment of relatively large volumes of slightly contaminated groundwater), resin materials may be advantageous. Therefore, we investigated the ability of several sorbents to remove soluble ^{99}Tc in the HMS filtrate.

Cement fixation of chemically toxic heavy metals is often regarded as the "best available technology" for reducing their potential for environmental impact. The sludge filter cake from the raffinate waste treatment may be stabilized in a cement-based grout waste form for final disposal. Use of granulated blast furnace slag (BFS) or other reductive admixtures in the grout greatly decreases the leachability of ^{99}Tc from the waste form (**9**).

Table I. Analysis of PORTS raffinate and filtrate from HMS precipitation of raffinate

Constituent or characteristic	Concentration (mg/L)		
	Raw raffinate		HMS filtrate[b]
	Typical range[a]	Sample as tested	
Aluminum		11,500	8.5
Barium		13.5	0.47
Cadmium		9.8	<0.03
Chromium		66	<0.1
Copper	15 to 360	160	<0.04
Iron	700 to 7,100	5,300	<0.04
Lead		56.5	<0.5
Manganese		105	<0.01
Nickel	80 to 620	635	<0.1
Uranium	2 to 1,450	45	<0.3
Zinc	15 to 210	150	<0.01
Mercury		2.32	0.92
Nitrate	250,000 to 400,000	250,000	94,500
Sulfate		2,600	330
Technetium	0.1 to 120	34	13
Gross alpha, pCi/L		410,000	<4,500
pH	<1	0.11	8.5
Specific gravity	1.05 to 1.25	1.20	1.15

[a]Raffinate characteristics reported by Acox (7).
[b]HMS filtrate prepared from PORTS raffinate sample (as received) by dilution, neutralization, and filtration (1.5 L of raw raffinate yields 4.0 L of slurry, before filtration).

Experimental

Chemical Analysis

To quantify ^{99}Tc activity, an aqueous sample (up to 4 mL) is mixed with 10 mL of scintillation cocktail (Aquasol) and counted in an LKB-Wallac Model 1211 Rackbeta liquid scintillation counter (**11,12**). Data are corrected for background as assessed by a distilled water sample in cocktail. Nitrate ion activity is assessed with use of an Orion Research model 93-07 nitrate ion selective electrode and an Orion model 901 ionalyzer. Heavy metals concentrations are surveyed by inductively coupled plasma spectroscopy (EPA method 6010), and mercury is analyzed by atomic adsorption spectroscopy (EPA method 245-1).

Sorbent Batch Testing

Weighed amounts of sorbent are added to known volumes of aqueous phase contained in scintillation vials or disposable plastic centrifuge cones and shaken at room temperature for ~24 h, after which they are separated by filtration or centrifugation.

A surrogate raffinate solution was prepared in ~2.3 *N* HNO_3 to contain Al (11,900 mg/L), Fe (5500 mg/L), Na (1250 mg/L), NO_3^- (245,000 mg/L), and SO_4^{2-} (2570 mg/L), thus approximating the composition of the major constituents in the raffinate sample from PORTS (cf. Table I). Aliquots of surrogate were adjusted to pH values between 4 and 9 by the addition of NaOH and centrifuged to remove hydrolyzed iron and aluminum. Samples of the clarified, pH-adjusted solutions were spiked with NH_4TcO_4 standard to a concentration of about 0.2 mg/L ^{99}Tc.

Detailed information on sorbents used in this study is given in Reference 8. Sorbent dosages refer to the mass of sorbent "as received" from the vendor, with no preconditioning. Dowex SRB-OH (hydroxide form of Dowex-1X8, Bio-Rad Laboratories) is the Type 1, strongly basic, beaded, porous polystyrene matrix resin historically used by the PORTS facility for ^{99}Tc removal. The as-received resin has an approximate Gaussian particle size distribution (mean diameter 620 ± 180 μm) and a moisture content of 54% (w/w). The as-received Reillex 402 resin (Reilly Industries, Inc.) is a granular material with a lognormal particle size distribution (median diameter 120 μm and spread factor 1.6) and a moisture content of 8.2% (w/w).

Several iron metal powders were tested and shown to efficiently reduce ^{99}Tc. The results reported here are for degreased iron filings, about 40 mesh (Fisher Scientific product no. I-57).

Packed Column Solute Breakthrough Studies

Aqueous slurries of resin are dispensed into disposable plastic chromatographic columns (Kontes Scientific, product no. 420161) until a nominal 5-mL bed volume is attained [0.76-cm ID by 11-cm bed height]. The column is rinsed with distilled water and allowed to drain by gravity. Then 5- or 10-mL aliquots (one or two nominal bed volumes) of HMS filtrate are sequentially applied to the column and allowed to drain into a collection vessel. Aliquots of the effluent are counted to determine the breakthrough of ^{99}Tc.

Grout Leach Testing Protocols

Specific waste form testing protocols and analytical and quality assurance procedures used in this investigation are detailed in documents by Gilliam and coworkers at the Oak Ridge National Laboratory (**13,14**).

The EPA Extraction Procedure Toxicity (EP-Tox) Test [described in 40 CFR 261, Appendix II (**15**)] is a laboratory-scale procedure designed to simulate the leaching a waste would undergo if disposed of in an improperly designed sanitary landfill. A representative waste sample is extracted for a 24-h interval with distilled water at a targeted solution pH of 5.0 ± 0.2, using 0.5 *N* acetic acid to maintain the pH. The extract is then analyzed using established analytical procedures for selected constituents [EPA/SW-846 (ref. **16**)]. If any of the designated contaminants exceed a threshold value (100 times the National Interim Primary Drinking Water Standard) (**17**), the waste is a candidate to be listed as a hazardous waste. The waste may be excluded as a hazardous waste by petitioning the EPA and providing proof that the waste was treated prior to disposal and is not hazardous [40 CFR 260.22(d) (**18**)]. Thus, if a raw waste fails the EP-Tox Test procedure, delisting may still be possible if the immobilized waste form passes the procedure (**5**). Waste that has been delisted as nonhazardous may be a candidate for relatively economical storage, disposal, and transportation options not available for the original nonstabilized hazardous waste (**5**).

The ANSI/ANS-16.1-1986 standard, "Measurement of the Leachability of Solidified Low-Level Radioactive Wastes by a Short-Term Test Procedure" (**19**), defines a leachability index as a parameter that characterizes the resistance of the solidified waste toward release of radioactive species. It expresses the leaching data in terms of basic mass transport theory (viz., diffusion from a semi-infinite medium). The index may serve as a figure of merit for the evaluation and comparison of different solidification systems and for the optimization of the associated processes. The standard requires that values of "effective diffusivity" for constituents of interest be determined for different leaching times in replacement leachant (demineralized water) and averaged:

$$D = \pi \left[\frac{a_n/A_o}{(\Delta t)_n}\right]^2 \left[\frac{V}{S}\right]^2 (T) \quad (1)$$

where

a_n = activity of a nuclide released from the sample during leaching interval n, corrected for radioactive decay,

A_o = total activity of a given radionuclide in the sample at the beginning of the leach test (i.e., after the initial 30-s rinse),

$(\Delta t)_n$ = $t_n - t_{n-1}$, duration of the *n*th leaching interval, s,

D = effective diffusivity, cm^2/s,

V = volume of sample, cm^3,

S = geometric surface area of the sample as calculated from measured dimensions, cm^2,

T = $[0.5\,(t_n^{0.5} + t_{n-1}^{0.5})]^2$, leaching time, representing the "mean time" of the leaching interval, s.

The negative logarithm of this average diffusivity in square centimeters per second is the defined leachability index.

$$L_i = \frac{1}{n} \sum_{1}^{n} [\log (\beta/D_i)]_n \tag{2}$$

where β is a defined constant (1 cm^2/s) and D_i is the effective diffusivity of nuclide i calculated from the test data. Values of L_i become greater as the diffusion rate D_i values become smaller. In this study, large values of L_i are desirable, and since L_i is a logarithm term, small increases in the values may be significant.

For purposes of quality assessment, we employed the abbreviated test (i.e., 5 days of leach testing) (**19**).

Results and Discussion

Sorbent Batch Testing Using HMS Filtrate

Trace element removal by sorbents may be influenced by (1) the element oxidation state and initial concentration, (2) sorbent dosage, (3) pH, and (4) general solution conditions (e.g., by the presence of competing ions or complexing ligands) (**20**). For the waste samples in this study, ^{99}Tc exists primarily as pertechnetate ion, and the nitrate ion is the principal competing species.

Removal of Soluble ^{99}Tc. In our initial scoping studies, we used a surrogate raffinate solution (see experimental section), adjusted to a range of pH values between 4 and 9, a single sorbent dosage (10 g/L), and a 24-h contact time to allow an approach to equilibrium. Results are illustrated in Table II. In general, we noted little or no significant effect of surrogate pH upon ^{99}Tc sorption efficiency under the defined experimental conditions. In general, the materials most efficient in removing soluble ^{99}Tc were inorganic solid-phase reducing agents (iron and ferrous sulfide), polyvinylpyridine (PVP) resins, and strong base anion exchange resins. Iron oxyhydroxides (hematite and magnetite) were not effective for Tc removal in surrogate media under oxic conditions [although Tc(VII) may be reduced and Tc(IV) may be chemisorbed by iron-containing minerals under anoxic conditions (**21**)]. Alumina and weak-base anion exchange resin demonstrated no selective sorption of Tc under the present experimental conditions, similar to results reported by Palmer and Meyer (**22**). Blast furnace slag, used to enhance the retention of Tc in grout host materials, was not effective in reducing Tc under conditions of short-term oxic exposure. Indian red pottery clay, an illite used to sorb radiocesium, did not affect soluble Tc.

Figure 1 illustrates sorption isotherms for removal of soluble ^{99}Tc in HMS filtrate (**8**), using several of the more effective sorbents that were identified in testing with surrogate raffinate. Dowex SRB-OH removes ^{99}Tc from the HMS filtrate, along with a significant amount of nitrate ion (about 76 mg nitrate per gram of resin) (**8**), but was less efficient than Reillex 402.

Reillex 402-I, an industrial grade of Reillex 402, was significantly less effective for removal of soluble Tc than is the purified resin. These PVP resins have pK_a values of

Table II. Technetium-99 removal efficiency for selected sorbents using surrogate solutions[a]

Sorbent	Type	pH range[b]	Percent Tc removed
Iron metal[c]	Inorganic	7 to 9	99.7 (±0.7)
"Ferrous Sulfide" (Fe_3S_4)[d]	Inorganic	8.1 to 8.3	92.1 (±3.6)
Reillex 402[e]	PVP resin	5.3 to 10.4	75.7 (±1.7)
Dowex SRB-OH[f]	Strong-base anion-exchange resin	4.3 to 11.3	46.8 (±1.0)
Reillex 425[g]	PVP resin	6.3 to 7.6	39.9 (±5.4)
Amberlyte A-26[h]	Strong-base anion-exchange resin	4.0 to 7.3	37.3 (±2.4)
Reillex 202[i]	PVP resin	5.0 to 8.0	31.8 (±2.1)
Fe_2O_3[j] (hematite)	Inorganic	3.9 to 6.9	5.0 (±3.1)
Fe_3O_4[j] (magnetite)	Inorganic	3.9 to 8.8	3.2
Blast furnace slag	Inorganic	11	3.0 (±3.9)
Alumina[k]	Inorganic	8.8	2.0 (±1.6)
Amberlyte IRA-68[h]	Weak-base anion exchange resin	4.7 to 10.3	0
Indian red pottery clay[l]	Inorganic	3.9 to 8.0	0
Zeolites[m]	Inorganic	4 to 9	0

[a]Initial (precontact) nominal pH values for surrogate ranged from 4 to 9. Surrogate contained 0.2 mg ^{99}Tc and 210,000 mg nitrate ion per liter. Sorbent dosage was 10 g/L. Sorbent and surrogate are contacted for 24-h before phase separation.

[b]Reported pH range is for solution phase after sorbent contact.

[c]Fisher Scientific Company, product I-57.

[d]Greigite (prepared in house).

[e]Poly-4-vinylpyridine cross-linked with 2% divinylbenzene (Reilly Industries).

[f]Bio-Rad Laboratories.

[g]Poly-4-vinylpyridine cross-linked with 25% divinylbenzene (Reilly Industries).

[h]Rohm & Haas.

[i]Poly-2-vinylpyridine cross-linked with 2% divinylbenzene (Reilly Industries).

[j]Thiokol/Ventron Division.

[k]Fisher Scientific Company, product no. A-540.

[l]American Art Clay Company.

[m]Materials tested include PDZ-300, PDZ-140D, and PDZ-150D (Tennessee Specialty Minerals); CH clinoptilite (Chem Nuclear); and IE-96 (Linde Division, Union Carbide Corp.).

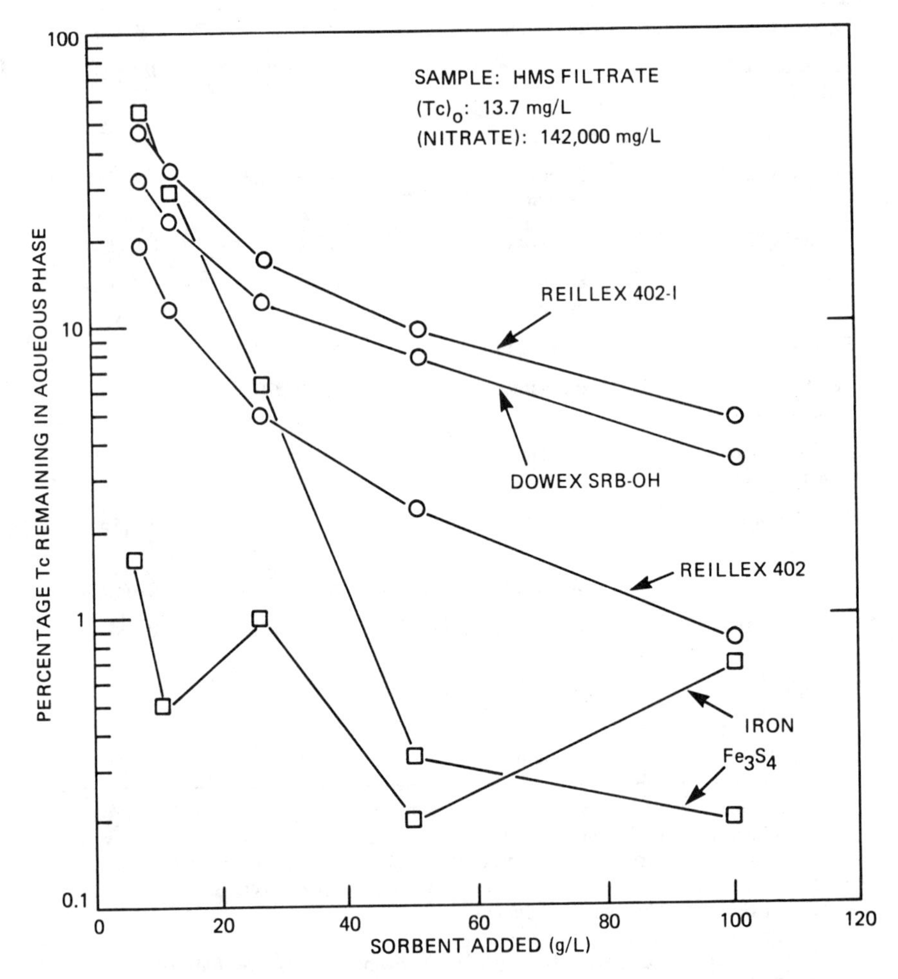

Figure 1. Technetium removal from heavy metals sludge filtrate by selected sorbents (24-h equilibration time).

~3 to 4 and thus may act as weak base anion exchangers in acidic media. In the slightly alkaline HMS filtrate, the pyridine moiety is essentially a free base, and the resin demonstrates little uptake of nitrate or other anions. The mechanism for uptake of pertechnetate is thus believed to be predominantly sorptive (**23,24**). As indicated in Table II, the poly-4-vinylpyridine resin was superior to poly-2-vinylpyridine for ^{99}Tc removal.

Several ferrous sulfide preparations were tested for removal of soluble Tc [predominantly by reduction to Tc(IV) compounds, although Tc_2S_7 is also sparingly soluble] (**2**). In general, the ^{99}Tc removal efficiency correlated with the redox potential (Eh) of the treated sample, with Greigite (Fe_3S_4) being the most efficient sulfide preparation tested. Ferrous sulfide (**25,26**) and iron metal (**25,27**), especially scrap iron (**28**), are inexpensive and very efficient for the removal of soluble pertechnetate by sorption and reduction to less soluble forms of Tc. Preliminary investigations also indicate that iron blast furnace flue dust waste is also a relatively effective sorbent or reductant for soluble ^{99}Tc. Because of their relatively high bulk density, these inorganic sorbents are capable of a high loading of ^{99}Tc per unit volume of spent sorbent, and the spent sorbent can be disposed of in a stabilized grout waste form (**9**). Reduction of Tc by these inorganic sorbents is relatively slow kinetically (**9,27**). Thus, they are best suited for batch equilibration applications.

Removal of Soluble Complexed Mercury Ion. Sorbents effective in removing soluble ^{99}Tc in the HMS filtrate were also effective in removing complexed mercury ion (**9**), with iron metal being most effective (Figure 2).

Removal of Soluble Cr(VI). Scrap iron powder is an efficient and economical reagent for the treatment of chromate-containing metal plating wastewaters (**28**). Soluble Cr(VI) is reduced by iron, in acidic medium, to form hydrolyzable Cr(III), which is removed by co-precipitation with iron oxyhydroxide upon solution neutralization. The HMS filtrate does not contain a significant concentration of soluble Cr(VI) (Table I). Therefore, to test the efficacy of iron metal treatment, we first added a small amount of $K_2Cr_2O_7$ to the filtrate, then treated the solution with 100 g iron filings per liter of solution. In our study, soluble Cr and ^{99}Tc concentrations were each reduced by over 99% (i.e., from 13.4 to 0.097 mg/L for ^{99}Tc and from 82.1 to 0.087 mg/L for Cr). The treatment with iron was performed without any prior pH adjustment.

^{99}Tc Breakthrough in Resin Columns

The use of iron or ferrous sulfide for ^{99}Tc removal is advantageous in terms of economy and waste minimization, but the relatively slow kinetics may be a disadvantage for treatment of relatively large volumes of slightly contaminated water in a continuous mode. We tested the ability of Dowex SRB-OH and Reillex 402 resins (5-mL bed volume columns) to overcome this problem by gravity flow elution with aliquots of HMS filtrate (Figure 3 and Tables III and IV). There was significant breakthrough of ^{99}Tc from the Dowex-SRB column after the first nominal bed volume of HMS filtrate applied, whereas the Reillex 402 column showed little breakthrough until the column had sorbed about 1 mg ^{99}Tc per gram resin (Figure 3 and Table IV). As HMS filtrate was applied to the Reillex resin, a yellow stain appeared on the off-white resin; the stain extended to the entire length of the column after application of ~30 bed volumes of HMS filtrate, at which time there was a corresponding increase in effluent concentration of ^{99}Tc (Figure 3).

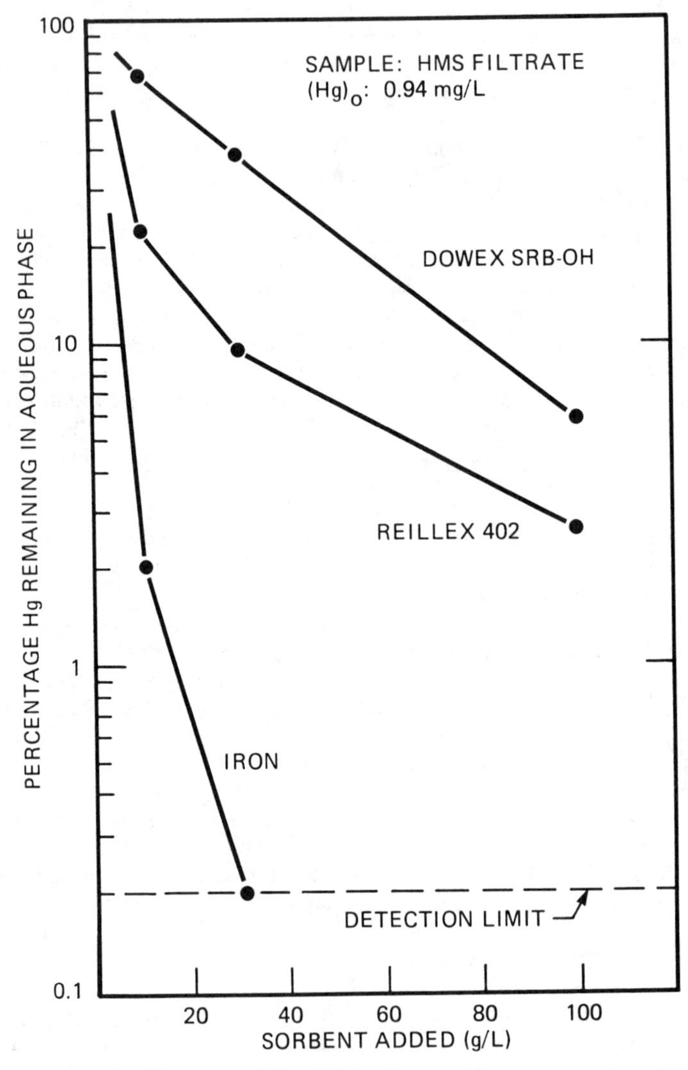

Figure 2. Soluble mercury removal from heavy metals sludge filtrate by selected sorbents (24-h equilibration time).

In the present study, the PVP resin demonstrated superior ^{99}Tc loading characteristics compared with a strong-base anion exchange resin; that is, more ^{99}Tc was sorbed with less breakthrough for a given volume of contaminated influent. This likely reflects the greater selectivity for Tc by the PVP sorbent. The Reillex resin used in this study is granular, and flow is less rapid than for the beaded Dowex resin; under experimental conditions, average flow as 1.3 to 2.9 mL/min/cm^2 for Reillex and 8.8 to 13 mL/min/cm^2 for Dowex. The granular PVP resin may be used in an upflow mode as

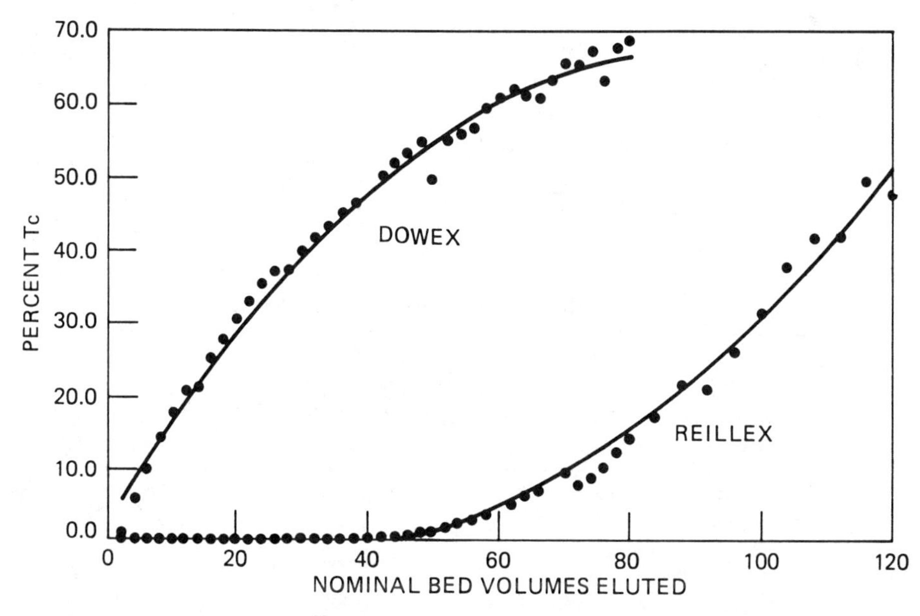

Figure 3. Breakthrough of ^{99}Tc from heavy metals sludge filtrate applied to 5-mL bed volume resin columns. Dowex resin refers to Dowex SRB-OH, and Reillex resin refers to Reillex 402.

Table III. Dowex SRB-OH column breakthrough data

Nominal bed volumes eluted[a]	Effluent ^{99}Tc (C/C_o)	Cumulative Total ^{99}Tc retained (mg)	Cumulative fraction total ^{99}Tc retained	Average ^{99}Tc loading on resin (mg ^{99}Tc/g, as received)
10	0.175	0.604	0.902	0.245
20	0.307	1.105	0.823	0.447
30	0.400	1.528	0.760	0.619
40	0.388	1.907	0.712	0.772
50	0.503	2.224	0.664	0.901
60	0.614	2.504	0.623	1.014
70	0.662	2.750	0.587	1.114
80	0.695	2.971	0.555	1.203

[a]Nominal slurry-packed resin bed volume is 5.0 mL [0.76-cm diam by 11-cm column height]. Column is eluted with HMS filtrate, containing 13.2 mg (228 μCi) ^{99}Tc and 94.5 g nitrate ion per liter. Average eluent flow was 4 to 6 mL/min. Drained, packed column contains 3.36 g total water and 1.127 g resin (dry-weight basis), equivalent to 2.47 g as-received resin.

Table IV. Reillex 402 column breakthrough data

Nominal bed volumes eluted[a]	Effluent ^{99}Tc (C/C_o)	Cumulative Total ^{99}Tc retained (mg)	Cumulative fraction total ^{99}Tc retained	Average ^{99}Tc loading on resin (mg ^{99}Tc/g, as received)
10	0.000	0.670	1.000	0.325
20	0.000	1.339	1.000	0.650
30	0.002	2.008	1.000	0.975
40	0.005	2.675	0.999	1.299
50	0.016	3.337	0.997	1.620
60	0.040	3.985	0.992	1.935
70	0.096	4.605	0.982	2.235
80	0.144	5.201	0.971	2.525
90	0.220	5.742	0.953	2.787
100	0.317	6.234	0.931	3.026
110	0.383	6.642	0.902	3.224
120	0.482	7.002	0.871	3.399

[a]Nominal slurry-packed resin bed volume is 5.0 mL [0.76-cm diam by 11-cm column height]. Column is eluted with HMS filtrate, containing 13.2 mg (228 μCi) ^{99}Tc and 94.5 g nitrate ion per liter. Average eluant flow rate was 0.6 to 1.3 mL/min. Drained, packed column contains 3.46 g total water and 1.89 g resin (dry-weight basis), equivalent to 2.06 g as-received resin.

a fore-column to minimize ^{99}Tc contamination of a less selective sorbent (e.g., ion-exchange resin or activated charcoal). Alternately, a beaded form of PVP (e.g., Reillex 425 resin) could be used.

Stabilization of Heavy Metals Sludge

Concentrations for selected constituents in a "representative" sample of heavy metals sludge from PORTS and the EP-Tox Test leachate from this sample are given in Table V. The sludge contains high concentrations of hydrolyzed toxic heavy metals, appreciable quantities of radionuclides from low-enriched uranium (at 5.77 wt % ^{235}U isotope) and ^{99}Tc, and a high concentration of soluble nitrate ion. Leachate from the raw HMS fails the EP-Tox criteria for cadmium and lead. Thus, it is a candidate to be listed as a hazardous waste. In addition, nitrate ion and ^{99}Tc are quantitatively leached by this procedure.

The HMS must be stabilized or placed in containers for long-term storage or disposal. Most heavy metals, including cadmium and lead, are effectively immobilized in the high-pH pore water within cement-based grouts (**9,29,30**), although anionic species (such as TcO_4^-) remain relatively mobile. Ground BFS has been shown to improve the leach performance of cement-based waste forms, particularly with respect to Tc (**9,31,33,34**). This improved performance has been attributed to the formation of smaller, more tortuous pores in the solidified waste form. An additional specific interaction for Tc(VII) and other redox-sensitive solutes is the residual reductive potential of pore water in cement-BFS blends (**35,36**).

Leach Testing Using HMS Filtrate as Surrogate Waste. Reducing Tc(VII) to hydrated TcO_2 should reduce the solubility and the migration rate of ^{99}Tc. Brodda reports that the leachability of electrolytically reduced ^{99}Tc (i.e., TcO_2) from aluminous or portland cement grouts is a factor of 10 to 100 lower than the corresponding leachability of ^{99}Tc as pertechnetate ion (**36**). In a similar study we immobilized HMS filtrate (containing a relatively high activity of TcO_4^-, for accurate determination of leach rate), with and without pretreatment with iron filings (Table VI for formulations). As discussed previously, batch treatment with iron at dosage of 100 g/L removes >99% of the soluble ^{99}Tc by reduction to Tc(IV) hydroxides which sorb to the metal surface. Representative data for ^{99}Tc leaching from OPC–fly ash grout are presented graphically in Figure 4. Since a plot of cumulative fraction contaminant leached (A_n/A_o) versus square root of time (Equation 1) should be linear for a simple diffusion-limited release, we can conclude that this is diffusion-limited process. Under the testing conditions, the average ANS-16.1 leachability indices are 8.6 ± 0.4 for Tc(VII) and 10.4 ± 0.3 for Tc(IV) (i.e., a decrease in effective diffusivity of ~70 for reduced ^{99}Tc).

Table VI also summarizes leach data for untreated HMS filtrate immobilized in a grout containing BFS as a component. Results using BFS from several sources, with and without trace amounts of gypsum or air-entrainment admixture (**9,37**), yielded average ANS-16.1 leachability indices for ^{99}Tc of 10.5 ± 0.5, a value equivalent to the mean leach index for Tc(IV) in a cement paste without BFS (Table VI). The quantity of Tc leached from the grout with time (Figure 5) is biphasic, with a greater release occurring during the initial 24-h interval. This may reflect mobilization of the near-surface technetium or the leaching of two oxidation states of ^{99}Tc at significantly different rates (**32**) (cf. Figure 4).

Table V. Characterization of PORTS HMS and its EPA EP-Tox test leachate

	Concentration of constituent		
	In raw HMS[a]	In EP-Tox Test leachate (mg/L)	
Constituent	(μg/g)	From raw HMS	Regulatory limit[b]
Arsenic	4.5	0.005	5.0
Barium	35	0.77	100
Cadmium	970	16	1.0
Chromium	490	0.42	5.0
Lead	3,550	28	5.0
Mercury	<1	0.0013	0.2
Selenium	<0.5	<0.05	1.0
Silver	5.7	0.019	5.0
Aluminum	96,000	170	
Copper	1,900	9.7	
Iron	36,000	0.51	
Nickel	6,000	23	
Uranium	310	3.84	
Zinc	1,300	8.5	
Nitrate	117,000	5,560	
Technetium	5.13	0.29	

[a]Material received from PORTS (can 994497).
[b]*Code of Federal Regulations*, 40, Part 261.24, U.S. Environmental Protection Agency.

Table VI. Immobilization of HMS filtrate in grout

	HMS filtrate		
	Pretreated with iron[a] (no BFS)	Raw, no BFS	Raw, with BFS
Constituent added to as-poured grout, wt %			
HMS filtrate[b]	38.5	40.0	40.0
Iron filings (about 40 mesh)[c]	3.9		
Portland cement (Type I-II)	28.9	30.0	20.0
Fly ash (Type F)	28.9	30.0	20.0
Blast furnace slag			20.0
ANS-16.1 leachability index (30-day cure)[d]			
^{99}Tc	10.4 ± 0.3	8.6 ± 0.4	10.5 ± 0.5[e]
NO_3^-	ND[f]	ND	7.3 ± 0.1[e]

[a]12-g iron filings and 120-g HMS filtrate were shaken overnight, and both iron and treated filtrate were added to grout dry solids mix (with no BFS component).
[b]Composition described in Table I.
[c]Product I-57, Fisher Scientific Company.
[d]Grout specimens are right cylinders [typical dimensions: 2.5 cm diam and 4.1 cm high (V/S about 0.48 cm), with a mass of about 35 g].
[e]Mean results for six blends, using four different sources of BFS (see Ref. 9).
[f]Not determined.

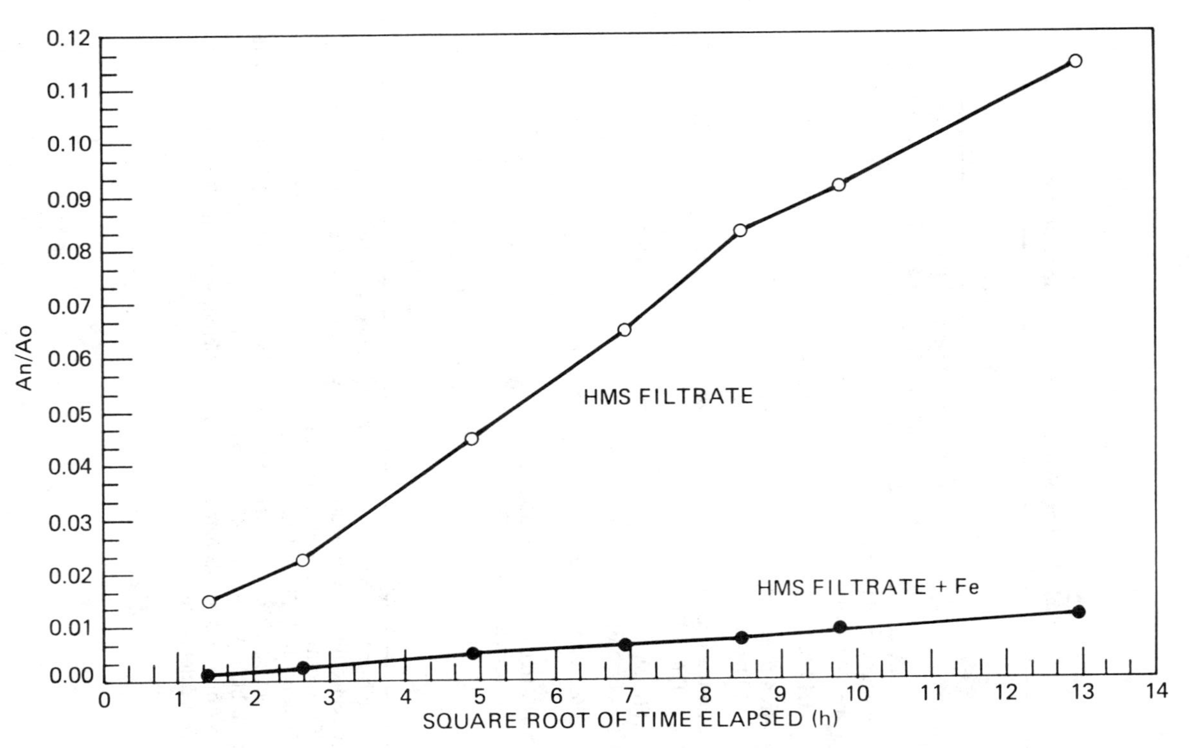

Figure 4. Cumulative fraction of contaminant leached from OPC–fly ash grout versus the square root of time (see Table VI for grout formulation).

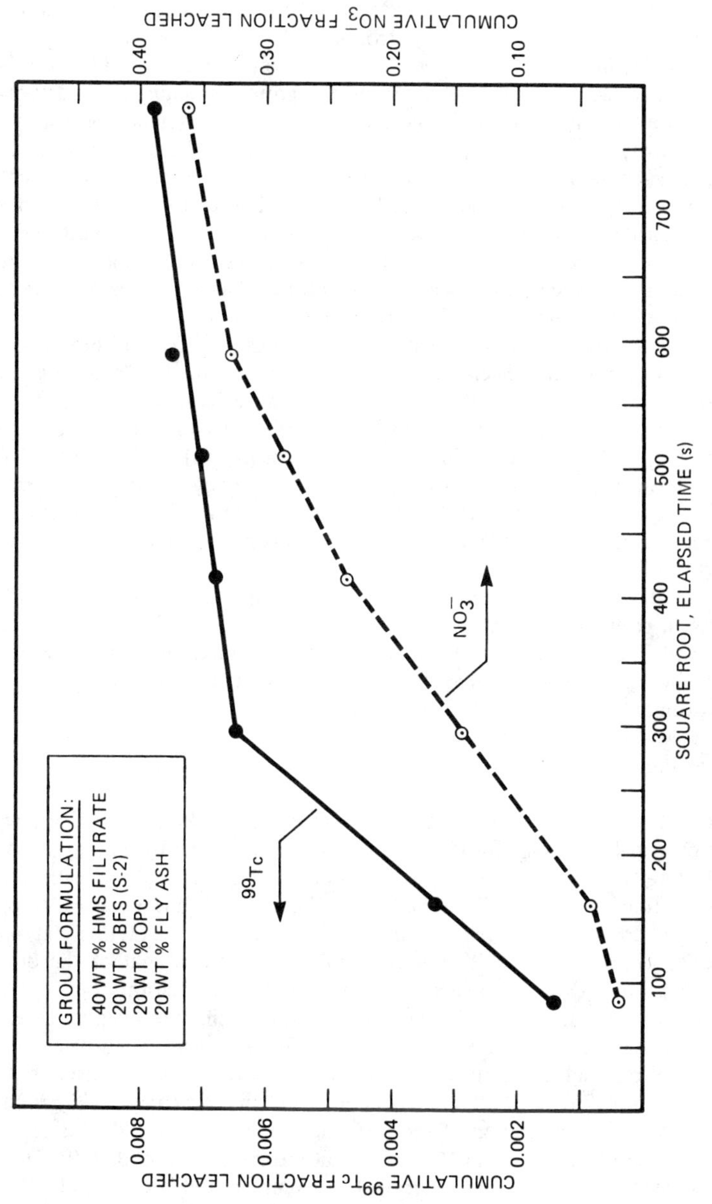

Figure 5. Cumulative fraction of contaminant leached from slag-cement grout versus the square root of time. (Note change in ordinate scale, as indicated by arrows.)

Leach Testing for Waste Forms Containing HMS Waste. Leachability test results for ^{99}Tc and nitrate ion in stabilized HMS are presented in Table VII for grouts prepared with two different dry solid blends. The reference dry solid blend, as used previously for heavy metals immobilization (**30**), consisted of 50 wt % each OPC and fly ash, whereas the comparison blend consisted of 33 wt % each of OPC, fly ash, and BFS. As indicated in Table VII, there is a pronounced effect of waste loading on the ANS-16.1 leachability index for ^{99}Tc, especially for the blend containing BFS. Also there is a dramatic improvement in retention of ^{99}Tc in the grout containing BFS component (i.e., as much as a five-unit increase in the ANS-16.1 leachability index or a decrease in effective diffusivity by as much as five orders of magnitude). The enhanced retention of ^{99}Tc in the slag-containing grout likely reflects the ability of the BFS component to effect reduction of Tc(VII). In contrast, the leachability indices for nitrate ion show a corresponding maximum increase in leach index of less than two units for the slag-containing grout (Table VII). Since nitrate ion is readily soluble in the grout matrix, it is presumed that the increase in its leachability index in slag-containing grout reflects a reduction in waste form porosity or increase in pore tortuosity.

Tallent and coworkers have investigated the retention of ^{99}Tc and nitrate ion in immobilized surrogate Double-Shell Slurry Feed waste for the Hanford Operations facility (**31,32**). They report that retention of ^{99}Tc in grout is generally improved by (1) an increase in the grout-mix ratio (the weight of dry solids blend mixed with a given volume of waste), (2) an increase in grout fluid density, (3) an increase in BFS component in the dry solid blend, and (4) a decrease in relative waste concentration. Glasser and coworkers suggested that the reductive potential of BFS in grout (as assessed by the redox potential of grout pore water) becomes dominant for >75 wt % BFS in the dry solid blend (**35,36**).

In Table VIII, we compare leachability for ^{99}Tc and nitrate ion for a relatively high sludge loading in grout prepared with dry mixes containing 33 wt % and 85 wt % BFS (grouts PSA and PSB, respectively). Retention of ^{99}Tc is significantly improved in the grout with higher slag content, whereas the leachability index for nitrate shows no further improvement.

Direct Calcination of Raffinate

Alderfer and Bundy have reported on the direct calcination of PORTS raffinate waste (**38,39**). The principal advantage for this approach is waste volume reduction; ~5 to 8% of the original raffinate mass remains as solid waste after calcination, with a bulk volumetric reduction of about 95% (excluding scrubber liquids and chemical trapping agents) (**39**). [In contrast, processing raffinate via precipitation of HMS (at about 30 wt % solids) yields a sludge with a volume of about 60% of that of the original raffinate treated, and HMS immobilized in grout yields a wasteform comparable in volume to the original volume of raffinate treated (**9**).] Calcination evaporates excess water and distills some of the nitric acid from the raffinate waste. At sufficiently high temperatures, the residual nitrate in the solids can be thermally decomposed, but the risk of volatilization of ^{99}Tc as $HTcO_4$ (pK_a about 0.1) from the strongly acidic raffinate is greatly increased (**3**). Calcination of the raffinate waste at about 380°C is reported to retain about 90% of the ^{99}Tc and 10% of the nitrate ion in the solids (**39**). The raw calciner ash failed the EP-Tox test criterion for cadmium, and nitrate and ^{99}Tc are

Table VII. ANS-16.1 leachability indices for technetium-99 and nitrate ion as a function of sludge waste loading in grouts prepared with two dry mix blends

Constituent added to as-poured wet grout (wt %)			ANS-16.1 Leachability Index[a] (30-day grout cure)			
			Dry blend 1[b] (no BFS)		Dry blend 2[c] (with BFS)	
Raw HMS[d]	Diluent water	Dry blend	^{99}Tc	NO_3	^{99}Tc	NO_3
36.0	11.2	52.8			7.0	6.2
38.3	11.7	50.0	6.1	5.9		
30.6	19.4	50.0			9.6 ± 0.7[e]	7.2 ± 0.0[e]
30.0	20.0	50.0	6.6 ± 0.1[e] 6.8 ± 0.3[e]	5.9 ± 0.2[e] 6.9 ± 0.2[e]		
13.9	36.1	50.0	7.0 ± 0.1[e] 7.2 ± 0.2[e]	6.6 ± 0.1[e] 7.5 ± 0.3[e]	12.4 ± 0.2[f]	8.0 ± 0.1[f]

[a]Abbreviated test (5 days of leaching).
[b]Dry blend is equal parts (by weight) ordinary portland cement (Type I-II) and class F fly ash.
[c]Dry blend is equal parts (by weight) ordinary portland cement (Type I-II), class F fly ash, and granulated blast furnace slag.
[d]Raw heavy metals sludge contains 32.6% solids and 67.4% water (weight basis).
[e]Result is mean plus or minus standard deviation for three replicate leach specimens.
[f]Result is mean plus or minus one-half the range for duplicate leach specimens.

readily leached (9). The ash may need to be containerized or immobilized for ultimate disposal. Calciner ash immobilized in a cement host at about a 10% waste loading yielded a wasteform volume equivalent to about 30% of the original volume of raffinate treated (9).

Conclusions

Several solid sorbents were investigated for their ability to remove soluble ^{99}Tc in decontamination and uranium recovery wastewaters. PVP resin was found to be more effective than strong-base anion exchange resin for removing TcO_4^- in the presence of competing anions. Under the conditions of testing, the PVP resin demonstrated relatively little breakthrough of Tc until an average resin loading of about 1 mg Tc per gram of resin.

Inorganic sorbents or reductants, especially elemental iron filings or powder, offer an inexpensive means to remove soluble ^{99}Tc and certain other metals of potential regulatory concern [e.g., Hg and Cr(VI)]. Batch equilibration times of several hours are recommended for this procedure. The spent sorbent incorporates a high Tc waste loading per unit volume and is shown to be easily incorporated into a grout matrix, with a relatively low leach rate (Table VI).

Table VIII. Effect of slag content on immobilization of HMS in slag-cement group

Constituent	Blend designation	
	"Low slag" formulation, blend PSA	"High slag" formulation, blend PSB
Constituent added to as-poured wet grout, wt %		
HMS[a]		
As total	30.6	30.6
As solids	10.0	10.0
Water		
As diluent	19.4	19.4
As total[b]	40.0	40.0
Portland cement (Type I-II)	16.7	7.5
Fly ash (Class F)	16.7	0.0
Blast furnace slag	16.7	42.3
Lime [$Ca(OH)_2$]	0.0	0.3
ANS-16.1 leachability index (42-d grout cure)		
^{99}Tc	9.6 ± 0.7	11.4 ± 0.2
NO_3^-	7.2 ± 0.0	6.8 ± 0.1

[a]Raw HMS contains 32.6 wt % solids and 67.4 wt % water.

[b]Total water is the sum of that added from the sludge and the diluent.

Soluble ^{99}Tc in a wet heavy metals treatment sludge is also amenable to stabilization in grout. Blast furnace slag, a by-product obtained in the manufacture of pig iron, possesses latent hydraulic cement properties and is generally available at a cost comparable to that of OPC (9). Addition of this slag to cement dry solids blend significantly enhances the retention of ^{99}Tc in a cement-based grout, presumably because of the reductive potential of the BFS.

Acknowledgments

T. M. Gilliam, R. D. Spence, I. L. Morgan, E. W. McDaniel, and O. K. Tallent of the Oak Ridge National Laboratory provided information and guidance for the development of grout formulations for the immobilization of Tc-contaminated sludges. The Oak Ridge Gaseous Diffusion Plant is operated by Martin Marietta Energy Systems, Inc., under contract DE-AC05-OR21400 with the U.S. Department of Energy.

Legend of Symbols

a_n = activity of a nuclide released from the specimen during leaching interval n, corrected for radioactive decay

A_o = total activity of a given radionuclide in the specimen at the beginning of the leach test (i.e., after the initial 30-s rinse)

D = effective diffusivity, cm^2/s

$(\Delta t)_n$ = $t_n - t_{n-1}$, duration of the *n*th leaching interval, s

S = geometric surface area of the specimen as calculated from measured dimensions, cm^2

T = $[0.5\ (t_n^{0.5} + t_{n-1}^{0.5})]^2$, leaching time, representing the "mean time" of the leaching interval, s

V = volume of specimen, cm^3

β = a defined constant (1 cm^2/s)

D_i = the effective diffusivity of nuclide i calculated from the test data

Literature Cited

1. Garten, C. T. Jr. *Environ. Int.* 1987, **13**, 311.
2. Wildung, R. E.; McFadden, K. M.; Garland, T. R. *J. Environ. Qual.* 1979, **8**, 156.
3. Pourbaix, M. *Atlas of Electrochemical Equilibria*; Pergamon Press: Oxford, 1966; pp 294–299.
4. Meyer, R. E.; Arnold, W. D.; Case, F. I.; O'Kelley, G. D. *Thermodynamic Properties of Tc(IV) Oxides: Solubilities and the Electrode Potential of the Tc(VII)/Tc(IV)-Oxide Couple*; NUREG/CR-5108 (ORNL-6480); Martin Marietta Energy Systems, Inc., Oak Ridge National Laboratory: Oak Ridge, Tenn., May 1988.

5. Mattus, A. J.; Gilliam, T. M.; Dole, L. R. *Review of EPA, DOE, and NRC Regulations on Establishing Solid Waste Performance Criteria*; ORNL/TM-9322; Martin Marietta Energy Systems, Inc., Oak Ridge National Laboratory: Oak Ridge, Tenn., July 1988.
6. Saraceno, A. J. *Control of Technetium at the Portsmouth Gaseous Diffusion Plant*; GAT-2010; Goodyear Atomic Corp.: Piketon, Ohio, November 1981.
7. Acox, T. A. *Proc. Fourth DOE Environmental Protection Information Meeting*; CONF-821215, August 1983, pp 77–85.
8. Bostick, W. D.; Evans-Brown, B. S. *Sorptive Removal of Technetium from Heavy Metals Sludge Filtrate Containing Nitrate Ion*; K/QT-160; Martin Marietta Energy Systems, Inc., Oak Ridge Gaseous Diffusion Plant: Oak Ridge, Tenn., January 1988; available from NTIS.
9. Bostick, W. D.; Shoemaker, J. L.; Fellows, R. L.; Spence, R. D.; Gilliam, T. M.; McDaniel, E. W.; Evans-Brown, B. S. *Blast Furnace Slag-Cement Blends for the Immobilization of Technetium-Containing Wastes*; K/QT-203; Martin Marietta Energy Systems, Inc., Oak Ridge Gaseous Diffusion Plant: Oak Ridge, Tenn., November 1988; available from NTIS.
10. Gangwer, T. E.; Goldstein, M.; Pillay, K. K. S. *Radiation Effects on Ion Exchange Materials*; BNL-50781; Brookhaven National Laboratory, November 1977.
11. Rucker, T. L.; Mullins, W. T. *Proc. Conf. Anal. Chem. Energy Technol.* 1980, 23rd, pp 95–100.
12. Pacer, R. A. *Int. J. Appl. Radiat. Isot.* 1980, **31**, 731.
13. Gilliam, T. M.; Loflin, J. A. *Leachability Studies of Hydrofracture Grouts*; ORNL/TM-9879; Martin Marietta Energy Systems, Inc., Oak Ridge National Laboratory: Oak Ridge, Tenn., November 1986.
14. T. M. Gilliam et al. *Summary Report on the Development of a Cement-Based Formula to Immobilize Hanford Facility Waste*; ORNL/TM-10141; Martin Marietta Energy Systems, Inc., Oak Ridge National Laboratory: Oak Ridge, Tenn., September 1987.
15. *Code of Federal Regulations*, 40, Part 261, Appendix II, "Identification and Listing of Hazardous Waste," "EP Toxicity Test Procedures," Office of the Federal Register, July 1, 1987.
16. *Test Methods for Evaluating Solid Waste*; EPA/SW-846, 2nd ed.; Environmental Protection Agency, Office of Solid Waste and Emergency Response: Washington, D.C., July 1982.
17. *Code of Federal Regulations*, 40, Part 261.24, "Characteristics of EP toxicity," Office of the Federal Register, July 1, 1987.
18. *Code of Federal Regulations*, 40, Part 260.22(d), "Hazardous Waste Management System: General," "Petitions to amend Part 261 to exclude a waste produced at a particular facility," Office of the Federal Register, July 1, 1987.
19. American Nuclear Society, "Measurement of the Leachability of Solidified Low-Level Radioactive Wastes by a Short-Term Test Procedure," ANSI/ANS-16.1-1986.
20. Merrill, D. T.; Maroney, P. M.; Parket, D. S. *Trace Element Removal by Coprecipitation with Amorphous Iron Oxyhydroxide: Engineering Evaluation*; EPRI Report CS-4087; Electric Power Research Institute: Palo Alto, Calif., July 1985.
21. Walton, F. B.; Paquette, J.; Ross, J. P. M.; Lawrence, W. E. *Nuclear Chem. Waste Mgt.* 1986, **6**, 121.
22. Palmer, D. A.; Meyer, R. E. *J. Inorg. Nucl. Chem.* 1981, **43**, 2979.
23. Boyd, G. E.; Larson, Q. V. *J. Phys. Chem.* 1960, **64**, 988.

24. Schwochau, K. *Radiochim. Acta* 1983, **32**, 139.
25. Strickert, R.; Friedman, A. M.; Fried, S. *Nucl. Technol.* 1980, **49**, 283.
26. Lee, S. Y.; Bondietti, E. A. *Mat. Res. Soc. Symp. Proc.* 1981, **15**, 315.
27. Vandergraaf, T. T.; Ticknor, K. V.; George, I. M. In *Geochemical Behavior of Disposed Radioactive Waste*; G. S. Barney, J. D. Navratil, and W. W. Schultz (eds.); Chap. 2, ACS Symposium Ser. 246, American Chemical Society: Washington, D.C., 1984.
28. Bowers, A. R.; Ortiz, C. A.; Cardozo, R. J. *Metal Finishing* 1986, **84**(1), 37.
29. Gilliam, T. M. *J. Underground Injection Practices Council* 1986, **1**, 192.
30. Shoemaker, J. L.; Bostick, W. D. *Support for Characterization, Formulations, and Stabilization of K-1407-B and K-1407-C Pond Sludges*; K/QT-199; Martin Marietta Energy Systems, Inc., Oak Ridge Gaseous Diffusion Plant: Oak Ridge, Tenn., September 1988.
31. Tallent, O. K.; McDaniel, E. W.; Del Cul, D. G.; Dodson, K. E.; Trotter, D. R. *Mat. Res. Soc. Symp. Proc.* 1988, **112**, 23.
32. Tallent, O. K.; McDaniel, E. W.; Del Cul, G. D.; Dodson, K.E.; Trotter, D. R. *Development of Immobilization Technology for Hanford Double-Shell Slurry Waste Feed*; ORNL/TM-10906; Martin Marietta Energy Systems, Inc., Oak Ridge National Laboratory: Oak Ridge, Tenn., August 1989.
33. Malek, R. I. A.; Roy, D. M.; Barnes, M. W.; Langton, C. A. *Slag Cement-Low Level Waste Forms at the Savannah River Plant*; DP-MS-85-9; Savannah River Laboratory: Aiken, S.C., 1985.
34. Brodda, B. G. *Science of the Total Environment* (Netherlands) 1988, **69**, 319.
35. Angus, M. J.; Glasser, F. P. *Mat. Res. Soc. Symp. Proc.* 1986, **50**, 547.
36. Rahman, A. A.; Glasser, F. P. *Cements in Radioactive Waste Management. Characterization Requirements of Cement Products for Acceptance and Quality Assurance Purposes*; EUR-10803-EN; Commission of the European Communities, 1987.
37. Gilliam, T. M.; Spence, R. D.; Evans-Brown, B. S.; Morgan, I. L.; Shoemaker, J. L.; and Bostick, W. D. *Proc. Inter. Top. Mtg. Nucl. and Haz. Waste Mgt. (Spectrum '88)*, American Nuclear Society: La Grange Park, Ill., 1988, pp 109–111.
38. Bundy, R. D.; Alderfer, R. B. *Bench-Scale Study of Direct Calcination of Raffinate Waste*; K/QT-105; Martin Marietta Energy Systems, Inc., Oak Ridge Gaseous Diffusion Plant: Oak Ridge, Tenn., September 1987.
39. Alderfer, R. B. *Disposal of Portsmouth Raffinate Waste by Rotary Calcination*; K/QT-196; Martin Marietta Energy Systems, Inc., Oak Ridge Gaseous Diffusion Plant: Oak Ridge, Tenn., October 1988.
40. Rimshaw, S. J.; Case, F. N.; Tompkins, J. A. *Volatility of Ruthenium-106, Technetium-99, and Iodine-129, and the Evolution of Nitrogen Oxide Compounds During the Calcination of High-Level, Radioactive Nitric Acid Waste*; ORNL-5562; Union Carbide Corp., Oak Ridge National Laboratory: Oak Ridge, Tenn., February 1980.

RECEIVED November 10, 1989

Chapter 21

Extraction of Plutonium from Lean Residues by Room-Temperature Fluoride Volatility

G. M. Campbell, J. Foropoulos, R. C. Kennedy, B. A. Dye, and R. G. Behrens

Nuclear Materials Division, Los Alamos National Laboratory, Los Alamos, NM 87545

The use of dioxygen difluoride and krypton difluoride for the recovery of plutonium from lean residues by conversion to gaseous plutonium hexafluoride is being investigated. The synthesis of dioxygen difluoride in practical quantity has been demonstrated. Fluorination of plutonium compounds under ideal conditions supports the contention that a viable process can be developed. Application of the method to lean plutonium residues is in the early stage of development.

The high cost and political sensitivity associated with the disposal of radioactive waste makes it imperative that the quantity of waste generated be reduced to the lowest reasonable level. The processing of plutonium residues by direct conversion to plutonium hexafluoride, which can be easily separated as a gas, was examined many years ago (1). The process was limited then to reaction of residues with fluorine at elevated temperature.

Because of the corrosive nature of hot fluorine, there was a significant materials compatibility problem. Also, because plutonium hexafluoride becomes increasingly unstable at temperatures above 470° K, there was reduced reaction efficiency, and plutonium hexafluoride decomposition product deposits outside of the reaction zone. To some, the advantages of the process, including the reduction of the number of processing steps, space requirements, and the need for fewer chemical additives, outweighed the disadvantages. As the cost of waste disposal has sharply increased in recent times, the fluoride volatility process has gained new significance.

To complement the high temperature process and overcome some of the perceived shortcomings, LANL has been examining the possibility of room temperature fluoride volatility. The compound dioxygen difluoride and its gas phase equilibrium product, the dioxygen monofluoride radical, have had the most emphasis in this study because of the potential for making it in sufficient quantity economically. They have been shown to be powerful fluorinating agents for the actinides (2, 3). Krypton difluoride is believed to have desirable chemical properties (4), but is more difficult to produce in sufficient quantity at this time. LANL has found that the reaction with plutonium residues has a half time greater than 5 hours and a reaction efficiency at least as good as dioxygen difluoride. At room temperature the half time of dioxygen difluoride is a few seconds.

0097–6156/90/0422–0368$06.00/0

The chemical reactions of interest in the dioxygen difluoride process are:

$$F_2(g) + O_2(g) \rightarrow FOOF(g) \quad (1)$$

$$3\ FOOF(g) + PuR(s) \rightarrow PuF_6(g) + 3O_2(g) + R(s,g) \quad (2)$$

$$6\ FOO(g) + PuR(s) \rightarrow PuF_6(g) + 6O_2(g) + R(s,g) \quad (3)$$

where R(s) is a solid residue. The product R(g) represents gaseous compounds such as phosphorus pentafluoride, silicon tetrafluoride, and carbon tetrafluoride. The first reaction requires the input of energy either by photolysis, microwave excitation, or thermal heating. The others proceed spontaneously at room temperature.

Kinetic studies (5, 6) of the dioxygen difluoride, dioxygen monofluoride, oxygen system showed that there was an equilibrium,

$$2\ FOO(g) \rightleftharpoons FOOF(g) + O_2(g)\ , \quad (4)$$

that produced dioxygen monofluoride in the gas phase. The reaction rate was very temperature sensitive with an activation energy of 13 kcal per mole. It was found that dioxygen monofluoride reacted to produce plutonium hexafluoide much more efficiently when the plutonium residue was spread over a metal surface.

The first gas circulating loop was designed to optimize the fluorination reaction in light of the information gained from the kinetic studies. It was operated at a high flow rate so that dioxygen difluoride passed quickly from the supply reservoir to the gas-solid reactor. The gas-solid reactor had a large volume so that the gas reactant remained in contact with the solid residue for several seconds. The solid residue was spread over a metal surface (metal matrix).

After demonstrating that the metal matrix reactor operated efficiently, the information obtained was used to adapt the reaction to a fluidized bed. It was believed that a fluidized bed would be more convenient for use in a production mode. To compensate for the slower flow rate used in fluidization, and the lack of a metal catalyst, the amount of dioxygen difluoride (as opposed to dioxygen monofluoride) reaching the reaction zone was optimized. This required additional cooling of the gas stream and control of the oxygen pressure.

A severe test of plutonium extraction was made by removing it as plutonium hexafluoride from incinerator ash. Fluorination of the ash at elevated temperature was shown to result in the formation of nonvolatile plutonium fluorides. When the untreated ash was fluorinated at room temperature, volatile plutonium hexafluoride was formed.

Experimental.

The apparatus used in carrying out the gas-solid reactions involving the fluorination of plutonium residues was enclosed in a glove box designed for the safe handling of plutonium. Figure 1 is a simplified schematic of the equipment used. Although nickel or aluminum were the preferred materials of construction for handling fluorinating agents, stainless steel was found to be perfectly adequate for many uses at room temperature. Type 316 stainless steel was used in this case.

The dioxygen difluoride was made outside of the glove box and cryopumped to a receiving reservoir inside the glove box. The receiving reservoir was one component of a gas circulating loop.

Metal Matrix Reactor.

To take advantage of the effect of a metal surface, the first reactor used was a stainless steel cylinder filled with compacted aluminum foil balls, Figure 2. The cylinder containing the aluminum balls accepted 13 liters of gas when filled. The balls formed a matrix for support of the solid reactant.

The solid reactant was distributed evenly throughout the reactor. The 13 liter volume of the reactor was the largest component of the 18 liter gas circulating system. At a circulation rate of one liter per second, the gas entered the reactor about 2 seconds after vaporization in the dioxygen difluoride reservoir and then spent 13 seconds in the reactor. The circulation rate was controlled by throttling a bypass valve around the bank of three model 601 Metal Bellows compressors. The rate of dioxygen difluoride addition to the fluorine and oxygen carrier gas mix was controlled by adjusting the temperature of the dioxygen difluoride reservoir.

Fluidized Bed Reactor.

The fluidized bed reactor consisted of a tapered aluminum cylinder with a 1.9-cm-diam base opening to a 3.8-cm-diam at the top of the 30.5-cm-long reaction cylinder. The reaction cylinder opened into a 10-cm-diam filter assembly. The gas entered through a nickel frit at the base of the reactor, mixed with the solid reactant in a fluidized state and exited after passing through particulate filters. The optimum circulation rate was about 1 standard liter per minute. The linear gas velocity through the bottom frit was 30 cm/s. The pressure drop across the bed was nominally the weight of the solid reactant per unit cross sectional area. The measured pressure drop across the bottom frit and fluidized bed was about 60 torr in most of these experiments.

The dioxygen difluoride reservoir was located as close as possible to the entrance of the fluidized bed to minimize the gas travel time. At liquid nitrogen temperature no dioxygen monofluoride radical was present under the conditions used, but was formed in the gas phase after evaporation of dioxygen difluoride. The rate at which dioxygen monofluoride was formed increased with temperature. To maintain a high ratio of dioxygen difluoride to dioxygen monofluoride in the fluidized bed reactor, the carrier gas stream was precooled by a second liquid nitrogen trap before it entered the dioxygen difluoride reservoir. In addition to retarding the formation of dioxygen monofluoride from dioxygen difluoride (as a consequence of the lower temperature), precooling the gas limited the oxygen pressure in the gas stream. An equal mixture of oxygen and fluorine has a vapor pressure of only about 160 torr at liquid nitrogen temperature. This favorably effects the equilibrium [Eq. (4)].

Chemical Reactants.

The dioxygen difluoride was prepared in a separate operation by reacting thermally excited fluorine atoms with oxygen at a cold interface. Within a year LANL expects to have the capability of producing a kilogram of dioxygen difluoride per day by this method* (T. R. Mills, personal communication, April 1989).

The plutonium tetrafluoride used in these experiments was the unreacted portion of the material used for the thermal generation of plutonium hexafluoride for another project. The powder density was about 1.3 g/cc. Particle size ranged from 25 to 125 microns.

The plutonium dioxide (not generated by the incineration of contaminated waste) was produced by the calcination of plutonium oxalate precipitate.

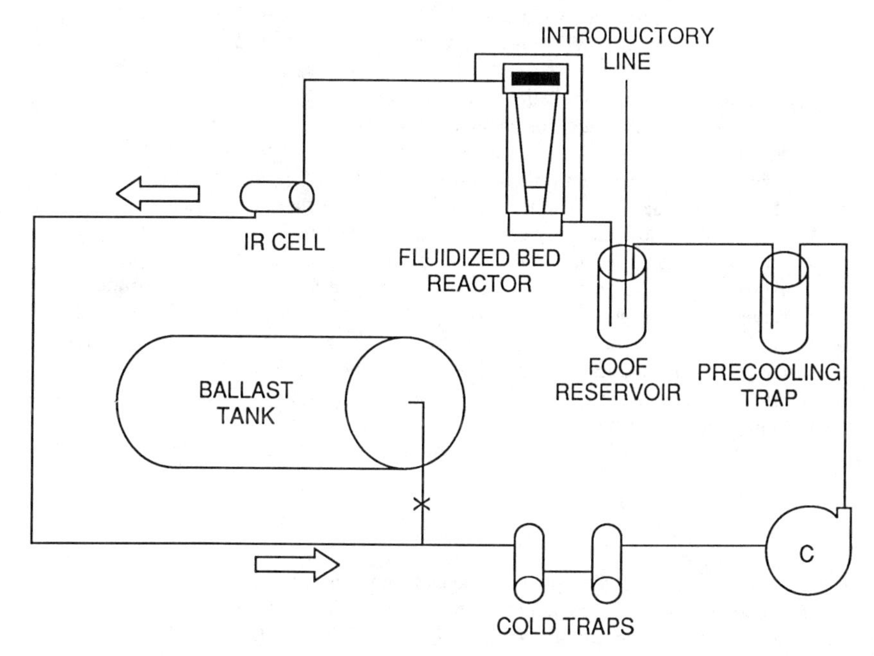

Figure 1. Simplified Fluorination Loop Schematic.

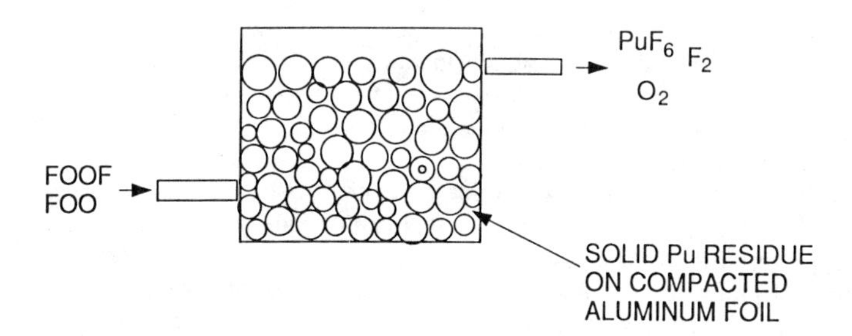

Figure 2. Metal Matrix Gas-Solid Reactor.

Procedure.

At liquid nitrogen temperature the dioxygen difluoride is a solid with very low vapor pressure. The physical properties of dioxygen difluoride have been reviewed by Streng (7). A carrier gas was introduced to the gas circulating part of the loop not including the ballast tank. The carrier gas was fluorine initially, but became a mixture of fluorine and oxygen as dioxygen difluoride was introduced. This gas was circulated so that the vapor from the dioxygen difluoride receiving reservoir was carried first to the gas-solid reactor, then through the infrared diagnostic cell, cold traps, compressor, then back to the reservoir.

The plutonium hexafluoride generated in the gas-solid reactor was condensed in the cold traps at a temperature of about 190° K. The vapor pressure of dioxygen difluoride in the reservoir, and therefore the rate of addition, was controlled by adjusting the temperature of the dioxygen difluoride reservoir. Adjusting the height of the liquid nitrogen Dewar provided adequate temperature control for these experiments.

The pressure in the circulating portion of the loop could be controlled by bleeding gas into the ballast tank. In the fluidized bed experiments the circulating gas pressure was maintained by condensing the gas (at liquid nitrogen temperature) into a trap located between the compressor and the dioxygen difluoride reservoir. The flow rate was monitored by measuring the calibrated pressure drop across an orifice before the gas entered the dioxygen difluoride reservoir.

All experiments were done in a batch mode. The dioxygen difluoride was maintained at liquid nitrogen temperature until introduction to the gas-solid reaction loop. The progress of the reaction was followed by monitoring pressures and by Fourier transform infrared spectroscopy (FTIR).

Results of the Metal Matrix Reactor Experiments.

The chemical reaction efficiency of the oxygen fluorides (dioxygen difluoride and dioxygen nonofluoride) with plutonium tetrafluoride varied from about 12 percent at a loading of 82 g plutonium tetrafluoride, to 24 percent with a loading of 400 g. Figure 3 shows the amount of plutonium hexafluoride generated as a function of dioxygen difluoride used.

At the lower loading, the presence of dioxygen monofluoride at the FTIR cell, downstream from the reactor, could easily be observed at moderate addition rates. At loadings of 400 g, only a trace was observed at the highest rate that could be achieved by allowing the dioxygen difluoride reservoir to warm in ambient air after having removed the liquid nitrogen Dewar. In one experiment 51 g of plutonium hexafluoride was generated by the addition of 44 g of dioxygen difluoride in about 20 minutes. This demonstrates that a practical reaction rate is possible.

A graph of the plutonium hexafluoride generated versus the amount of dioxygen difluoride used during the fluorination of plutonium dioxide is shown in Figure 4. In this case the curve is concave upward rather than downward, as was the case during the fluorination of plutonium tetrafluoride. This reflects the fact that there are intermediate stable compounds such as plutonium oxyfluoride generated before plutonium hexafluoride can be produced. By using ultraviolet light absorption to monitor the amount of fluorine used during the run, it was determined that the heat generated by the reaction of dioxygen difluoride with plutonium dioxide was sufficient to activate a reaction of molecular fluorine with plutonium dioxide during the early stages of the fluorination. When this factor was excluded, the reaction efficiency was found to approach that of plutonium tetrafluoride with dioxygen difluoride.

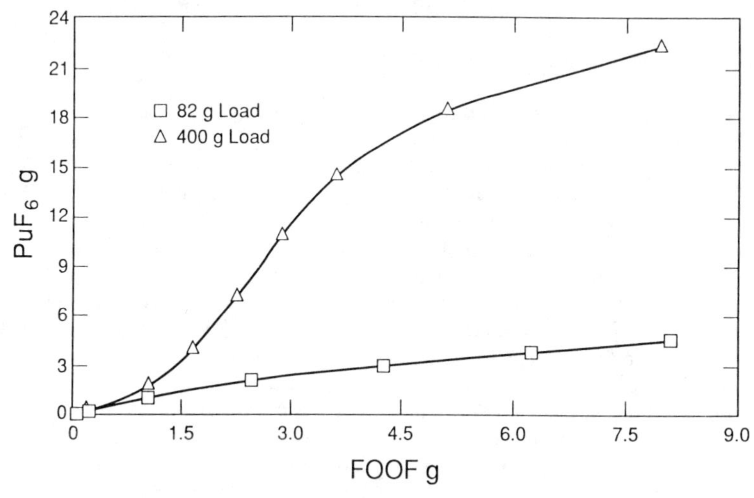

Figure 3. The Effect of Loading on Fluorination in the Aluminum Matrix Reactor.

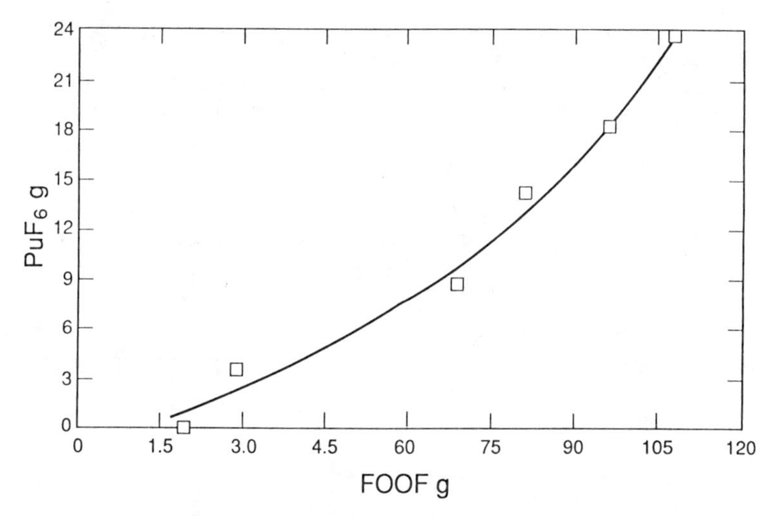

Figure 4. Fluorination of PuO_2 in the Aluminum Matrix Reactor.

A comparison of the fluorination of plutonium tetrafluoride when the supporting material was Teflon instead of aluminum is shown in Figure 5. Although the plutonium tetrafluoride loading, as well as the total surface area was slightly greater when Teflon was used, it is clear that the reaction efficiency has greatly decreased. It is felt at this time that, at room temperature, the metal surface enhances the reaction of dioxygen monofluoride but may not be necessary for the dioxygen difluoride reaction. It is more chemically reactive.

Results of Fluidized Bed Reactor Experiments.

The fluidized bed reactor has been the gas-solid reactor of choice for many chemical reactions in industry. Heat is usually added to drive the reaction. The solid powder can be easily loaded and unloaded. A convenient way to use a metal catalyst to assist the reaction of dioxygen monofluoride in the fluidized bed has not been found. To improve the reaction efficiency, the effort has been concentrated toward increasing the ratio of dioxygen difluoride to dioxygen monofluoride. This has been accomplished by precooling the gas, minimizing the travel volume between the dioxygen difluoride reservoir and the fluid bed reactor, and by controlling the oxygen pressure. Another alternative would be to increase the oxygen pressure to drive the equilibrium [Eq. (4)] to increased concentration of the more stable dioxygen monofluoride. The temperature of the fluid bed could then be raised slightly to activate the dioxygen monofluoride reaction.

A comparison of the plutonium hexafluoride generated versus the dioxygen difluoride used, in the fluid bed and in the aluminum matrix reactor is shown in Figure 6. The conditions used (except for the plutonium tetrafluoride loading) were chosen to optimize the reaction efficiency in each case. In this example, the flow rate of the carrier gas through the fluidized bed was about 5 actual liter per minute. The carrier gas flow rate through the aluminum matrix reactor was about 60 actual liter per minute. The operating pressure was about 150 torr in both experiments. The rate that the reservoir temperature was increased was about the same in each case. The overall reaction efficiency was slightly higher (15% *vs* 12%) in the aluminum matrix reactor. A second generation fluidized bed that will allow for higher flow rates and larger sample loading is in the design stage* (Newman, H .J.; Martinez, H. E., personal communication, June 1989).

Fluorination of Plutonium Contaminated Incinerator Ash.

One of the plutonium containing residues being stored at DOE plutonium facilities is the ash generated from the incineration of combustible materials that accumulate in the process glove boxes. The typical composition of such ash is given in Table 1.

Table 1. Typical Composition of Incinerator Ash

Element	Percent Abundance	Element	Percent Abundance
Potassium	20	Chromium	2
Calcium	20	Silicon	2
Chlorine	17	Nickel	1
Plutonium	13	Carbon	0.9
Sodium	5	Hydrogen	0.5
Magnesium	5	Copper	0.5
Iron	4	Aluminum	0.5
Phosphorus	3		

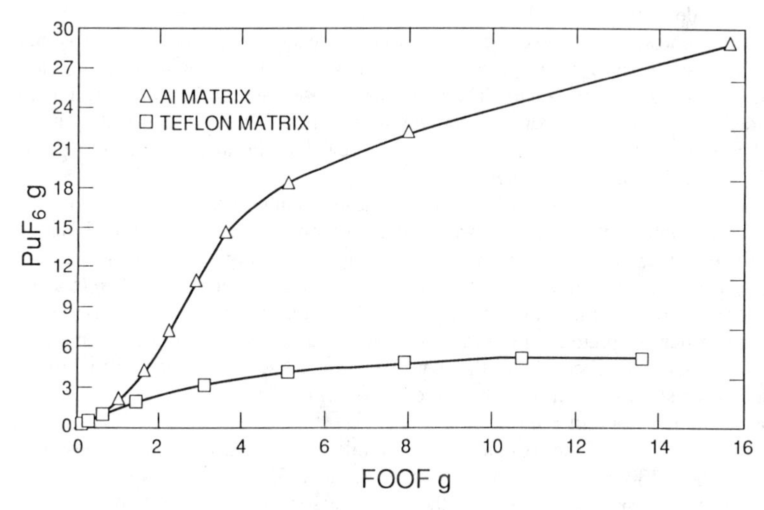

Figure 5. Fluorination of 400 g PuF_4 on an Aluminum Matrix Compared to Fluorination of 446 g PuF_4 on a Teflon Matrix.

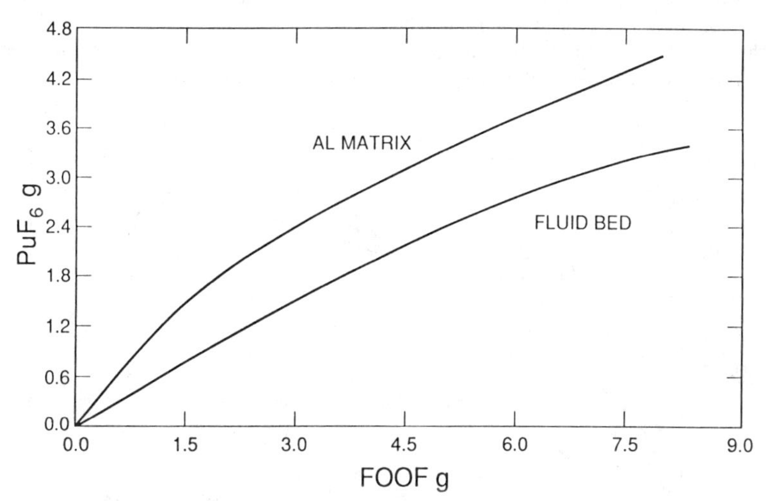

Figure 6. Comparison of the Fluorination of PuF_4 in the Aluminum Matrix and the Fluid Bed Reactor.

The literature indicates that alkali metal fluorides can form double salts with plutonium fluorides (8,9). If this happens it could prevent the formation of gaseous plutonium hexafluoride. To test the severity of this at room temperature, a preliminary experiment was run in which 15 g of plutonium hexafluoride was mixed with 85 g of sodium fluoride in the fluidized bed.

Figure 7 shows a comparison of the plutonium hexafluoride produced versus the dioxygen difluoride used when fluorinating the mixture, and when fluorinating a load of 15 g neat plutonium tetrafluoride. (The word neat is used here to indicate that nothing was added to the laboratory grade plutonium tetrafluoride.) It was found that double salt formation was not a factor in this relatively mild reaction. Further study of this problem will be made using other surrogates.

To conserve dioxygen difluoride and prevent high temperatures from heat released by the fluorination reaction, the ash was pretreated with fluorine. It was found that ash prefluorinated at temperatures of 520° K did not produce plutonium hexafluoride. An experiment in which the untreated ash was added to the fluidized bed, and pretreated with fluorine at room temperature, did produce plutonium hexafluoride however. Figure 8 shows the amount of plutonium hexafluoried generated versus the dioxygen difluoride used. X-ray analysis indicated that the plutonium was present in the form of plutonium dioxide. Other substances in the ash may occlude some of the plutonium dioxide, decreasing the reaction efficiency.

At the start, 67% of the particles in the sample were larger than 90 microns, 43% were larger than 180 microns and 25% were larger than 350 microns. No attempt was made to reduce the particle size except the mixing associated with the fluidization process in the bed. Simple ways of pretreating the sample to reduce particle size, as well as fluorinate most of the oxides, before using dioxygen difluoride are currently being examined. Preliminary experiments indicate that doubling the amount of ash in the bed increases the reaction efficiency by a factor of 3. Nothing has been found that would preclude obtaining practical reaction efficiencies.

The ash represents the most stringent application of the process. There seem to be no severe problems associated with processing residues rich in plutonium.

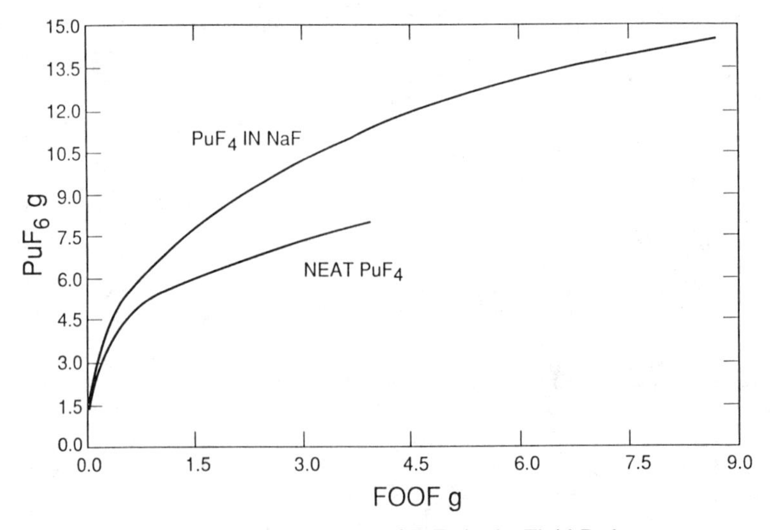

Figure 7. Fluorination of PuF_4 in the Fluid Bed.

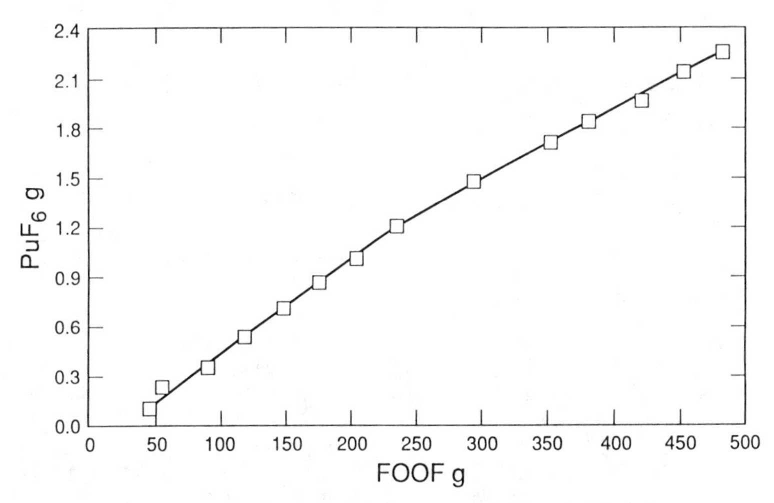

Figure 8. Fluorination of Incinerator Ash Lean in Plutonium.

Acknowledgments.

We wish to acknowledge the contribution of H. J. Newman and H. E. Martinez for the design and fabrication of the fluid bed. E. A. Trujillo and R. M. de Duval were very helpful in materials preparation and characterization. T. R. Mills and R. J. Kissane produced most of the dioxygen difluoride used in these experiments.

Performed under the auspices of the U.S. Dept. of Energy.

Literature Cited.

1. Jonke, A. A.; "Reprocessing of Nuclear Reactor Fuels by Processes Based on Volatilization, Fractional Distillation and Selective Adsorption"; *At. Energy Rev.*, **March 1965**, vol. 3, p. 1.
2. Malm, J. G.; Eller, P. G.; Asprey, L. B.; "Low-Temperature Syntheses of Plutonium Hexafluoride"; *J. Amer. Chem. Soc.* **1984**, vol. 106, p. 2726.
3. Kim, K. C.; Campbell, G. M.; "Fourier Transform Infrared Spectrometry Using a Very-Long-Pathlength Cell: Dioxygen Difluoride Stability and Reactions with Plutonium Compounds"; *Appl. Spect.*, **1985**, vol. 39, p. 625.
4. Asprey, L. B.; Eller, P. G.; Kinkead, S. A.; "Formation of Actinide Hexafluorides at Ambient Temperatures with Krypton Difluoride"; *Inorg. Chem.*, **1986**, vol. 25, p. 670.
5. Campbell, G. M.; "The Infrared Vibrational Intensity of Gaseous Dioxygen Monofluoride"; *J. Mol. Structure*, **1988**, vol. 189, p. 301.
6. Campbell, G. M.; "A Kinetic Study of the Equilibrium Between Dioxygen Monofluoride and Dioxygen Difluoride"; *J. Fluorine Chem.*, in press.
7. Streng, A. J.; "The Oxygen Fluorides"; *Chem. Rev.*, **1963**, vol. 63, p. 607.
8. Katz, S.; Cathers, G. I.; "A Gas Sorption-Desorption Method for Recovering Plutonium Hexafluoride"; *Nuclear Applications*, **July 1968**, vol. 5, p. 5.
9. Galkin, N. P.; Veryatin, U. D.; Bagryantsev, V. F.; Gusev, V. F.; "Interaction of Perovskite-Type Ternary Elements with Elementary Fluorine"; *Radiokhimiya*, **1975**, vol. 17, p. 604.

RECEIVED November 10, 1989

Chapter 22

Peptide Synthesis Waste Reduction and Reclamation of Dichloromethane

Richard V. Joao, Ilona Linins, and Edward L. Gershey

Rockefeller University, 1230 York Avenue, New York, NY 10021

Waste management has two important benefits. Obviously, reducing volume decreases costs and liabilities and it may also effect generator status. Reclaiming chemicals is a cost effective means of minimizing waste.

Peptide synthesis, an important process in academic research and industrial production laboratories, produces large amounts of chemical waste, the primary component of which is methylene chloride (dichloromethane, DCM). Other significant components are ethanol, methanol, dimethylformamide, and trifluoroacetic acid. Due to the high halogen content of the mixture, disposing of this is costly. At this university reclaiming DCM reduces the volume of chemicals shipped as waste by 25%.

Separating and purifying DCM is a four part process involving distillation, aqueous extraction, sieving, and analysis. Recovery of the distillate between 37 and 40.5°C results in an azeotropic mixture which is approximately 95% DCM, 5% methanol, and trace amounts of other components. Most of the methanol is removed by two 1:1 water washes. The remaining trace amounts of methanol and water are removed by 4 angstrom molecular sieves. High sensitivity capillary gas chromatography and flame ionization detection indicate that the product is greater than 99.9% pure.

Reclamation and recycling of solid phase peptide synthesis effluents are important means of waste reduction. In recent years automated peptide synthesis has become an important and widely practiced biochemical technique. However, a single investigator working with an automated peptide synthesizer can generate 20 to 40 liters of waste per week. The waste produced is approximately 60% dichloromethane (DCM). Assuming that 75% of the DCM can be recovered, a yearly waste reduction of 936L of halogenated solvent and a sixteen hundred dollar decrease in waste disposal costs per investigator can be realized. Also, at a cost of approximately eight dollars per liter for HPLC grade DCM, a recycling program could save approximately seven thousand dollars per investigator, annually.

Besides decreasing costs, reducing wastes also lowers liabilities. RCRA places a "cradle to grave" responsibility and provides for penalties of up to $50,000 per day and/or by imprisonment for up to two years (1). The liabilities associated with waste disposal include not only the potential criminal penalties but also the cost of negative

0097–6156/90/0422–0378$06.00/0

public perception of waste generators. DCM has been found in 11% of the waste disposal sites on the National Priorities List (2). Media attention to improper waste disposal practices has resulted in a public hesitant to accept waste handling facilities. This has led to costlier disposal solutions, higher transportation costs and restricted access to these facilities.

Peptide synthesis (3, 4) produces a complex waste stream that is primarily dichloromethane (DCM). It also contains N,N-dimethylformamide (DMF), methanol (MeOH), ethanolamine trifluoroacetate (EATFA), and diisopropylethylamine (DiEA). DCM and DMF are used throughout the peptide synthesis process. DiEA is used to accelerate the esterification step. MeOH is used to wash away excess reactants between steps. TFA is used to remove protecting groups from the amino acids which are being added to the peptide chain. Ethanolamine is used to neutralize the TFA to produce a free base which can be coupled with the next amino acid. This neutralization results in the production of EATFA. For recycling to a peptide synthesizer, it is crucial that the purified DCM be anhydrous and uncontaminated with these reagents.

EXPERIMENTAL

DISTILLATION. Material was distilled with a spinning band type still (B/R model 8400, Pasadena, MD). This distillation apparatus has a separation capability of 30 theoretical plates. The mantle rate was set at 75% of the maximum. An equilibrium time of thirty minutes was selected. Test samples were taken over a temperature range from 35°C to 45°C at 10 minute intervals. The head and pot temperatures were also recorded. The still was then programmed to begin collecting material at 37°C and finish at 41°C (the boiling point of DCM is 40.1°C). The reflux ratio was set at 10. A maximum of 10L of peptide synthesis waste may be distilled at a time and requires approximately 12 hours.

AQUEOUS EXTRACTION. Much of the MeOH and other water-soluble contaminants were removed by aqueous extraction. The distillate was extracted twice at a ratio of 1:1 (distillate:water) by shaking in 4L separatory funnels. Aqueous extraction was tried before and after distillation, the number of extractions was varied from one to three, and the ratios of 1:2, 1:1, and 2:1 (distillate:water) were investigated.

DESICCATION. Trace water and MeOH were removed by batch adsorption with 4Å molecular sieves (Aldrich, Milwaukee, Wi). Desiccation using 3Å molecular sieves and anhydrous calcium chloride was also examined.

GAS CHROMATOGRAPHIC ANALYSIS. All samples were run on a Gas Chromatograph (Varian Model 3700, Sugar Land, TX) modified for use with capillary columns; data were analyzed with an automated integrator (Shimadzu Chromatopac C-R3A, Kyoto, Japan).

Two protocols were used. A screening protocol was used to analyze peptide synthesis waste and monitor the product throughout the recycling process, 0.2 microliters of sample were introduced into a direct injector at 250°C connected to a capillary column (Supelco SPB-35 Glass Capillary, 60m x 0.75mm x 1.0 micron film). Helium was used as a carrier gas (Matheson Ultra High Purity - 99.999%). A Thermal Conductivity Detector (TCD) was used, with its temperature set at 270°C, the TCD range was set at 0.5mV. The filament temperature was set at 290°C. The oven temperature was held at ambient for 5 minutes and then increased to 250°C at a rate of 15°C/min. In the second protocol, for quality control of the final product,

0.5 microliters (unsplit) of sample were introduced into an injector set at 150°C and onto a column (Hewlett-Packard Ultra 2 Capillary column, 25m x 0.32mm x 0.52 micron film). Helium was used as a carrier gas. A Flame Ionization Detector (FID) was used, its temperature set at 150°C and its range set at 10^{-12} amp/mV.

GAS CHROMATOGRAPHY-MASS SPECTROMETRY (GC-MS). Peptide synthesis waste samples were analyzed and components identified on a Hewlett-Packard 5890 Gas Chromatograph, outfitted with a column described in GC protocol 2 above, coupled with a VG Analytical Quadrupole Mass Spectrometer. This work was performed by the Rockefeller University Mass Spectrometry Service Laboratory.

RESULTS AND DISCUSSION

ANALYSIS OF PEPTIDE SYNTHESIS WASTE. Figure 1a illustrates the composition of a typical sample of peptide synthesis waste. DCM accounts for 56.7%, DMF constitutes 35.9%, MeOH 6.7%, DiEA 0.5%, EATFA 0.2%. Due to the nature of the peptide synthesis process, the quantity of a given solvent can vary significantly from one waste sample to the next. The specific solvents used also vary depending on the synthesis chemistry used. Analysis of many samples from different sources of peptide synthesis waste show that certain generalizations can be made. DCM is consistently the largest component of all peptide synthesis wastes, varying from 50% to 80%. The second largest component, DMF makes up between 20% and 35% of the waste volume. The amount of methanol varies from 3% to 15% and the next significant component, EATFA, comprises from <1% to 10%. Other components which are frequently found are diisopropylethylamine, dicyclohexylurea, hydroxybenzotriazole, dimethylsulfide (DMS) and ethanol. Butyloxycarbonyl amino acids are always found. Generally these other components are present in trace amounts and rarely comprise more than 1% of the waste.

Initially, GC-MS was used to identify the components of the peptide synthesis waste. In order to confirm the information given by the GC-MS analysis, samples of peptide synthesis waste, to which known standards were added, were analyzed on our gas chromatograph. Samples were mixed with known amounts of ethanol, propanol, butanol, DMS, acetonitrile and the major waste constituents. All of these chemicals have been found occasionally in peptide synthesis waste. These added standards were seen by increase in peak size to co-migrate with material in the sample being identified.

Figure 2 shows the relationship between time, temperature, and the changing composition of the distillate. Analysis of samples of the distillate taken over time indicated that a stable ratio of DCM to MeOH existed between 37°C and 40.5°C. Distillate collected in this temperature range was found to contain 95% DCM and 5% MeOH (Figure 1b). Analysis of the still bottom (Figure 1c) showed that it contained 12% DCM as well as the other components of the original mixture and had to be disposed as hazardous waste. Attempts to recover more DCM resulted in the distillate being contaminated with DiEA and/or DMF.

Redistillation of the product failed to significantly reduce the concentration of MeOH, suggesting that MeOH and DCM were co-distilling as an azeotrope. Slowing down the distillation by increasing the reflux ratio and equilibrium time decreased the amount of methanol only slightly and it slowed the reclamation process excessively. To remove this MeOH, aqueous extraction of the starting material (Figure 1a) was performed before distillation.

Aqueous extraction of the starting material prior to distillation reduced the MeOH to 0.3% (Figure 3a), but resulted in an aqueous layer of significant volume, heavily contaminated with DMF (Figure 3c) which would have had to have been

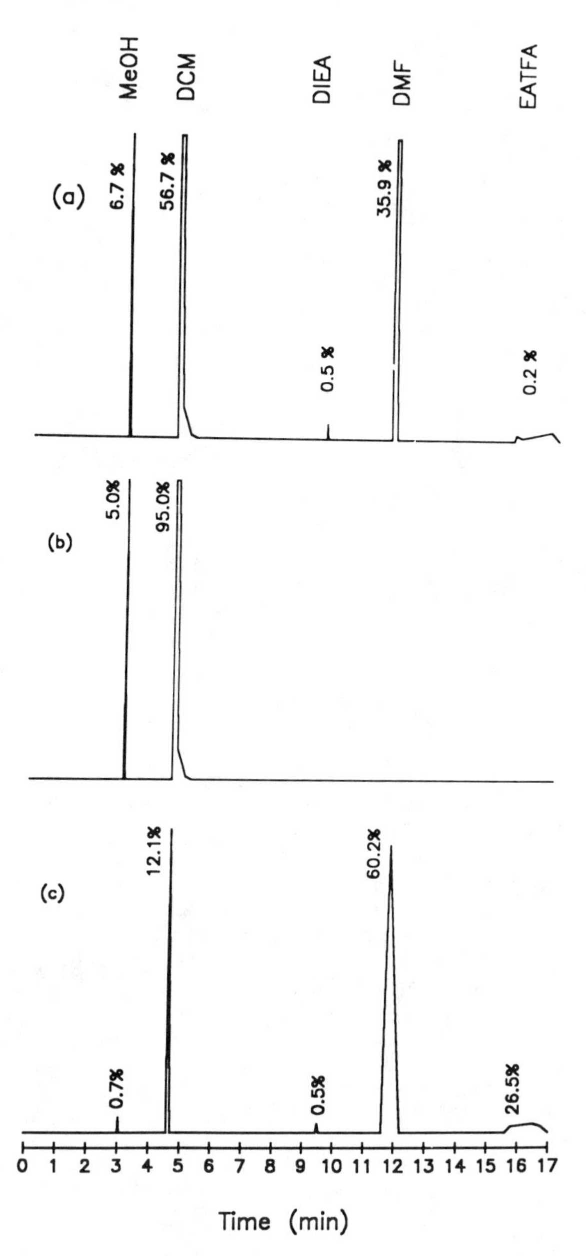

Figure 1: Gas chromatographs of a typical sample of peptide synthesis waste. (a) The waste usually contains methanol (MeOH), dichloromethane (DCM), N,N-dimethylformamide (DMF), ethanolamine trifluoroacetate (EATFA), and diisopropylethylamine (DiEA). (b) Peptide synthesis waste after distillation. (c) Material remaining in still bottom after distillation of peptide synthesis waste. Attempts to recover more DCM resulted in the distillate being contaminated with DiEA and/or DMF.

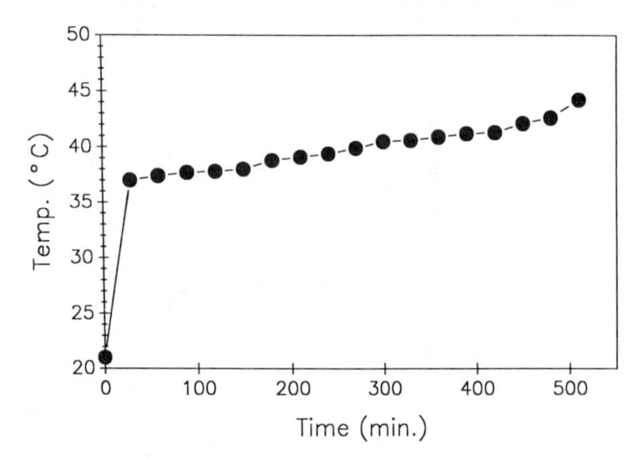

Figure 2: Relationship between time, temperature, and the changing composition of the distillate. To determine the temperatures at which to take an appropriate cut, head temperature was recorded and samples of distillate were taken at 10 minute intervals. Distillate collected in the plateau region (37°C to 41°C) was primarily DCM.

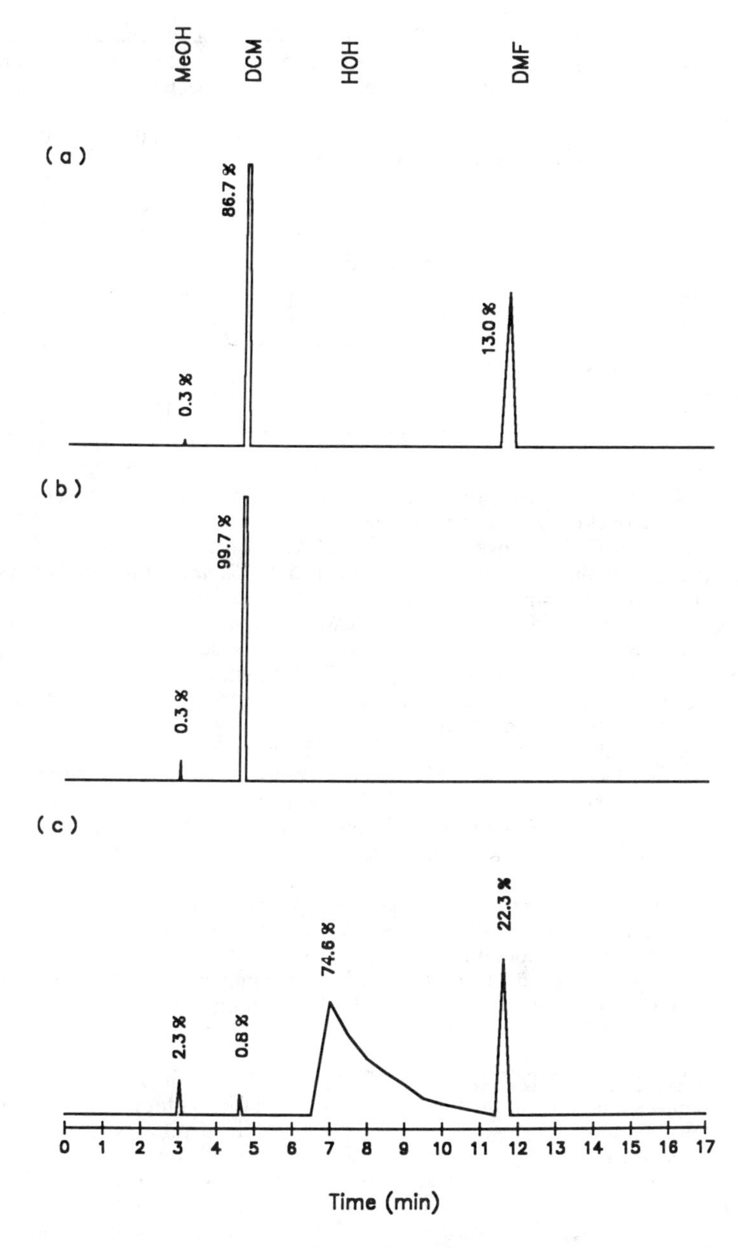

Figure 3: Gas chromatographs illustrating the effect of aqueous extraction of peptide synthesis waste before distillation. (a) Organic layer after extraction. The concentrations of MeOH and DMF are greatly decreased after a single 1:1 aqueous extraction. (b) Distilled organic layer. DMF has been completely eliminated but 0.3% MeOH remains in the distillate. (c) Aqueous layer after extraction. Due to the high concentration of DMF, it must be disposed of as hazardous waste.

disposed of as hazardous waste. Distillation of the organic layer removed the DMF and other high-boiling contaminants more effectively (Figure 3b). If this sequence had been followed, additional extractions would have been necessary, in turn contributing more DMF contaminated waste.

The effect of varying the water:product ratio and the number of extractions following distillation, are summarized in Table I.

Table I. Water Extraction of Methanol

Water:Product Ratio	1:2	1:1			2:1
Number of Washes	1	1	2	3	1
% MeOH	0.5	0.2	0.03	0.02	0.2
% DCM	99.5	99.8	99.97	99.98	99.8

Extraction with either 2:1 or 1:1 ratio produced a product with approximately 0.2% MeOH. The product of the 1:2 ratio however contained 0.4% MeOH. The 1:1 ratio was chosen to because it removed most of the MeOH without excessive handling. A second extraction reduced the amount of MeOH to 0.03% and a third extraction did not further purify the product.

Because water is slightly soluble in DCM, it was necessary to desiccate the product. Desiccation using anhydrous Calcium Chloride produced too much particulate matter in the final product and did not remove residual MeOH. 3Å molecular sieves removed the water but were less effective than the 4Å molecular sieves at removing any residual MeOH.

Quality control of the final product was done on a gas chromatograph using a Flame Ionization Detector (FID). The detector was set to its most sensitive setting (10^{-12}Amp/mV). A series of standard solutions were made up to test the sensitivity of the detector. The solutions ranged in concentration from 1 to 10,000ppm MeOH in DCM. The detector, on its most sensitive setting, was able to detect 1ppm MeOH in DCM.

The final product was analyzed and compared with several commercially available brands of HPLC Grade DCM. All were >99.9% pure, although several extremely small (<0.01%) contaminant peaks were detected by gas chromatography. The purity of the final product, 99.96% DCM was comparable to the best HPLC grade DCM. GC-MS analysis of the final product failed to identify any contamination.

PRE-SCREENING OF STARTING MATERIAL. If the initial waste contained certain chemicals the purity of the final product was insufficient for peptide synthesis. Many of these contaminating agents were not a result of the peptide synthesis process but of careless waste handling on the part of laboratory personnel. To detect the presence of these solvents it is necessary to screen the waste before distillation. Figure 4 illustrates the effect of distillation on a purposely contaminated sample of peptide synthesis waste. Ethanol, propanol, DMS, acetonitrile, and butanol were added to peptide synthesis waste and the resulting mixture was analyzed. The contaminant peaks were clearly resolved (Figure 4a). After distillation acetonitrile and butanol were removed, but MeOH, ethanol, propanol, and DMS remained (Figure 4b). With a sufficient number of water extractions, it was possible to remove the alcohols (Figure 5), but the DMS remained. It is likely that other organic non-

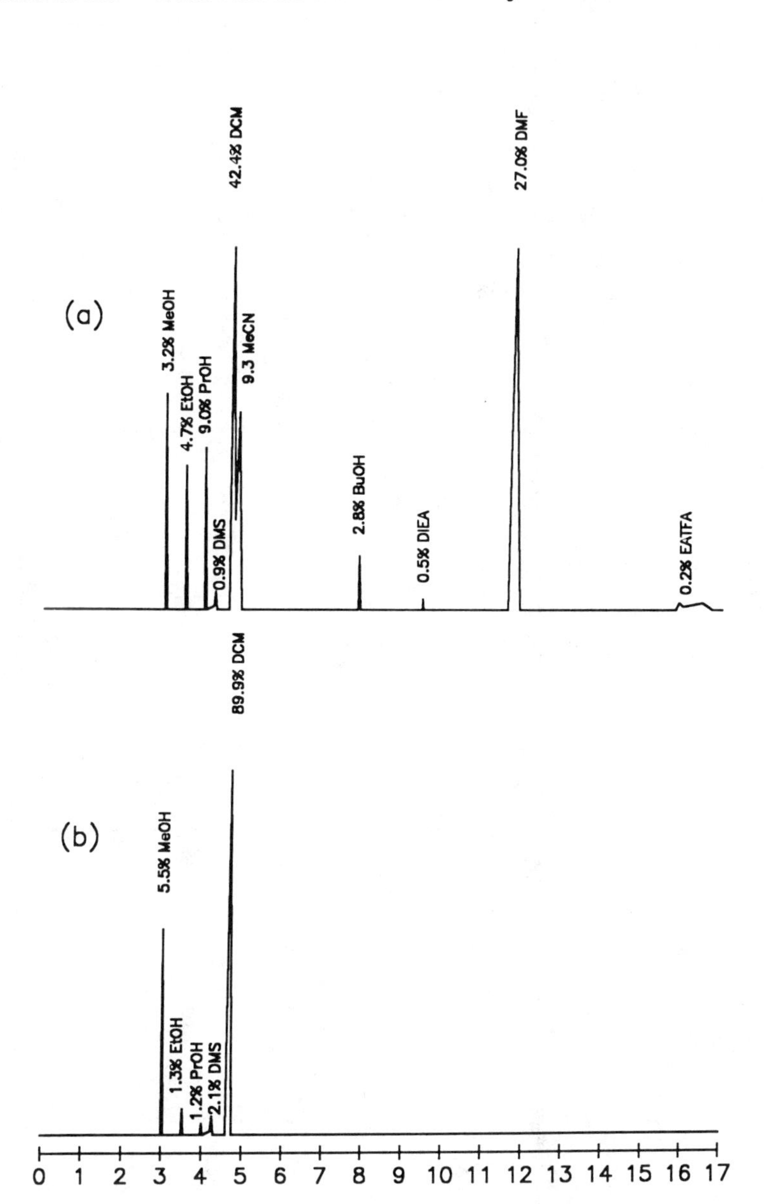

Figure 4: Gas chromatographs of contaminated peptide synthesis waste not suitable for distillation. (a) Before distillation. (b) After distillation. Distillation removed all acetonitrile (MeCN), butanol (BuOH), DiEA, DMF, and EATFA but could not remove MeOH, ethanol (EtOH), propanol (PrOH) or dimethylsulfide (DMS).

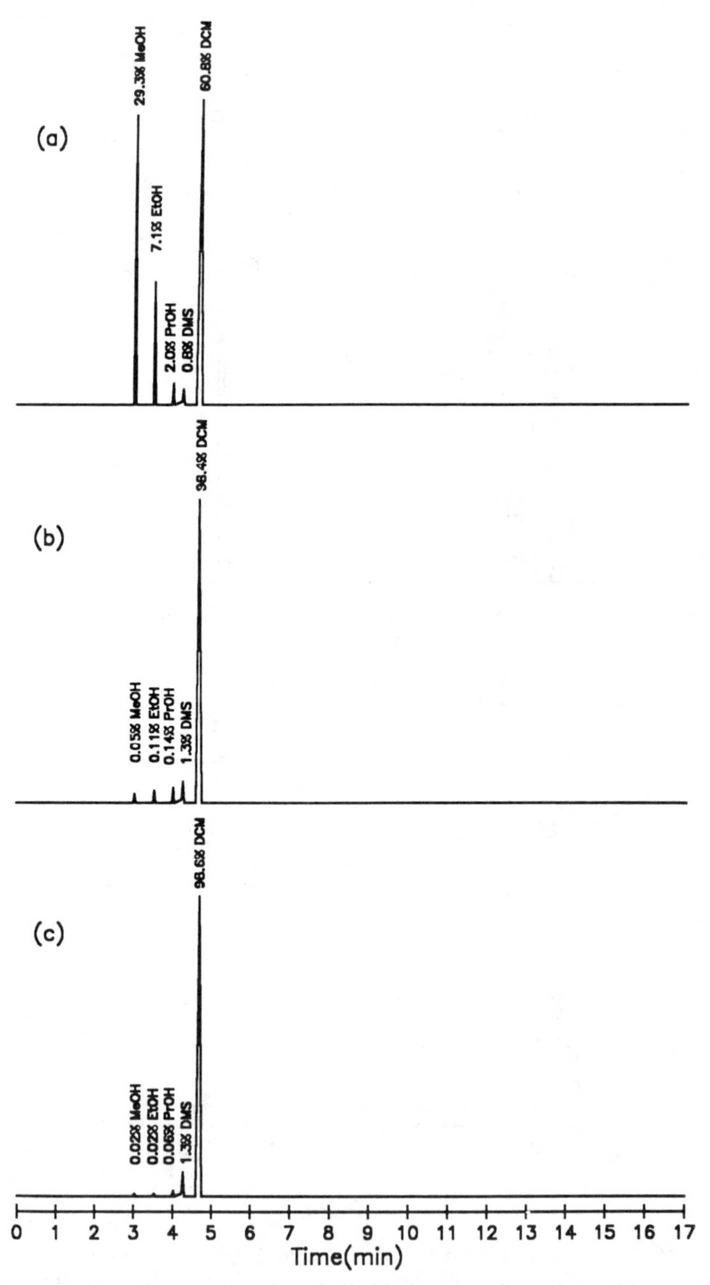

Figure 5: Gas chromatographs of distilled contaminated peptide synthesis waste. (a) Before distillation. (b) After one 1:1 aqueous extraction. Concentrations of alcohols were greatly reduced, but DMS remained. (c) After two 1:1 aqueous extractions. Alcohols were removed almost entirely but because DMS is insoluble in water, it remains in the product.

polar solvents with boiling points in the range of 40°C to 60°C will also compromise the quality of the final product and make it unacceptable for peptide synthesis.

In order to reclaim and recycle DCM from peptide synthesis waste, starting materials must be pre-screened to be sure that DMS and other difficult to remove contaminants have been segregated from the synthesis waste. The labor involved in distillation, water extraction and desiccation can be automated to some degree. However, it is possible that at 95% purity, without aqueous extraction and desiccation, DCM could be recycled for purposes other than peptide synthesis, e.g. paint stripping and degreasing. It is also likely that this reclamation protocol will be applicable to other waste streams. In research environments the primary use of DCM is for solid phase peptide and DNA synthesis, which both require the highest purity. However, from a liability standpoint it is advantageous to recycle chemicals on-site. A well documented quality control approach, for example, with a well-standardized and calibrated GC analysis, is most essential to the acceptance of reclaimed products.

LITERATURE CITED

1. Watson, T., Hall, R.M. Jr., Davidson, J.J., Case, D.R., RCRA Hazardous Wastes Handbook, Government Institutes, Washington, D.C. 1980.
2. Upton, A.C.; Kneip, T.; Toniolo, P. In Annual Review of Public Health; Breslow, L.; Fielding, J.E.; Lave, L.B., Eds.; Annual Reviews Inc.: Palo Alto, 1989; Vol.10, p 8.
3. Merrifield, R.B., J. Am. Chem. Soc. 1963, 85, 2149-54.
4. Stewart, J.M., Young, J.D., Solid Phase Peptide Synthesis, (Second Edition), Pierce Chemical Company: Rockford, IL 1984.

RECEIVED November 10, 1989

INDEXES

Author Index

Affiliation Index

Subject Index

D

T

Production: Donna Lucas
Indexing: Deborah H. Steiner
Acquisition: Cheryl J. Shanks

Elements typeset by Hot Type Ltd., Washington, DC
Printed and bound by Maple Press, York, PA

Paper meets minimum requirements of American National Standard for Information Sciences—Permanence of Paper for Printed Library Materials, ANSI Z39.48–1984 ∞

Other ACS Books

Chemical Structure Software for Personal Computers
Edited by Daniel E. Meyer, Wendy A. Warr, and Richard A. Love
ACS Professional Reference Book; 107 pp;
clothbound, ISBN 0–8412–1538–3; paperback, ISBN 0–8412–1539–1

Personal Computers for Scientists: A Byte at a Time
By Glenn I. Ouchi
276 pp; clothbound, ISBN 0–8412–1000–4; paperback, ISBN 0–8412–1001–2

Biotechnology and Materials Science: Chemistry for the Future
Edited by Mary L. Good
160 pp; clothbound, ISBN 0–8412–1472–7; paperback, ISBN 0–8412–1473–5

Polymeric Materials: Chemistry for the Future
By Joseph Alper and Gordon L. Nelson
110 pp; clothbound, ISBN 0–8412–1622–3; paperback, ISBN 0–8412–1613–4

The Language of Biotechnology: A Dictionary of Terms
By John M. Walker and Michael Cox
ACS Professional Reference Book; 256 pp;
clothbound, ISBN 0–8412–1489–1; paperback, ISBN 0–8412–1490–5

Cancer: The Outlaw Cell, Second Edition
Edited by Richard E. LaFond
274 pp; clothbound, ISBN 0–8412–1419–0; paperback, ISBN 0–8412–1420–4

Practical Statistics for the Physical Sciences
By Larry L. Havlicek
ACS Professional Reference Book; 198 pp; clothbound; ISBN 0–8412–1453–0

The Basics of Technical Communicating
By B. Edward Cain
ACS Professional Reference Book; 198 pp;
clothbound, ISBN 0–8412–1451–4; paperback, ISBN 0–8412–1452–2

The ACS Style Guide: A Manual for Authors and Editors
Edited by Janet S. Dodd
264 pp; clothbound, ISBN 0–8412–0917–0; paperback, ISBN 0–8412–0943–X

Chemistry and Crime: From Sherlock Holmes to Today's Courtroom
Edited by Samuel M. Gerber
135 pp; clothbound, ISBN 0–8412–0784–4; paperback, ISBN 0–8412–0785–2

For further information and a free catalog of ACS books, contact:
American Chemical Society
Distribution Office, Department 225
1155 16th Street, NW, Washington, DC 20036
Telephone 800–227–5558